# Ecological Restoration Law

Ecological restoration is as essential as sustainable development for the health of the biosphere. Restoration, however, has been a low priority of most countries' environmental laws, which tend to focus narrowly on rehabilitation of small, discrete sites rather than the more ambitious recovery of entire ecosystems and landscapes. Through critical theoretical perspectives and topical cases studies, this book's diverse contributors explore a more ambitious agenda for ecological restoration law. Not only do they investigate current laws and other governance mechanisms; they also consider the philosophical and methodological bases for the law to take ecological restoration more seriously. Through exploration of themes relating to time, space, geography, semiotics, social justice, and scientific knowledge, this book offers innovative and critical insights into ecological restoration law.

**Professor Afshin Akhtar-Khavari** is at Queensland University of Technology, Australia.

**Professor Benjamin J. Richardson** is based at the University of Tasmania, Australia.

**Law, Justice and Ecology**

**Series editor**: Anna Grear, Law School, Cardiff University, UK

For information about the series and details of previous and forthcoming titles, see www.routledge.com/Law-Justice-and-Ecology/book-series/LAWJUSTECO

# Ecological Restoration Law

## Concepts and Case Studies

**Edited by**
**Afshin Akhtar-Khavari and**
**Benjamin J. Richardson**

Routledge
Taylor & Francis Group
a GlassHouse book

First published 2019
by Routledge
2 Park Square, Milton Park, Abingdon, Oxon OX14 4RN

and by Routledge
52 Vanderbilt Avenue, New York, NY 10017

A GlassHouse book

*Routledge is an imprint of the Taylor & Francis Group, an informa business*



*British Library Cataloguing in Publication Data*
A catalogue record for this book is available from the British Library

*Library of Congress Cataloging-in-Publication Data*
Names: Evans, Matthew, (Political scientist), editor.
Title: Transitional and transformative justice : critical and international perspectives / edited by Matthew Evans.
Description: Abingdon, Oxon [UK] ; New York, NY : Routledge, 2019. |
Series: Transitional justice | Includes bibliographical references and index.
Identifiers: LCCN 2018045125 | ISBN 9781138478510 (hbk)
Subjects: LCSH: Transitional justice. | Restorative justice. | Reparation (Criminal justice) | Political crimes and offenses.
Classification: LCC K5250 .T73 2019 | DDC 340/.115--dc23
LC record available at https://lccn.loc.gov/2018045125

ISBN: 978-1-138-60501-5 (hbk)
ISBN: 978-0-429-46831-5 (ebk)

Typeset in Galliard
by Taylor & Francis Books

# Contents

# Contributors

**Afshin Akhtar-Khavari** is a professor of international law and governance at the Queensland University of Technology, Australia, where he is also the director of a research program in international law and global governance. Afshin has published extensively in international environmental law; his most recent contributions to the restoration literature include a co-authored book published by Routledge titled *Ecological Restoration in International Environmental Law* and the 2017 special issue collection on 'Eco-Restoration and the Law' in the *Griffith Law Review*. His research examines and imagines new ethical foundations for international environmental law. He is currently working on a monograph examining the importance of the Ramsar Convention for conserving and restoring wetlands.

**Emily Barritt** is a lecturer at The Dickson Poon School of Law at King's College London, UK, where she teaches courses on tort, environmental law and public interest litigation. She has a BA in Jurisprudence from the University of Oxford, UK and a PhD from King's College London. She has previously taught environmental law at the University of Oxford. In 2015 she was a Research Fellow at the Centre for Environment, Energy and Natural Resource Governance, University of Cambridge, UK, working on a United Nations Environment Programme project developing legal options for a new implementing agreement on the protection of marine biodiversity for the UN Law of the Seas Convention. Emily's research considers the nature of environmental justice, rights, democracy and stewardship. Her forthcoming monograph is *The Foundations of the Aarhus Convention*.

**Robyn Bartel** is a multi-award-winning associate professor with wide-ranging expertise in geography, law and education. Known internationally for her contribution to legal geography, her research encompasses regulation, regulatory agencies and the regulated, as well as the social, institutional and natural landscape in which all are situated. Her work has been influential in environmental policy development, heavily cited in the scholarly literature, and handpicked for prestigious international collections and seminal texts in environmental law. Robyn designed and introduced the Bachelor of Sustainability at the University of New England, Australia, one of the first programmes of its kind. Robyn is a founding member of AELERT, the Australasian Environmental Law Enforcement and Regulators Network, and the Legal Geography Study Group of the Institute of Australian Geographers.

**An Cliquet** is a professor of international environmental and biodiversity law at Ghent University, Belgium. Her research is situated in the field of international, European and national biodiversity law, and encompasses both marine and terrestrial biodiversity law. She is involved in research or supervising PhD research on ecological restoration, climate change and nature conservation; ecological refugees; gender and biodiversity; protection of ecosystem services; a rights-based approach to conservation; the protection of the Congo Basin; health and biodiversity; wildlife trade; the legal protection of urban biodiversity; the protection of transboundary watercourses under biodiversity law; and the protection of the environment during armed conflict. She teaches courses on international and European environmental and biodiversity law, law and ethics on marine nature conservation, as well as some courses on public international law in general.

**Kris Decleer** is a senior researcher at the Research Institute for Nature and Forest, based in Brussels, Belgium. His research mainly encompasses studies on wetland management and restoration, and landscape ecology in general. For many years he has coordinated numerous consultancy projects in the field of ecological restoration, nature management, spatial planning and nature policy development for a variety of governmental organizations and other target groups. Since 2009 he has been guest lecturer at the University of Antwerp, the Netherlands, in the Master's training course on habitat conservation and restoration. He is co-founder (2006) and an active board member of the European Chapter of the Society for Ecological Restoration (SER) and has served as a member of the Board of Directors of SER International since 2011, where he attempts to contribute to bridging the gap between the scientific world on one hand and the world of practitioners and policy makers at the other hand. In his spare time, he is responsible for the management of a wetland nature reserve, where he build up restoration experience in all its practical aspects.

**Uzuazo Etemire** is a Senior Lecturer in the Faculty of Law, University of Port Harcourt, Nigeria and the Acting Head of the Faculty's Department of Jurisprudence and International Law. Dr Etemire holds an LLB, LLM and PhD in Law degrees from the University of Benin, Nigeria, University of Nottingham and University of Strathclyde, both UK, respectively. He is called to the Nigerian Bar, and is a Fellow of the Higher Education Academy, UK. In 2017, Dr Etemire was a Visiting Scholar at the University of Barcelona, Spain, where he furthered his primary research interest in the field of Environmental Law. In this area of the law, he has participated in several conferences as well as research projects and has published widely in major peer-reviewed journals. Dr Etemire is the recipient of several academic scholarships and awards, and is the author of the book *Law and Practice on Public Participation in Environmental Matters:*

*The Nigerian Example in Transnational Comparative Perspective*, published in 2016 by Routledge.

**Nicole Graham** is an Associate Professor of Law at the Sydney Law School, the University of Sydney, Australia. She teaches and researches in the fields of property law and theory, and legal geography. Nicole has written on the relationships between law, environment and culture with a particular focus on property rights, natural resource regulation and the concept and experience of place. Nicole has received teaching awards for her work teaching property law, is recognised as a highly effective first-year specialist, and has made significant contributions to educational development in embedding Indigenous laws and perspectives into the law curriculum, and in sustainability in legal education

**Phillipa McCormack** is a lecturer in administrative law and legal research methods at the University of Tasmania, Australia. Phillipa graduated from the University of Tasmania in 2007 with a combined degree in social ecology and law (first class honours) and a university medal. She was admitted to practice and worked as a solicitor at an international commercial law firm in government and environmental law, before taking up a position as an associate to the Honourable Justice Lasry at the Supreme Court of Victoria, Australia. She was recently awarded a doctorate for her thesis, 'Australia's Legal Frameworks for Biodiversity Conservation: Facilitating Adaptation in a Rapidly Changing World'. Her work is cross-disciplinary and she has published with lawyers, scientists and policy makers in a range of scientific and legal journals on the challenges of climate change for biodiversity conservation.

**Menes Abinami Muzan** is a Lecturer in the Department of Public Law, Faculty of Law, University of Port Harcourt, Nigeria, where he also doubles as Research Editor in the *UNIPORT Journal of Public Law*, an annual peer-reviewed journal published under the auspices of the Department of Public Law. He holds an LLB from the Rivers State University, Port Harcourt and an LLM in Environmental Law from the School of Oriental and African Studies (SOAS), University of London, UK. He is a Barrister and Solicitor of the Supreme Court of Nigeria, and is currently a Doctoral Researcher at the Birmingham Law School, University of Birmingham, UK where he is studying for a PhD in Environmental Law under the Tertiary Education Trust Fund. He has published several articles in leading peer-reviewed journals on critical aspects of environmental law and is the recipient of several academic scholarships and awards.

**Froukje Maria Platjouw** is a researcher at the Norwegian Institute for Water Research and a post-doctoral research fellow at the University of Oslo, Scandinavian Institute for Maritime Law, Norway. She graduated from the University of Groeningen, the Netherlands in 2006 and the University of Oslo in 2008 with degrees in public international law. Currently, she works on research projects in the interface between environmental and societal dynamics, and has a main interest in the governance of transboundary marine ecosystems. In 2016 she published *Environmental Law and the Ecosystem Approach – Maintaining*

*Ecological Integrity through Consistency in Law* (Routledge). Her work is interdisciplinary, integrating elements of ecological sciences and the economics of ecosystem services into legal research.

**Benjamin J. Richardson** is Professor of Environmental Law at the University of Tasmania, Australia. His international academic career over two decades has included positions in law faculties in Canada, the UK, the Netherlands and New Zealand. He also collaborates with a variety of academic groups, including the IUCN Academy of Environmental Law and the Australian Panel of Experts on Environmental Law. He practises environmental stewardship on his Tasmanian eco-sanctuary, Blue Mountain View.

**Francine Rochford** is an Associate Professor at La Trobe University, Victoria, Australia. Her research often traverses the intersection of sociology and the law; her master's thesis was an application of Habermas' discourse theory to common law, and her doctoral thesis considered the causes and effects of litigation in higher education. More recently, she has been researching Victorian water policy and its effects, including semiotic approaches to restoration of water to the environment. She coordinates the regional operations of the La Trobe Law School and convenes subjects in water law, the law of education and the law of torts, and teaches contract law, consumer law and constitutional law.

**Hendrik Schoukens** graduated as a Master in Laws at the Catholic University of Leuven, Belgium in 2005. In 2007 he graduated as Master in Environmental Law at Ghent University, Belgium. In 2010 he received his complementary Master's Degree in Environmental Law at the Facultés Universitaires Saint-Louis and the Catholic University of Louvain, Belgium. Since January 2006 Hendrik has been working as an environmental lawyer at LDR Advocaten, where he focuses on biodiversity-related cases. In 2011 he co-authored the first edition of the *Handbook of Biodiversity Law* in the Flemish Region, and in 2016 he edited the first edition of a comprehensive handbook on EIA law in the region. At the Department of European, Public and International Law, Hendrik wrote his doctoral thesis on the legal aspects of ecological restoration, which he successfully defended in 2017. He currently works as a postdoctoral researcher at Ghent University.

**Anastasia Telesetsky** is a Professor of Law at the University of Idaho College of Law, US, where she teaches the Natural Resources and Environmental Law program. She is a member of the World Commission on Environmental Law and the IUCN Academy of Environmental Law's Research Committee. She is a former Fulbright Fellow (the Philippines), Bosch Fellow (Germany) and Axford Fellow (New Zealand). She has lectured on international environmental law in Japan, China and New Zealand. Her primary research focuses are on ecological restoration, fisheries governance and disaster law.

# 1 Ecological restoration and the Anthropocene

*Afshin Akhtar-Khavari and Benjamin J. Richardson*

## 1.1 Importance of ecological recovery

Ecological recovery has never been more important yet incongruously remains a low priority in environmental law. Most policy-makers perceive the intensifying upheavals of the Anthropocene as reasons to pay ever more attention to the future so as to forestall further degradation. Climate change, species extinctions, oceans of plastic debris and other ecological tolls loom on the horizon as an ever-real dystopia. We cannot ignore the urgency to halt dissipation of the life-sustaining biosphere, yet equally we should heal past losses in order to make sustaining what remains more viable.[1] The Anthropocene is not a recent phenomenon but derives from a long history of anthropogenic environmental change that began at least with the onset of industrialisation two centuries ago and possibly earlier with the advent of agriculture.[2] Under the aegis of the philosophy of sustainable development, which provides environmental law's conceptual ballast, regulators dwell on forestalling future adversity rather than addressing past follies.[3] The legal priority is commonly to avert, mitigate or adapt to new ecological impacts rather than to repair past damage. This stance may also emotionally and culturally weaken people's sense of environmental stewardship on the presumption that nature has the capacity to passively restore itself through processes of ecological succession, species evolution and so forth. Damaged or degraded ecosystems sometimes can recover through their own processes, as evident in how nature rebounds after fires, floods or droughts; however, some recovery may be effectively impossible, such as when invasive species have fundamentally altered ecological equilibriums or toxic pollutants become embedded in land or water.

This book aims to deepen scholarship on ecological restoration (hereafter 'eco-restoration') law in order to better understand its methods, goals, achievements and other foundational dimensions. Providing neither an exegesis of current legal doctrine nor blueprints for law reform, the book instead offers critical and often conceptual insights into a variety of themes that permeate restoration governance, including its spatial and temporal properties, the role of science, social justice and

1 Garver (2017).
2 Ruddiman (2003).
3 Cordonier Segger and Khalfan (2004).

community participation, and other issues across a diversity of jurisdictions sampled. Building on a special edition of the *Griffith Law Review* in 2017 devoted to eco-restoration law,[4] this book includes additional authors and topics to enable a more comprehensive analysis of the subject, including additional insights on financing eco-restoration and climate change. As a multi-authored volume, the contributors bring diverse perspectives about eco-restoration law, but they share the leitmotif that the law should be more ambitious in helping nature to recover. In particular, lawmakers should go beyond the existing agenda that emphasises remediation of discrete sites, such as former mines and oil spills, to promote regeneration of entire landscapes and ecosystems. The difference between these approaches is epitomised in the terminological distinction this book makes between 'environmental' and 'ecological' restoration. The global practice of restoration ecology spans a diversity of methods, as evoked by the ubiquitous terminology of 'remediation', 'regeneration', 'restoration', 'reclamation' and so on, with these terms sometimes used interchangeably. In legal regimes, however, the broad distinction we make between 'environmental' and 'ecological' restoration is plausible because the law emphasises restoration actions towards the former spectrum while downplaying the latter. Governments occasionally intervene to 'remediate', 'reclaim' or 'regenerate' the most degraded enclaves but the widespread, incremental dissipation of the biosphere over centuries has gone largely unchecked and unrepaired.

A casual survey of international and regional environmental law and policy might suggest otherwise, however. The European Community's Habitats Directive[5] obliges its Member States 'to maintain or restore, at favourable conservation status, natural habitats and species of wild fauna and flora of Community interest'.[6] The Directive is supported by the European Biodiversity Strategy, which aspires by 2020 to restore 'at least 15% of degraded ecosystems'.[7] The European Landscape Convention also acknowledges restoration as one seminal method of 'landscape planning' as mandated by the treaty.[8] Of global reach, the Convention on Biological Diversity of 1992 obliges its current 196 parties, as far as possible, to 'rehabilitate and restore degraded ecosystems and promote the recovery of threatened species'; and to 'adopt measures for the recovery and rehabilitation of threatened species and for their reintroduction into their natural habitats under appropriate conditions'.[9] The Aichi Biodiversity Targets, the Convention's

4 Akhtar-Khavari and Richardson (2017). The chapters in this book that draw from that special edition of the *Griffith Law Review* are those authored by Francine Rochford (2017), Nicole Graham and Robyn Bartel (2017), An Cliquet and Kris Decleer (2017), Anastasia Telesetsky (2017), and Uzuazo Etemire and Menes Muzan (2017).
5 Council Directive 92/43/EEC of 21 May 1992.
6 Ibid, Article 2(2).
7 See http://ec.europa.eu/environment/nature/biodiversity/strategy.
8 European Treaty Series No. 176, 2000, article 1 ('Landscape planning', a term used in the Convention, is defined as 'strong forward-looking action to enhance, restore or create landscape').
9 (1992) 1760 UNTS 79; articles 8(f) and 9(c).

strategic plan for the decade to 2020, include the goal of restoring at least 15 per cent of degraded ecosystems.[10] Many further examples of international legal recognition of restoration are evaluated in the landmark book by Telesetsky, Cliquet and Akhtar-Khavari (2017), but they recognise that official edicts do not necessarily reflect actual practices. Recent research suggests these largely hortatory norms have limited traction in restoration governance within states for reasons that include their economic costs, community opposition, existing land-use entitlements and the lack of meaningful legal accountability for compliance.[11]

Consider the following examples that illustrate states' practices.[12] The United States has some legal precedents for eco-restoration, such as its Collaborative Landscape Restoration Program enacted in 2009 under the Omnibus Public Land Management Act.[13] The program prescribes ambitious mandates to restore forest ecosystems, although the rationales serve the forestry economy and control of destructive wildfires rather than recovery of biodiversity and ecological integrity for their own value.[14] New Zealand legal practice emphasises the latter goals more, and through the Conservation Act 1987[15] authorities have initiated projects to restore populations of the country's endemic and highly endangered birds,[16] and adopted a national strategy to eliminate entirely stipulated destructive exotic pests by 2050.[17] New Zealand's ambitions thus far await more extensive action, as so far its strategy is mainly directed to existing conservation-managed lands rather than any of the vast areas managed for farming or other economic activities. The governance of Australia's Murray-Darling Basin – as considered in some detail by Francine Rochford (Chapter 11 in this book) – presents another cautionary tale. The objects of the applicable Water Act 2007 (Cth) include to 'restore and provide for the ecological values and ecosystem services of Murray-Darling Basin': in practice, the managing authority has been hamstrung because of opposition from riparian towns and farmers wishing to retain their share of the river flows.[18]

Less ambitious forms of *environmental* restoration are more likely to be supported by governments because the private sector usually bears the costs; they deliver more tangible, near-term benefits; and they entail less complex governance given their more discrete temporal and spatial scales. Pollution regulation commonly has provisions to require landowners to remediate contaminated lands, and mining legislation similarly obliges operators to restore the amenity of their sites after mining ceases. Another avenue for environmental restoration is biodiversity and carbon offsets, in which, as a condition of regulatory approval, the developer

10 Target 15; see Jorgensen (2013).
11 Telesetsky, Cliquet and Akhtar-Khavari (2017).
12 As discussed in more detail in Richardson (2017), 226–235; Richardson (2016a).
13 Omnibus Public Land Management Act 2009 (US), Public Law 111–11, See www.fs.fed.us/restoration/CFLRP/index.shtml.
14 Ibid, s 4001.
15 See Conservation Act 1987 (NZ) s. 2(1).
16 Butler, Lindsay and Hunt (2014).
17 Predator Free 2050: www.doc.govt.nz/predator-free-2050.
18 Tsigas (2016).

must restore habitat in another location in exchange for the right to impair environmental values on their site being developed. Despite the benefits that some of these approaches can yield, they generally offer only ad hoc improvements to the environment, and with offsets may even worsen overall conditions.

The foregoing lacunae and weaknesses of official regulation are starting to be mitigated by community and environmental non-governmental organisations (NGOs) voluntarily restoring degraded landscapes.[19] The Yellowstone to Yukon Region Conservation Initiative (Y2Y), undertaken by a loose confederation of such organisations, has since 1997 being restoring and conserving a 3,200 km corridor of the Rocky Mountains from the United States to northern Canada that has been ecologically impaired by a history of development from roads, dams, forestry, mining and so forth.[20] In southwest Australia, the Gondwana Link project is similarly harnessing community resources for reconnecting fragmented landscapes that conserve imperilled fauna and flora in a global biodiversity hotspot. In Scotland, a group called Trees for Life is replanting 230,000 ha of the former Caledonian forest in the Scottish Highlands with a similar focus on biodiversity recovery.[21] Rewilding Europe is spearheading reintroduction of wolves to restore the ecological equilibrium that such apex predators can provide.[22] Some restoration projects are driven by business entrepreneurs, such as the former heads of clothing companies Patagonia and North Face who funded the acquisition and ecological restoration of former cattle ranches in Chile's Patagonia region, which have now become national parks.[23] Later in this book we will learn much more about the role of non-state actors in restoration governance, including to the level of individual landowners, as discussed in Chapter 5 by Robyn Bartel and Nicole Graham.

## 1.2 Terminology

What is 'eco-restoration'? The term has no authoritative definition in law, yet is widely used by academics and practitioners working on restoration issues. Restoration also has a long history, with antecedents in reforestation and game management in medieval Europe.[24] The Society for Ecological Restoration (SER), the leading global network of professionals in this field today, defines the term as 'an intentional activity that initiates or accelerates the recovery of an ecosystem with respect to its health, integrity and sustainability'.[25] The SER elaborates that an ecosystem can be 'restored' when it can 'sustain itself structurally and functionally', showing sufficient 'resilience to normal ranges of environmental stress and disturbance'.[26]

19 See further Richardson (2017), 235–249.
20 Wilcox and Aengst (1999); Levesque (2001).
21 See http://treesforlife.org.uk.
22 See https://rewildingeurope.com.
23 Bonnefoy (2018).
24 Baker (2014), 510.
25 Society for Ecological Restoration (2004).
26 Ibid, section 10.

The SER's approach thus emphasises three important ideas. First, restoration has a spatial dimension, the 'ecosystem', which exists in a huge variety of sizes from a small creek or decaying log to a large rainforest. Yet all have in common their quality as an environmental system, rather than being defined by anthropocentric spatial concepts such as property tenure or government jurisdiction. Second, restoration serves to improve and sustain the ecosystem's health and integrity rather than meeting only anthropocentric needs, such as returning polluted land to some economic purpose (the SER also acknowledges 'functionality' as an outcome, so that restoration can serve human needs in addition to, but not instead of, non-anthropocentric goals). Third, human intervention is necessary to assist damaged ecosystems to recover but with the expectation that they will eventually evolve and function on their own accord. In other words, damaged ecologies will not necessarily recover through their internal momentum.

The lexicon within ecological restoration studies encompasses several other terms. 'Rehabilitation' is defined by the SER as the 'reparation of ecosystem processes, productivity and services'.[27] These functional criteria could include the provision of portable water, fertile soils or harvestable forests. The related term 'reclamation', which, like rehabilitation, appears in some environmental laws and policies, commonly refers to the treatment of a former mine or industrial area in order to remove pollutants, control soil erosion or improve aesthetics. Reclamation can also sometimes denote environmentally pernicious practices, such as land reclamation programs that drain wetlands and build dikes. Another common term is 'remediation', involving removal of pollution or contaminants from soil, water and other environmental media, and it is commonly done on former industrial or urban sites. But rehabilitation can be construed differently, as feminist writer Germaine Greer showed in calling the project she initiated on her degraded rainforest property in southeast Queensland the 'Cave Creek Rainforest Rehabilitation Scheme'. Greer explained, 'restoration wasn't the right word … because it made the trees sound like furniture. I went for "rehabilitation" because it suggested the role that the forest would play in rebuilding itself'.[28] In fact, any of the foregoing terms might be open to interpretation in a variety of ways depending on their context and the participants' intentions.

Another term that does not come up in the SER discussion, but that should be distinguished from restoration, is 'geo-engineering'. It essentially means intervention to stabilise the Earth's climate system in order to mitigate adverse global warming, although geo-engineering could extend to other contexts, such as biotechnology projects to prevent species extinctions or even to revive extinct species.[29] Proposed methods of geo-engineering include iron fertilisation of the ocean to encourage algal growth and thereby carbon dioxide sequestration; and spraying sun-reflecting sulphates into the stratosphere.[30] Geo-engineering the climate does

27 Ibid.
28 Greer (2013), 104.
29 Brent, McGee and McDonald (2016).
30 Blackstock and Low (2018).

not repair past environmental damage but rather aims to forestall future adversity. This book acknowledges how eco-restoration governance relates to its alternatives, within and beyond the field of restoration, and contributes to understandings of how legal scholars and practitioners can contribute to debates about these options in the active recovery and management of the biosphere.

Some scientists believe 'renewal ecology' affords a better way to conceive the purpose of some restoration projects. David Bowman and others argue that we must not only restore degraded ecosystems but also *create* new ones that support biodiversity and provide ecosystem services for human communities in the context of rapid environmental change in the Anthropocene.[31] Sometimes this strategy will entail the deliberate creation of novel ecosystems into which threatened species can be relocated or migrate by themselves. This goal is likely to be most feasible in existing, degraded landscapes or areas suffering from declining economic productivity such as marginal farmlands that can be 'let go' for new environmental purposes. In this sense, 'renewal ecology' overlaps with existing approaches of restoration that aim to re-establish connectivity in landscapes between remnant wildlife habitat, in which new patches of habitat are nurtured to assist species on the move and to provide collateral productivity benefits to land owners, such as from reduced soil erosion and better water retention.

The mantra of 'rewilding' appeals to some environmentalists seeking large-scale restoration of landscapes while reducing human activity. The rewilding language originated in the 1980s out of the Wildlands Project with the goal of returning half of the United States to 'wildness'.[32] As the project's vision statement explains, '[r]ewilding means making our landscapes whole again … by protecting core reserves, reconnecting them via vast corridors of habitat, and restoring apex predators'.[33] It can have seemingly adverse consequences for people, as when infrastructure such as dams and roads must be removed for landscapes being rewilded.[34] In Europe, with its much greater human population density, the NGO network Rewilding Europe offers a somewhat different view of rewilding: '[it] helps landscapes become wilder, whilst also providing opportunities for modern society to reconnect with such wilder places for the benefit of all life'.[35] The language of rewilding thus far remains outside of official eco-restoration governance but retains appeal for non-state actors as a catchy slogan to attract public support for their cause.

This book does not attempt to resolve the complex terminological differences in the literature because restoration legitimately spans a range of strategies catering to diverse contexts: restoring a former mining site is not the same as rewilding an entire rainforest ecosystem. However, lawmakers must apply terminology accurately and consistently to ensure that the intended and appropriate form of

31 Bowman et al (2017).
32 Sandom et al (2013), 431.
33 See https://wildlandsnetwork.org/our-vision.
34 Blumm and Erickson (2012), 1047.
35 See www.rewildingeurope.com/about/what-is-rewilding.

restoration occurs. Legislation commonly fails here; the Canadian Environmental Protection Act 1999, to illustrate, mixes variable terminology such as 'remediate', 'repair', 'rehabilitate' and 'restore', and never defines them.[36] The lack of statutory definition of key terms and their admixture also undermines legal accountability for restoration obligations.

The emphasis on *ecological* restoration, the focus of this book, does not imply that smaller-scale approaches, such as remediation of brownfield sites or former mining sites, are not worthwhile. The distinction this book makes between 'eco-restoration' and other approaches, which we collectively label 'environmental restoration', might be rejected by some restoration practitioners as unjustified or unhelpful. Eco-restoration, after all, can serve to restore both ecosystems' structure and function. However, in the legal context this distinction has analytical traction because existing regulatory regimes for restoration focus heavily on *environmental* restoration. With few exceptions, governments generally ignore landscape-scale restoration as a legal requirement, with the initiative left primarily to non-state actors to undertake such projects, albeit sometimes with government seed funding. By highlighting 'eco-restoration law', this book seeks to emphasise deficiencies in current governance approaches that should be overcome.

Restoration ecology must not only emphasise recovery of biological and ecological systems, it must also accommodate social considerations, from gauging community opinion to active citizen participation in projects. Several other terms in the restoration literature capture this consideration more explicitly. One is 'reconciliation ecology', coined by Michael Rosenzweig[37] to describe restoration projects that benefit and engage people by drawing them closer to nature, including for the aesthetic and recreational benefits nature offers. This strategy is advocated particularly for urban denizens having limited access to relatively unmodified natural environments, which reconciliation ecology enables through creation of more urban greenery and wildlife habitat. In a sense, reconciliation ecology can be understood as a form of urban land-use planning with a stronger environmental overlay. Environmental 'stewardship', as explored by Emily Barritt (Chapter 4 in this book), is a ubiquitous term in the environmental studies literature,[38] and although it most commonly denotes steady, benign environmental management that serves to maintain the status quo, it also can encompass restoration to improve conditions. The stewardship ethos is particularly strong in some Indigenous people's cultures, such as Aboriginal Australians who call it 'caring for country', and New Zealand Maori, who express their stewardship values through the principle of *kaitiakitanga*, which translates roughly as 'guardianship and conservation of nature'.[39]

36 Canadian Environmental Protection Act 1999, S.C. 1999, c. 33, ss 2(1), 196, 205, 234(4), 294.1.
37 Rosenzweig (2003).
38 For an overview, see Berry (2006).
39 Whitt et al (2001), 9.

## 1.3 Recent legal developments

Environmental law has not kept pace with human beings' increasingly fast-paced reactions to being in the Anthropocene. Restoration science is one of these examples and, as discussed above, eco-restoration is often not provided for in legislative instruments in a range of countries. Often a turn to mitigation and remediation in environmental law instruments is likely because of the complexities and costs created by the kind of monitoring that goes on when self-sustaining ecosystems are being rebuilt. In this section, we briefly discuss some methods by which environmental laws and governance approaches to restoration can be influenced to cope with a range of increasing harms from climate change, biodiversity loss and extinction, and so on.

One such pressure on environmental law to develop has always been through new insights from science about what is possible and necessary to be done. Although restoration ecology has been around for a while, eco-restoration or the rewilding movement is dependent on more recent scientific movements. By not keeping pace with scientific developments, environmental laws have generally not deepened their responses to restoring ecosystems.[40] Palmer and Ruhl have argued, for instance, that the United States has not adequately addressed the recovery of self-sustaining ecosystems because scientific discourses on eco-restoration have not been used to build more precision into environmental laws that address restoration and remediation.[41] Cliquet and Decleer (Chapter 6 in this volume) put it differently and argue that lawyers can make decisions that are not informed by good ecological science, which can then hamper their efforts towards recovery efforts. They point to two concepts, extinction debt and ecological trap, as examples of the problems with potential legal solutions when they are not adequately grounded in ecological science.

Increasingly, significant scientific assessments are influencing developments internationally in relation to the science and law interface. The Intergovernmental Science-Policy Platform on Biodiversity and Ecosystem Services (IPBES) in March 2018 released a three-year study (hereafter the IPBES 2018 Study) on land degradation and restoration.[42] This report assesses the science and other reasons for the severity of land degradation globally. The assessment was conducted to provide more in-depth and comprehensive scientific support for the policy measures being pursued through the United Nations' international agreements and its Sustainable Development Goals in particular.[43] The IPBES assessment supports the idea that scientific work and assessments are necessary support for larger policy shifts globally.

Sometimes, however, drawing on popular scientific ideas and approaches to frame legal responses to recovery can create systemic problems. Akhtar-Khavari

40 eg, Palmer and Ruhl (2015); Cliquet and Decleer (2017); and Cliquet and Decleer (Chapter 6 in this volume).
41 Palmer and Ruhl (2015).
42 IPBES (2018); see also van Leenders and Bor (2016).
43 Ibid, 4.

(Chapter 2 in this volume) argues that certain scientific ideas receive priority and are shaped in particular ways because of what human beings believe as real or facts about the world. He uses the example of plant intelligence to highlight how scientific discourses have ignored something important in terms of appreciating the animism inherent in the world of living matter. This, he argues, has had important impacts on how human beings listen to the natural world and help with its recovery.

A second and alternative method and source of influence on the development of eco-restoration law and governance approaches comes from being able to compare practices across countries. The SER, which was established in 1988, now has 13 chapters where regional developments and initiatives drive norm developments and initiatives. For instance, the Australasian chapter of SER helped launch in 2017 the second edition of the 'National standards for the practice of ecological restoration in Australia', which took nearly five years to draft and was a collaboration between 12 different partner organisations.[44] The global arm of SER therefore builds its international standards drawing from a range of regional initiatives. However countries in the Association of Southeast Asian Nations do not appear to be involved in the activities of SER, despite, for instance, Japan having enacted back in 2002 one of the most well-developed laws that specifically focuses on the 'Promotion of Nature Restoration'.[45] Eco-restoration approaches are likely to vary among nations, given the significance of the variations between countries in terms of native plant availability, the nature of property ownership, food security and land degradation, as well the history of engaging in restoration activities. A significant influence on eco-restoration laws moving forward is likely to come from comparative studies from around the world.

Another important way to understand legal and governance developments and complexities is to draw on the range of knowledge systems and communities for the work that needs to be done. For instance, the IPBES 2018 Study, as a result of 21 case studies, was able to identify ways in which robust governance had to include Indigenous knowledge systems and communities in restoration activities.[46] The IPBES 2018 Study also identified that multi-stakeholder involvement, which includes anyone with potential interest and involvement in a landscape, was critical for effective restoration.[47] There are no surprises in these findings, but they indicate that legal developments in relation to eco-restoration will be successful only if they are inclusive and respectful of local conditions and communities that understand them and use them. The idea that a plurality of knowledge systems and actors is critical to the eco-restoration legal agenda is a feature of several chapters in this book, including in particular those by Benjamin Richardson (Chapter 3), Robyn Bartel and Nicole Graham (Chapter 5), and Uzuazo Etemire and Menes Muzan (Chapter 9). More than anything else, having to take into account the

44 See www.seraustralasia.com/standards/home.html.
45 See www.env.go.jp/en/laws/nature/law_pnr.pdf.
46 IPBES (2018), 33.
47 Ibid, 40.

range of knowledge systems and actors enriches decisions in relation to eco-restoration and makes them more complex than they already are.

International law and other intergovernmental agreements can influence the development of country responses to eco-restoration efforts. Eco-restoration is recognised as contributing to how certain goals within international agreements can be achieved. Prominent amongst these are the 1992 Convention on Biological Diversity, the 1992 Convention on Climate Change, the 1994 Convention to Combat Desertification, and the 1971 Convention on Wetlands of International Importance. These agreements all have Conferences of Parties that have subsequently developed the law relating to restoration efforts around the world.[48] The Aichi Biodiversity Targets, as an example, set goals for achieving the objective of the Convention on Biological Diversity, which include for instance that 15 per cent of all degraded lands have to be restored. Not all countries respond effectively to these targets and goals, and as Hendrik Schoukens (Chapter 8 in this volume) highlights, the lacklustre efforts of the European Union (EU), for instance, should be improved upon through strategic environmental litigation against it. Given the responses of the international community to dealing with climate change, the alternative is that restoration will be used as a mitigation strategy for non-compliance with current legal arrangements. Due to the early stage of international law's engagement with restoration, this area of legal development is likely to become an increasing feature of the work of international lawyers in the future.

## 1.4 Key themes

This book's collection of international perspectives critically evaluates the nature and impact of current laws and other governance mechanisms that address eco-restoration, advances theoretical understandings for governance reforms for eco-restoration, and contributes to critical and interdisciplinary insights into environmental law, as part of this Routledge monograph series on 'Law, Justice and Ecology'. Within these aspirations, the book engages with seven principal themes.

First, the book evaluates **eco-restoration in the context of its alternatives**, as the foregoing section on terminology already acknowledges. Eco-restoration is one way of capturing the idea of the recovery of ecosystems, going beyond the narrow approaches of 'environmental' restoration (eg, contaminated lands remediation). Among some environmental NGOs and activists, the term 'rewilding' is preferred to capture the former goal. Eco-restoration or rewilding is not necessarily incompatible with these other discrete approaches, which can contribute valuably to improving specific localities; yet eco-restoration focuses on rebuilding fundamental ecological and biological processes for their own value rather than primarily for their functional benefits for humankind. In 'novel ecosystems', a term coined to describe ecosystems that have fundamentally and perhaps irreparably changed due to extensive environmental degradation,[49] exact restoration of

48 For more detail see Telesetsky, Cliquet and Akhtar-Khavari (2017).
49 Hallett et al (2013).

former ecological conditions may not be possible but some environmental improvements may still be recoverable to generate functional benefits. Philippa McCormack (Chapter 12 in this volume), for instance, explores the idea of renewal ecology and how it can be used to justify novel ecosystems. She argues that legal responsiveness to climate change requires that conservationists take strategic actions to enable habits and species to respond to rising sea levels or habitat loss. It supports the idea that new habitats that are more likely to adapt to climate change will be important for restoring biodiversity.

Eco-restoration also should be analysed in the context of other ideas in the environmental governance and policy literature that express humankind's place in nature. In light of broader approaches to recovery of the natural world, restoration science can be seen as having limitations for imagining new governance arrangements or designing legal tools. Chapter 2, by Akhtar-Khavari, examines how the social life of trees and plants can emphasise the significance of the soft and animated dimensions of the natural world that provide for alternative insights to eco-restoration and alternatives that emphasise novel ecosystems. Homegardens and urban agriculture can enable human beings to engage more dynamically with the information that plants and trees receive, process, store and share. An appreciation of the animated life of plants will encourage and enable more 'listening', which is important in restoration efforts.

Another alternative context is the idea of 'stewardship', explored in Chapter 4 by Emily Barritt, with a long pedigree that entered the modern environmental movement in the wake of Aldo Leopold's advocacy for a land ethic to mediate people's relationship with nature.[50] Stewardship encompasses a wide variety of environmental management that may include restoration, although most commonly stewardship denotes a largely docile and benign relationship with natural systems that are allowed to flourish with minimal human interference. Restoration suggests a much more active role for managers, in which environments may be substantially transformed.

Secondly, eco-restoration governance has important **temporal and spatial dimensions**, as Benjamin Richardson explicitly examines in Chapter 3, that can help align human decision-making with nature's time scales and spatial ecology. Restoration in some forms involves conscious consideration of former environmental conditions as a 'baseline' to guide ecological recovery, which may reference the quite recent past, such as returning an abandoned mine to its pre-mine amenity, or restoration of much deeper historical periods. An example of the latter approach includes the restoration of Western Australia's Dirk Hartog Island to its condition in 1616 before the first Europeans brought substantial adverse ecological change.[51] Another example is the re-establishment of Scotland's Carrifran Wildwood to its splendour of 6,000 years ago.[52] In New Zealand's capital city, Wellington, the Zealandia sanctuary hosts a 225 ha ecosystem being returned to

50 Leopold (1949).
51 See www.sharkbay.org/restoration/dirk-hartog-island-return-1616.
52 See further www.carrifran.org.uk.

its rich biodiversity of 500 years ago.[53] Historical baselines nonetheless are sometimes problematic to ascertain or achieve, and thus some restoration project outcomes are judged primarily by reference to a forward-looking agenda of provision of future ecosystem services and preparing for climate change. Alagona and others contend that while use of strict historical baselines 'is fraught with challenges', some historical knowledge of nature is 'essential for conservation and restoration'.[54]

Space is another important denominator of eco-restoration in several ways, including the physical, geographical setting of restoration projects, ranging from highly discrete, small areas to large-scale landscapes; and the spatial organisation of legal governance such as property tenure and agency jurisdictions, which can fragment geographical space to the detriment of restoration. Furthermore, restoration can involve re-establishing healthier human emotional, aesthetic and spiritual relationships with nature, thereby restoring a 'sense of place'.[55] Restoration projects involving Indigenous peoples, for whom nature is often foremost a cultural landscape, are particularly relevant to this sense of place connection.[56] The marketisation of environmental space through carbon and biodiversity offsets is a new arena for restoration governance of some controversy, owing inter alia to the methodological reliability of making trade-offs that promise environmental gains in one space to compensate for losses in another.[57] The law may thus connect or fragment ecological spaces, with ramifications for the success of eco-restoration and nature stewardship.

The **science of eco-restoration** is a third theme addressed by some contributors to this book, especially in An Cliquet's and Kris Decleer's Chapter 6, as well as the ones by Afshin Akhtar-Khavari (Chapter 2) and Phillipa McCormack (Chapter 12). Restoration ecology has tended to be dominated by expertise from the natural sciences, such as zoology, botany and edaphology, which are indispensable to biologically feasible restoration. Less input into restoration projects usually comes from the social sciences or humanities. This does not mean that ecological restoration occurs without consultation with or assistance from affected local communities, which may transpire pursuant to legal obligations or because a restoration project is initiated by a local community. However, the dominance of the natural sciences in project design may create difficulties if relevant cultural issues are overlooked or downplayed, such as if a landscape of cultural heritage or aesthetic significance is altered.[58] By defining restoration primarily around the natural sciences, we risk losing sight of the importance of how people culturally and emotionally connect to nature and building ethical stewardship. At stake, in

53 See www.visitzealandia.com.
54 Alagona, Sandlos and Wiersma (2012), 49.
55 Drenthen (2009)
56 Hay (2009).
57 Gibbons and Lindenmayer (2007).
58 See especially for the tensions and synergies between science and public participation in eco-restoration: Egan, Hjerpe and Abrams (2011).

other words, are the types of knowledge that should inform restoration governance.

The interface between science and policy also matters, as scholars have already observed that environmental law can fail to utilise the appropriate scientific frameworks when it develops policy frameworks for restoration projects.[59] Environmental law is no stranger to referencing scientific knowledge in many of its domains, notably pollution standards, threatened species listings and environmental impact assessments, which in theory foster decisions based on scientific evidence and models rather than guesswork. Science is just as important to eco-restoration, although surprisingly legislation is silent about the scientific knowledge and methods that should inform restoration. Important decisions that require input from the natural sciences include ascertaining past environmental conditions that might serve as the baseline for restoration, choosing appropriate species to reintroduce, and modelling the impacts of global warming and other future environmental changes on places being restored. This book helps to inform readers about how scientific knowledge is incorporated into eco-restoration standards and procedures, and the obstacles and opportunities to a better synergy between science and policy.

Eco-restoration is often economically expensive to implement, requiring substantial financial outlays and in-kind contributions from the community. The costs of restoration may range from removal of contaminated soil and cleanup of oil spills, which can be extremely expensive, to replanting trees, controlling weeds and pests, and replenishing native fauna. **Financing eco-restoration** is thus a fourth theme of this book canvassed in some chapters, especially those by Froukje Platjouw (Chapter 7) and Anastasia Telesetsky (Chapter 10). In an era of declining public resources, the private sector has become more involved in restoration and remediation through legal compulsion or as a matter of corporate social responsibility.[60] Environmental liability rules can support restoration where there are clearly identifiable parties responsible for the pollution or other environmental damage to be repaired: landowners may be compelled to cleanse their contaminated properties and mining companies obliged to restore their sites to their former condition. The liability regime, however, fails when responsible parties are absent, such as for historical damage, or are insolvent. Landscape-scale eco-restoration rarely attracts regulators' interest, but NGOs and communities may voluntarily undertake it through their in-kind contributions (ie, time, labour, local expertise) as well as raise philanthropic funds from the general public and the private sector. Regulation can aid funding some eco-restoration indirectly through creation of offset markets, which developed in the 1980s in the United States through wetland mitigation banking.[61] While the costs of eco-restoration might seem high, they must be evaluated against the ecological, economic and social benefits, which over the long term can be very considerable; the 'restoration economy', as such investments are labelled by some, has been described by one

59 Palmer and Ruhl (2015).
60 Richardson (2016b).
61 Marsh, Porter and Salvesen (1996).

American commentator as the 'greatest new growth frontier'.[62] Restoration projects can also help to build communities' sense of place in nature with attendant cultural and emotional benefits.[63]

Monetising any form of restoration might be problematic for some because it risks framing our appreciation of nature through the capitalistic logic that itself has driven the environmental problems that now need to be repaired. The centrality of capitalism, especially its recent neo-liberal form, to global environmental decline has led some researchers to speak of the 'Capitalocene' rather than the Anthropocene.[64] They wish to focus attention on capitalism's responsibility for the crisis in the biosphere through the global expansion of markets and trade, the cult of economic growth, the advent of destructive technologies, and the weakening of public regulation of an increasingly unchecked private sector. Proponents of economic incentives, such as pollution taxes, and markets for ecosystem services and offsets, which utilise the architecture of capitalism, believe they can encourage more efficient use of natural resources and protection of natural 'capital'.[65] Yet this 'financialisation of nature', explains one NGO critic, commodifies nature into contrived, fungible units that violate the complex and interwoven properties of natural systems, and 'extends the damage done by a predatory and exclusive development model' controlled by the business sector.[66] Restoration, of course, comes in a variety of forms, some of which resemble this financialisation of nature, notably biodiversity offsets, while others do not, such as community-led eco-restoration projects that focus on provision of local social and ecological benefits.

The fifth overarching theme of this book acknowledges the **environmental humanities and the natural world**. Ecological restoration is not just a technical and scientific approach targeted at the recovery of ecosystems; it also has relevance for the attitudes and values of individuals and society that regulate behaviour, and it may connect people emotionally to elements of the natural world such as wildlife, forests, rivers, the weather and so forth. Community participation in eco-restoration projects may help nurture a 'sense of place' – an idea from human geography that encapsulates how people feel grounded and attached to specific localities.[67] The related concept of 'cultural landscape' expresses how some areas are viewed as containing both natural and cultural heritage, a relationship particularly important for Indigenous peoples with a long history etched in their traditional territories.[68] Having a sense of place helps to motivate people to care for nature beyond simply the functional benefits such as food and shelter that it can provide. Related theories of topophilia and biophilia (affinity to cultural landscapes and natural ecosystems, respectively) and ecological reconciliation[69] can help

62 Cunningham (2002).
63 Poe, Donatuto and Satterfield (2016).
64 Moore (2016); Demos (2017).
65 See eg, Helm (2015).
66 Kill (2015), 16.
67 Relph (1976).
68 Altman (2012).
69 Tuan (1977); Wilson (1984); Rosenzweig (2003).

inform how eco-restoration may emotionally and culturally engage human communities with their natural surroundings. This book's chapters from Emily Barritt (Chapter 4) and Robyn Bartel and Nicole Graham (Chapter 5) should help readers to understand whether and how environmental law can be used to shift people's mindset or emotional connections away from protecting or sustainably using nature to actively assisting with its recovery. Community participation rights, property tenure and cultural heritage protection legislation are among the legal mechanisms that might help align environmental governance with this agenda.

Another dimension of this discussion relates to how we create meaning in relation to a landscape when we engage in supposed restoration activities. While there are a variety of ideas that help us recognise the different ways that we can address problems, semiotics can explain why going in one direction may not actually work. Francine Rochford (Chapter 11) illustrates, for instance, how the single and pre-ordained measure taken to address water issues in a major landscape (the Murray-Darling Basin in Australia) disrupted established mappings of meanings. She argues that the single narrative created a semiotic re-regulation of the environment and a disruption of the relationship between inhabitants and the environment of the landscape that is being restored.

This role of communities in eco-restoration also feeds into the sixth key theme of the book, concerning **social justice and ecological acceptability**. The law has an important task to manage the trade-offs between restoration for ecological and biodiversity benefits on the one hand, and for delivery of functional benefits to human communities on the other. The focus on the former goal can be perceived by host communities to have adverse consequences for themselves. Local opposition may occur where residents have to coexist with reintroduced species such as carnivores, or where their economic opportunities to use natural resources are curtailed.[70] Some communities may also oppose eco-restoration for merely aesthetic reasons, because the changing character of the landscape interferes with their sense of place, as evident in reforestation projects in Scotland.[71] Conversely, local communities can be a significant resource for eco-restoration through citizen science projects and grassroots implementation efforts including tree planting and removal of pests. And in addition they benefit themselves from such participation, explain Lee and Hancock, through 'psychological, physiological, economic, and spiritual benefits, including learning new things, connecting with the natural environment, earning a living, doing something worthwhile' and so forth.[72] This nexus can be particularly strong for Indigenous communities, as acknowledged earlier in this chapter, and their participation in such projects can be crucial to both their social acceptability and ecological success, as evident in the Gondwana Link initiative in Western Australia.[73] How the law manages tensions between biological and social criteria in eco-restoration is addressed in a number of this

70 Nilsen et al (2007); Baker (2014).
71 Prior and Brady (2017).
72 Lee and Hancock (2011), 23.
73 Bradby, Keesing and Wardell-Johnson (2016).

book's chapters that touch on the interests of local communities, property owners, Indigenous peoples and scientific experts in project design and implementation. For instance, while Etemire and Muzan (Chapter 9) argue that general community participation made all the difference in the restoration initiatives in Ogoniland, Schoukens (Chapter 8) argues that strategic environmental litigation is the only vehicle that can really push the EU to achieve its restoration targets by 2020.

The final theme of this book considers the **modalities of eco-restoration governance**. The book's title refers to 'law', which is intended to encompass not only official regulations and legal policies in both the domestic and international spheres, but also a range of non-state, law-like governance arrangements. These might be voluntary agreements with landowners and local community activism and collaboration, as legal pluralism scholarship acknowledges.[74] Several important issues arise in the intersection between eco-restoration and governance, as other scholars have noted too.[75] First, we have the regulatory context to restoration, such as legislative duties on land managers, environmental damage liability rules, contractual arrangements, fiscal incentives including tax breaks and tradeable offsets, and dispute settlement. A second governance issue is its articulation at different scales – global, national, and local, and combinations of these, as conceptualised in multi-level governance theory.[76] A restoration project might draw on several levels of governance, beginning with the European Habitats Directive, the EU Member State's implementing national legislation, and then collaboration with local authorities and the impacted community, as evaluated recently by Telesetsky, Cliquet and Akhtar-Khavari,[77] and by Hendrik Schoukens in Chapter 8 of this volume.

Third, actors provide an important lens for understanding governance, which in eco-restoration commonly transpires through both state and non-state actors. The former may comprise government regulators and land managers as well as intergovernmental institutions such as the European Commission. The latter may comprise international donors, business corporations, environmental NGOs, local communities, Indigenous peoples and individual property owners, as Anastasia Telesetsky (Chapter 10) canvasses in some detail. The contributions of the non-state sector to eco-restoration can sometimes be so significant as to effectively displace state agencies. Actors can contribute in different roles, such as giving funding, setting policies, providing scientific expertise and implementing projects on the ground. Collaborations among different categories of non-state actors are also evident, such as BHP's philanthropic contributions to the 'Five Rivers Conservation Project' in Tasmania, partnering the global mining giant with the Tasmanian Land Conservancy and Conservation International.[78] The Y2Y (Yukon to Yellowstone) Conservation Initiative involves more than 300 partnering actors.

74 Peters et al (2015).
75 Mansourian (2017); Richardson (2016a); and Richardson (2016b).
76 Bache and Flinders (2004).
77 Telesetsky, Cliquet and Akhtar-Khavari (2017).
78 BHP Billiton 2014.

Fourth, eco-restoration can sometimes be fruitfully understood as 'governance regimes' or 'collaborative governance'. The former is a concept pioneered Oran Young to explain how different structures, rules and actors come together in a specific context to manage international environmental issues, but the concept can apply as well to the domestic context.[79] Theories of collaborative governance are also relevant to understanding the contributions and interactions among stakeholders in managing environmental issues, with collaboration increasingly a characteristic of environmental governance as states delegate or share authority with the community and private sector.[80] To understand eco-restoration governance regimes thus takes us beyond any focus on a single actor or legal instrument to consider the aggregation of a variety of such elements and their interactions. Collaborative and regime theory not only is a tool for analysing governance practices, but can have a normative slant as well in guiding how procedures, rules and institutions should coalesce to be most effective.[81]

## 1.5 Methodologies of the book

A variety of intellectual traditions have informed how we think about nature and its role in the Anthropocene.[82] These discussions are as old as any philosophical debates or subjects. Restoration ecology, while driven by scientific methodologies, is inherently infused with unacknowledged ideological positions about nature. This is why, as discussed above and in several chapters of this book, there are differences in approach as to whether rewilding, ecological restoration or renewal ecology is the optimal way to deal with environmental degradation. How human beings think about the natural world is likely to drive considerations by ecologists about different ways of restoring the environment.

This book is methodologically driven by, and frames its concern around, the notion of 'critique'. This critique is not just about understanding the place of humankind within nature, but more specifically with the overwhelming priority that appears to have emerged lately in relation to recovery of the natural world. While this critique can be enhanced and focused by drawing on one tradition, school of thought or idea, the literature on restoration and recovery in the law and environmental humanities is still rather undeveloped. We have drawn on the broad idea of critique to try and unpack the emerging connections between the ideas of restoration and recovery and the law.

This book is concerned with two forms of critique. The first, brought together in Part 1, exposes to critique the concept of restoration, and ecological restoration in particular. It does so by drawing on theories, ideas and doctrines to expose and evaluate the potential of the idea of restoration. For instance, Benjamin

79 Young (1989).
80 Margerum and Robinson (2016).
81 See especially Neil Gunningham's research on policy instrument combinations in the environmental context: Gunningham and Sinclair (1999).
82 See Philippopoulous-Mihalopoulos and Brooks (2017).

Richardson (Chapter 3) draws on theories and ideas about time and space to expose the potential of environmental law to address gaps in terms of how it has previously dealt with important different dimensions of the natural world. Afshin Akhtar-Khavari (Chapter 2) and also Robyn Bartel and Nicole Graham (Chapter 5) draw on new materialist approaches to expose restoration for its inability to genuinely involve human beings in the life of the natural world. Francine Rochford (Chapter 11) also picks on the lack of reflexivity in terms of governance policies for achieving restoration goals. She argues through eco-semiotics that single narratives can disrupt existing relationships between the inhabitants and the environment, which will in turn affect potential restoration ambitions for a landscape.

Emily Barritt (Chapter 4), on the other hand, argues that the eco-restoration framework on its own has limits as it does not focus directly enough on the human and nature relationship. Restoration, she argues, will achieve its cultural ambitions more effectively by embedding the scientific processes in the context of a richer humanities discourse around the notion of 'stewardship'. While these chapters focus on critiquing the concept of restoration, An Cliquet and Kris Decleer (Chapter 6) go to the heart of the disciplines of science and law, criticising their autopoetic tendencies and therefore their potential problems for managing the important agenda of recovery. However, their work also provokes them to come up with concrete ways in which science and law can inform each other to improve governance regimes for recovery.

Part 2 of this book, which is also informed by critique, aims to unpack the new modes of governance that have to be encouraged, identified and explored when thinking about and mobilising efforts to help with recovery of the natural world. The critique is focused on the problems that exist with the nation-state, which has always stood at the helm of environmental law developments. Hendrik Schoukens (Chapter 8) draws on the work being done by the EU to achieve certain restoration targets, and argues that strategic environmental litigation is needed to help achieve any kind of ambitious restoration targets. Etemire and Muzan (Chapter 9) instead argue that community participation by those in the Niger Delta is the key driver of the Ogoniland restoration initiative in Nigeria. The United Nations Environment Programme (UNEP) study of the horrendous oil contamination in this area identified a role for communities, which effectively mobilised their involvement in the restoration projects now under way. In Chapter 10, Anastasia Telesetsky identifies the significance of the role of private actors in taking restoration initiatives, and argues that the role of the state is rather limited given the political and legal status of private property and thus the power of private landholders. This part of the book also highlights and critiques the multiple scales of eco-restoration, from the local to the global realms, with implications for how close governance is, and should be, to affected human communities and the recovering ecosystems.

Several other elements of the methodologies of this book will be familiar to seasoned scholars who engage in critical environmental law research, including other contributors to this Routledge monograph series on Law, Justice and Ecology. One methodology is the interdisciplinary or multi-disciplinary enquiry used to understand eco-restoration in its intertwined scientific, cultural, political and governance contexts. To study eco-restoration law purely through the archaic lens of legal formalism would reveal little of relevance to its contemporary governance challenges. Another methodology is the comparative law perspectives that enable legal precedents and innovations around the world to be compared and contrasted, including recognising the value of non-state contributions to environmental governance as highlighted by the legal pluralism literature.[83] While most of the contributors draw on Western jurisdictions, Chapter 9 considers eco-restoration governance in Nigeria, and several canvass the role of non-state actors. And finally, some of the research for this volume draws directly or indirectly on fieldwork and empirical research, such as Chapter 5 from Bartel and Graham that includes extensive interviews with Australian farmers, and Chapter 9 from Etemire and Muzan in which a major UNEP community study informs their Nigerian case study. In sum, the analyses of eco-restoration governance in this book draw on a rich smorgasbord of research methodologies from law and other disciplines.

## 1.6 Chapter synopses

Many readers appreciate a road map that provides a succinct overview of the various chapters in a volume with multiple contributors and topics, and we conclude this chapter by catering to this demand. This book has two parts. Part 1 discusses some key foundational issues that underpin the concept of eco-restoration and its governance. Part 2 explores some specific examples or contexts of restoration, in order to illustrate some of these broader themes, and which in themselves raise insights for the development of environmental law in this field.

In Part 1, Akhtar-Khavari, Richardson, Barritt, and Bartel and Graham assess the relevance of eco-restoration for law and human environmental behaviour more broadly. In Chapter 2, Afshin Akhtar-Khavari argues that while environmental law must deepen its engagement with eco-restoration science, this alone will not help with the recovery of the natural world. Drawing on the idea of the 'natural contract', he argues that people should reorient their lives so that the social contract does not continue to remain as the only significant foundation for their governance. Plants and trees are more than matter, and his chapter argues that if we examine their existence in terms of the kind of information that they receive, process, store and transmit, human beings will understand more about their animated life and develop the capacity to 'listen' more actively to the natural world. Science, unfortunately, has generally ignored that plants and trees are intelligent and socially oriented. Once we recognise that the world is not just made up of matter that we can recreate, move, engineer or destroy, then we are compelled to reimagine how we will govern and behave accordingly. Akhtar-Khavari argues that a new 'legal imagination' has to embrace the idea that the social life of plants and trees is significant, an idea that eco-restoration activities can help realise.

83 Peters et al (2015).

In Chapter 3, Benjamin J. Richardson draws on the concepts of time and space to explain how eco-restoration aims to shift degraded ecosystems into different times and spatial arrangements. However, these 'timescapes' of eco-restoration face significant governance challenges. One spatial challenge for eco-restoration is the fragmentation of space created by property tenure and the allocation of governmental authority. Time also brings governance challenges: for instance, at what point in time can we judge that a restoration project is successful; and does a project ever have a completion date, given the dynamic quality of ecosystems? Such questions are relevant for legal accountability of those obliged to undertake restoration. The foregoing timescapes of eco-restoration become particularly complicated when we introduce the impact of *globalisation*, which is reconfiguring time and space in dramatic ways with important repercussions for environmental law. Globalisation is shifting the boundaries and frontiers of governance and changing the altered scales of action, and also accelerating and compressing time, with the consequence that a long-term and patient commitment to eco-restoration is less likely to attract support from nation-states. Richardson's chapter concludes with advice for better timescapes for eco-restoration governance, including with regard to public participation and land tenure regimes.

Chapter 4, from Emily Barritt, explores eco-restoration in its broader philosophical framework, and she argues that human relationships with the natural world matter more than meticulously replicating a degraded ecosystem. Barritt believes that restoration should focus on lost relationships and their characteristics. Stewardship will help achieve this goal because of its deeply entrenched and cross-cultural ethical values that focus human beings on collective participation in the task of caring for the natural world. Stewardship, argues Barritt, is a value that can be embraced by communities and lawmakers at multiple scales from local landowners to international institutions, and it aligns their thinking with both nature's past and future. In infusing stewardship ideals into restoration ambitions and practices, she argues that more will be done in the long run by encouraging less environmentally exploitative relationships.

While the foregoing chapters explore the potential and challenges with eco-restoration as an idea, Robyn Bartel and Nicole Graham in Chapter 5 argue that 'reconciliation' of the human and nature relationship is more likely than eco-restoration to de-centre the human. Drawing on regenerative agricultural experiences in Australia, they argue that private landholdings are not necessarily the problem, as is often suggested, and can in fact support pro-environmental ends. Bartel and Graham believe that place-based approaches that recognise non-human agency, and reconcile landholders with the dynamic qualities of nature, can help overcome problems with property tenure systems that historically have been complicit in significant environmental transformation and degradation. These experiences of reconciliation through innovative agricultural practices have depended heavily on the initiatives of particular landholders, whose attitudes and practices are meticulously documented in Chapter 5. The question of how such reconciliation practices can be disseminated more broadly remains a governance challenge yet to be resolved in Australia or other jurisdictions.

The final two chapters in this first part, from An Cliquet and Kris Decleer (Chapter 6) and Froukje Maria Platjouw (Chapter 7), also explore the conceptual significance of eco-restoration for law and governance by unpacking different dimensions of the bigger picture. Cliquet and Decleer, in Chapter 6, discuss the importance of lawyers cooperating with ecologists to enable the further development of restoration law. They explore three issues that require attention. The first is the conflict amongst restoration ecologists and scientists themselves, which makes it difficult for the legal community to act as the arbiter of what is the right or best thing to do. The second is the lack of understanding amongst lawyers of the science of restoration ecology, and they argue that while the concepts of 'extinction debt' and 'ecological trap' are significant, they are not explicitly recognised in legal instruments. Lastly, they make the point that legal expertise needs to be communicated through scientific publications to ensure that ecologists are also well informed about what is needed by the legal community.

Platjouw, in Chapter 7, enriches the conceptual debate by discussing how effective green financing arrangements can make a big difference in ensuring that restoration initiatives do not stall. While offsetting schemes (eg, biodiversity offsets, wetland banking, payments for ecosystem services) and cap-and-trade mechanisms have been around for some years, they can also be mobilised for restoration efforts. Platjouw's chapter explores and gives examples of these and other green financing opportunities from a range of jurisdictions. She argues that multi-actor regimes that combine states and non-state institutions are more likely to support the financing restoration initiatives. However, she also argues against a one-size-fits-all mechanism for securing proper financing for restoration initiatives. Green financing is also not without risks and limitations, believes Platjouw, if we fail to address the underlying economic drivers associated with the financial system that degrades and overconsumes natural resources.

Part 2, comprising five chapters, explores a range of case studies of governance of eco-restoration, from existing practices to emerging threats to be reckoned with.

In Chapter 8, Hendrik Schoukens discusses EU initiatives and argues that it has established significant governance strategies to restore 15 percent of degraded ecosystems in the region. He argues that the EU has created solid restoration responses in conservation policies and governance frameworks, but that it has not effectively extended the scope and depth of its reactions to achieve the targets it has set. For instance, restoration goals have not been extended effectively into the agriculture and fisheries sectors. While the EU can enact a new directive to deal with eco-restoration more broadly, Schoukens argues that strategic environmental litigation to enforce existing law is more likely to broaden and deepen the EU initiatives in this space.

Uzuazo Etemire and Menes Muzan, in Chapter 9, identify the significance of involving stakeholders in decision-making and the socio-economic improvements that can come for local communities from restoration of degraded ecosystems. They use the UNEP study of 2011 relating to petroleum hydrocarbon contamination in Ogoniland, Nigeria, and argue that its acceptability and adoption came from involving stakeholders as participants in developing the broader eco-restoration plans for the degraded ecosystems in that region. The community's acceptance of the report and recommended initiatives ensured that it could holistically address restoration goals. This, they argue, was far more effective than any command-and-control-regulation techniques that Nigerian authorities could have devised for the Niger Delta.

In Chapter 10, Anastasia Telesetsky picks up on this theme of broadening participation, and focuses on the most dominant actors who can engage in restoration activities, namely private actors. These include individual philanthropists, Indigenous communities, corporations, NGOs and blended groups of citizens. Her chapter focuses on what she calls the 'residual private land owners' who can restore environments but often are not as effectively mobilised for those ends. With analysis of some existing practices, mainly from Australia, New Zealand and the United States, she proposes 'special districts' as an option to create and build relationships amongst non-state actors at a landscape scale for eco-restoration. The special districts, she argues, will create opportunities for greater participation from amongst private landowners.

Francine Rochford (Chapter 11) and Phillipa McCormack (Chapter 12) approach their case studies from different angles, and instead of focusing on the nature and strength of participation of actors in restoration, they deal with the changes of thinking through the eco-restoration lens to achieve outcomes and goals. In Chapter 11, Francine Rochford analyses the governance of Australia's Murray-Darling Basin to illustrate the formidable problems with eco-restoration in an area the size of France. Drawing on eco-semiotics, she argues how the single narrative around water consumption, created for the Murray-Darling Basin governance regime, disrupted the multiple meanings and values that had once been in place between people, communities and the natural environment. The single narrative created a kind of semiotic re-regulation of the environment and disrupted relationships between inhabitants and the environment of the water basin. In effect, her chapter is a wake-up call for what might sometimes appear as straightforward and simple solutions to restoration goals and ambition.

While climate change looms as a major threat to the integrity of the Murray-Darling Basin, McCormack in Chapter 12 focuses specifically on climate adaptation as her case study. She explores how Australian legal frameworks have failed to develop climate-adapted laws to support species and ecological communities more generally. Drawing on the philosophy of 'renewal ecology', McCormack looks closely at a range of Australian laws to identify areas where reform could ensue to achieve conservation and restoration of ecosystems in the context of climate adaptation. She argues that the idea of restoring ecosystems back to their historical trajectory is not enough if Australia is going to cope with conserving biodiversity in a warmer planet. She argues for a range of more strategic governance changes to create and conserve new habitats that will cope with climate-driven redistributions of species, and to enable threatened species to relocate to such areas.

This book does not cover all aspects of ecological or environmental restoration law, nor does it need to, given the existing literature. The recently published

*Routledge Handbook of Ecological and Environmental Restoration* includes several chapters devoted to 'management and policy issues' that consider the role of community participation and market instruments, although these perspectives are written primarily by non-law scholars.[84] Our book gives only limited consideration to international law on restoration, as this subject is already covered comprehensively by another recent volume authored by three contributors featured in this book.[85] The subject of environmental restoration law, covering remediation, reclamation and other spatially and temporally narrow forms of restoration, has attracted considerable literature, much of which is framed around environmental damage liability regulation.[86] Substantial literature also provides case studies of specific projects or jurisdictions,[87] an approach also found in Part 2 of our book but linked to the conceptual and multi-jurisdictional themes canvassed in Part 1. This book also should be read in conjunction with literature that delves into the scientific, ethical and social dimensions of restoration practice, such as Dave Egan et al's volume on the *Human Dimensions of Ecological Restoration*, and Francisco Comín's *Ecological Restoration: A Global Challenge*.[88] While our book hopes to engage the reader in detailed exploration of a variety of themes that explore and critique eco-restoration and its governance, we also hope that it will contribute to this broader field of literature to facilitate further conversations around the idea of recovery of the natural world in the Anthropocene.

## References

### *Primary sources (law sources)*

Canadian Environmental Protection Act 1999, S.C. 1999, c. 33.

Conservation Act 1987 (NZ).

Convention on Biological Diversity (1992) 1760 UNTS 79.

Convention on *Wetlands* of International Importance especially as Waterfowl Habitat (1972), 11 ILM 963.

European Council Directive 92/43/EEC of 21 May 1992 on the conservation of natural habitats and of wild fauna and flora.

European Landscape Convention, European Treaty Series No. 176, 2000.

Omnibus Public Land Management Act 2009 (US), Public Law 111–111.

Water Act 2007 (Cth).

Intergovernmental Science-Policy Platform on Biodiversity and Ecosystem Services (IPBES) (2018), *Thematic Assessment of Land Degradation and Restoration* IPBES/6/INF/1/Rev.1

United Nations Convention to Combat Desertification (1994) 33 ILM 1328.

United Nations Framework Convention on Climate Change (1992) 31 ILM 849.

84 Allison and Murphy (2017).
85 Telesetsky, Cliquet and Akhtar-Khavari (2017).
86 eg, Larsson (1999); Winter et al (2008).
87 eg, Richardson (2014).
88 Egan, Hjerpe and Abrams (2011); Comín (2010).

### *Secondary sources*

Afshin Akhtar-Khavari and Benjamin J. Richardson (2017) 'Ecological Restoration and the Law: Recovering Nature's Past for the Future' 26(2) *Griffith Law Review* 1487.

Peter Alagona, John Sandlos and Yolanda Wiersma (2012) 'Past Imperfect: Using Historical Ecology and Baseline Data for Conservation and Restoration Projects in North America' 9(1) *Environmental Philosophy* 49.

Stuart K. Allison and Stephen D. Murphy (eds) (2017) *Routledge Handbook of Ecological and Environmental Restoration*, Routledge.

Jon Altman (2012) *People on Country: Vital Landscapes, Indigenous Futures*, Federation Press.

Ian Bache and Matthew Flinders (2004) *Multi-level Governance*, Oxford University Press.

Susan Baker (2014) 'Political Science and Ecological Restoration' 23(3) *Environmental Politics* 509.

Richard J. Berry (ed.) (2006) *Environmental Stewardship: Critical Perspectives – Past and Present*, T&T Clark.

BHP Billiton (2014) 'BHP Billiton Launches Five Rivers Conservation Project', www.bhpbilliton.com/investors/news/BHP-Billiton-Launches-Five-Rivers-Conservation-Project.

Jason Blackstock and Sean Low (eds) (2018) *Geoengineering our Climate? Ethics, Politics and Governance*, Routledge.

Michael Blumm and Andrew Erickson (2012) 'Dam Removal in the Pacific Northwest: Lessons for the Nation' 42(4) *Environmental Law* 1043.

Pascale Bonnefoy (2018) 'With 10 Million Acres in Patagonia a National Park System is Born', *The New York Times*, 19 February.

David Bowman et al (2017) 'Renewal Ecology: Conservation for the Anthropocene' 25(5) *Restoration Ecology* 674.

Keith Bradby, Amanda Keesing and Grant Wardell-Johnson (2016) 'Gondwana Link: Connecting People, Landscapes, and Livelihoods Across Southwestern Australia' 24(6) *Restoration Ecology* 827.

Kerryn Brent, Jeffrey McGee and Jan McDonald (2016) 'The Governance of Geoengineering: An Emerging Challenge for International and Domestic Legal Systems?' 24(1) *Journal of Law, Information and Science* 1.

Dave Butler, Tony Lindsay and James Hunt (2014) *Paradise Saved*, Random House New Zealand.

Marie-Claire Cordonier Segger and Ashfaq Khalfan (2004) *Sustainable Development Law: Principles, Practices and Prospects*, Oxford University Press.

An Cliquet and Kris Decleer (2017) 'Halting and Restoring Species Loss: Incorporating the Concepts of Extinction Debt, Ecological Trap and Dark Diversity into Conservation and Restoration Law' 26(2) *Griffith Law Review* 178.

Francisco Comín (ed.) (2010) *Ecological Restoration: A Global Challenge*, Cambridge University Press.

Storm Cunningham (2002) *The Restoration Economy: The Greatest New Frontier*, Berrett-Koehler Publishers.

T.J. Demos (2017) *Against the Anthropocene: Visual Culture and Environment Today*, Sternberg Press.

Martin Drenthen (2009) 'Ecological Restoration and Place Attachment: Emplacing Non-Places' 18 *Environmental Values* 285.

Dave Egan, Evan E. Hjerpe and Jesse Abrams (eds) (2011) *Human Dimensions of Ecological Restoration*, Island Press.

Uzuazo Etemire and Menes A. Muzan (2017) 'Governance and Regulatory Strategies Beyond the State: Stakeholder Participation and the Ecological Restoration of Ogoniland' 26(2) *Griffith Law Review* 275.

Geoffrey Garver (2017) 'Ecological Integrity in the Anthropocene: Lessons for Law from Ecological Restoration and Beyond' in L. Westra, J. Gray and F.T. Gottwald (eds) *The Role of Integrity in the Governance of the Commons*, Springer, 191.

Philip Gibbons and David Lindenmayer (2007) 'Offsets for Land Clearing: No Net Loss or the Tail Wagging the Dog?' 8(1) *Ecological Management and Restoration* 26.

Nicole Graham and Robyn Bartel (2017) 'Farmscapes: Property, Ecological Restoration and the Reconciliation of Human and Nature in Australian Agriculture' 26(2) *Griffith Law Review* 221.

Germaine Greer (2013) *White Beach: The Rainforest Years*, Bloomsbury.

Neil Gunningham and Darren Sinclair (1999), 'Regulatory Pluralism: Designing Policy Mixes for Environmental Protection' 21(1) *Law and Policy* 49.

Lauren M. Hallett et al (2013) 'Towards a Conceptual Framework for Novel Ecosystems' in R.J. Hobbs, E. Higgs and C. Hall (eds) *Novel Ecosystems: Intervening in the New Ecological World Order*, Wiley-Blackwell.

Robert Hay (2009) 'A Rooted Sense of Place in Cross-Cultural Perspective' 42(3) *Canadian Geographer* 245.

Dieter Helm (2015) *Natural Capital: Valuing the Planet*, Yale University Press.

Dolly Jorgensen (2013) 'Ecological Restoration in the Convention on Biological Diversity Targets' 22 *Ecological Restoration* 2077.

Jutta Kill (2015) *The Financialization of Nature: Creating a New Definition of Nature*, Friends of the Earth.

Marie-Louise Larsson (1999) *The Law of Environmental Damage: Liability and Reparation*, Kluwer Law.

Marty Lee and Paul Hancock (2011) 'Restoration and Stewardship Volunteerism' in D. Egan, E.E. Hjerpe and J. Abrams (eds) *Human Dimensions of Ecological Restoration*, Island Press, 23.

Aldo Leopold (1949), *A Sand County Almanac*, Oxford University Press.

Suzanne Levesque (2001) 'The Yellowstone to Yukon Conservation Initiative: Reconstructing Boundaries, Biodiversity, and Beliefs' in J. Blatter and H.M. Ingram (eds) *Reflections on Water: New Approaches to Transboundary Conflicts and Cooperation*, MIT Press.

Stephanie Mansourian (2017) 'Governance and Restoration' in S.K. Allison and S.D. Murphy (eds) *Routledge Handbook of Ecological and Environmental Restoration*, Routledge.

Richard Margerum and Cathy Robinson (2016) *The Challenges of Collaboration in Environmental Governance: Barriers and Responses*, Edward Elgar Publishing, 401.

Lindell Marsh, Douglas Porter and David Salvesen (1996) *Mitigation Banking: Theory and Practice*, Island Press.

Jason W. Moore (ed.) (2016) *Anthropocene or Capitalocene? Nature, History, and the Crisis of Capitalism*, PM Press.

Erlend Nilsen et al (2007) 'Wolf Reintroduction to Scotland: Public Attitudes and Consequences for Red Deer Management' 274(1612) *Proceedings: Biological Science* 995.

Margaret A. Palmer and J.B. Ruhl (2015) 'Aligning Restoration Science and the Law to Sustain Ecological Infrastructure for the Future' 13(9) *Frontiers in Ecology and the Environment* 512.

Anne Peters et al (eds) (2015) *Non-State Actors as Standard Setters*, Cambridge University Press.

Andreas Philippopoulous-Mihalopoulos and Victoria Brooks (2017) 'Introduction' in A. Philippopoulous-Mihalopoulos and V. Brooks (eds) *Research Methods in Environmental Law*, Edward Elgar, x.

Melissa Poe, Jamie Donatuto and Terre Satterfield (2016) '"Sense of Place": Human Wellbeing Considerations for Ecological Restoration in Puget Sound' 44(5) *Coastal Management* 409.

Jonathan Prior and Emily Brady (2017) 'Environmental Aesthetics and Rewilding' 26(1) *Environmental Values* 31.

Edward Relph (1976) *Place and Placelessness*, Sage.

Benjamin J. Richardson (2014) 'Rewilding Tasmania's Lake Pedder: Past Loss as Nature's Lex Ferenda' 33(2) *University of Tasmania Law Review* 194.

Benjamin J. Richardson (2016a) 'The Emerging Age of Ecological Restoration Law' 25(3) *Review of European, Comparative and International Environmental Law* 277.

Benjamin J. Richardson (2016b) 'Resourcing Ecological Restoration: The Legal Context for Commercial Initiatives' 24(5) *Restoration Ecology* 686.

Benjamin J. Richardson (2017) *Time and Environmental Law: Telling Nature's Time*, Cambridge University Press.

Francine Rochford (2017) 'Designing the Environment – the Paradox of Eco-restoration' 26(2) *Griffith Law Review* 202.

Michael Rosenzweig (2003) *Win-Win Ecology*, Oxford University Press.

William F. Ruddiman (2003) 'The Anthropogenic Greenhouse Era Began Thousands of Years Ago' 61 *Climatic Change* 261.

David Salvesen, Lindell L. Marsh and Douglas R. Porter (eds) (1996) *Mitigation Banking: Theory and Practice*, Island Press.

Chris Sandom et al (2013) 'Rewilding' in D. MacDonald and K. Wills (eds) *Key Topics in Conservation Biology II*, John Wiley and Sons, 430.

Society for Ecological Restoration (2004) *The SER International Primer on Ecological Restoration*, Society for Ecological Restoration International Science & Policy Working Group.

Anastasia Telesetsky (2017) 'Eco-restoration, Private Landowners and Overcoming the Status Quo Bias' 26(2) *Griffith Law Review* 248.

Anastasia Telesetsky, An Cliquet and Afshin Akhtar-Khavari (2017) *Ecological Restoration in International Environmental Law*, Routledge.

Manny Tsigas (2016) 'Farmers, Business Groups Cry Foul Over Murray Darling Basin Plan' *SBS News*, 20 February.

Yi-Fu Tuan (1977) *Space and Place*, University of Minnesota Press.

Laurie Whitt et al (2001) 'Indigenous Perspectives' in D. Jamieson (ed.) *A Companion to Environmental Philosophy*, Blackwell.

Stuart Whitten et al (2011) *A Compendium of Existing and Planned Australian Wildlife Corridor Projects and Initiatives, and Case Study Analysis of Operational Experience*, CSIRO.

Louisa Wilcox and Peter Aengst (1999) 'Yellowstone to Yukon: Romantic Dream or Realistic Vision of the Future?' 9(3) *Parks* 17.

Gerd Winter et al (2008) 'Weighing up the EC Environmental Liability Directive' 20(2) *Journal of Environmental Law* 163.

Edward O. Wilson (1984) *Biophilia*, Harvard University Press.

Oran R. Young (1989) *International Cooperation: Building Regimes for Natural Resources and the Environment*, Cornell University Press.

Caroline van Leenders and Anne-Marie Bor (2016). *Finance for One Planet 2016*, CoP Financial Institutions and Natural Capital, Netherlands Enterprise Agency.

# Part 1

# Concepts of ecological restoration law

# 2 The social life of plants and trees and the limits of environmental law's recovery imagination

*Afshin Akhtar-Khavari*

## 2.1 Introduction

In *The Natural Contract*,[1] and also *Biogea*,[2] Michel Serres inspires a critique of the social contract that has underpinned concepts like that of the nation-state, and also law and jurisprudence more generally.[3] Being wide-ranging in its scope and allegorical in style, Serres' work develops the idea that a rational order of any kind emerging from human politics has potential to do violence to objects, matter or nature because of the way that our concepts, language and power prioritise human concern.[4] He explores ways in which language, concepts, knowledge and other human artefacts capture the world and restrict how we engage with matter.[5] In particular, his work in *The Natural Contract* and also *Biogea* features the idea that human beings have to help the natural world to recover from the violence which, by not listening to it, we inflict upon it. In this sense, I do not use the word 'violence' to refer to actual harm that may be caused by cutting down a tree. It is used as a metaphor for the kind of damage done, or which eventuates from, structuring and giving order to our lives through language, or other activities, that we think are rational, scientific or appropriate. This violence comes from shutting out the 'noise' that exists everywhere, but which is ignored by us, as human beings, because we allow ourselves to have power over inert or living matter.[6] It is larger than the violence that is done by capitalism and consumption, because it is suggested that in our post-linguistic world, our inward-looking human concern makes all human beings culpable for the way that we engage with the natural world.

In *The Natural Contract*, Serres explores how the actions of human beings now have a global significance, as opposed to just being locally important.[7] In this sense, he suggests that human beings must define 'new responsibilities *towards* the

1 Serres (1995).
2 Serres (2012).
3 See Hobbes (1929 [1651]); Rousseau (2003 [1762]).
4 See Simons (2017).
5 See Serres (1995); Serres (2000); Serres (2007); Serres (2011)Serres (2012).
6 Serres (2007), 121–128.
7 Zournazi and Serres (2003), 196.

world'.[8] In *Biogea*, he introduces the even wider concept that human beings have an obligation to listen to the natural world:

> To amaze the crowd and get himself talked about, an artist wrapped bridges, buildings, statues in public squares. I see, I hear, I know the world wrapped with words, sentences, images. We put birds in cages, fish in aquariums, plants in pots, children in schools, adults in factories and offices, women under veils or in houses, God under the low crushing masses of stone in the churches of the countryside and the naves of the cathedrals, our love letters in envelopes, lastly, for settlement in full, all the things in the world in prison under words, locked up behind their bars. This so-called artist gives expression to this general wrapping.[9]

It is not that we must remove ourselves from the natural world but rather that we must consider the 'destructive and invasive tendencies' that can shape how we view the role of nature in the world. As Zournazi has noted, Serres 'asks us to consider the natural world outside of the usual human appropriation of it'.[10] In *Biogea,* Serres discusses this shift in human understanding in relation to our organic roots:

> Once again, we cannot claim to be subjects in the midst of a world of objects, for our behaviour resembles that of other insects, other rodents or poisonous plants. Not separated, but plunged, immersed in the Biogea, in cousin company. I want to think like that company, in it, by it, with it, for it.
>
> Warning! We can no longer resolve ourselves to this war of every man against every man, fatal in the final analysis to the entire Biogea and consequently to us. Peace. Would a new Eden emerge if we agree to a Natural Contract? Which? Drawn up in what language?[11]

By positioning human beings as actors involved in the natural world, Serres shifts from viewing 'matter' as inert, to something else that has energy and is capable of transmitting information.[12] This new appreciation of matter allows for the natural world to establish itself as the subject of the natural contract. This does not suggest that the natural world must be party to this contract, but rather that human beings must rethink how we 'listen to the world'.[13] In *Biogea*, Serres illustrates this assumption behind the natural contract and the importance of the concepts that underpin his theories by concluding:

8 Ibid.
9 Serres (2012) 37–38.
10 Zournazi (2012) 1.
11 Serres (2012), 170–171.
12 Connor (2009).
13 Zournazi (2012).

> And so, instead of always seeking temporary victories that can quickly be overturned into definitive defeats, instead of wanting to kill this rival microbe that, mutating as many times as necessary, will almost certainly kill ten great-grandsons of the child recently cured, I'd rather try to decipher its language: the signals that it emits, that it stores, processes and receives, since – it as well as me – we give ourselves over to these four operations. To attempt to open talks with it and negotiate together, thanks to the codes shared in this way, a mutual aid and benefit pact, so that we can pass from parasitism to symbiosis together. That's why I want to listen to the voices of the Biogea while comparing them with ours. Communication, interferences, translation, distribution, passages and bridges. How can the invasive order become a reciprocal dialogue? How can the object become subject? In what language does this mute world speak?[14]

The idea of listening to the world requires that we do more than what science has previously achieved on its own. In *The Natural Contract* and *Biogea*, Serres emphasises the significance of the soft and fluid dimensions of the work required to effectively listen to the natural world. Serres criticises science for ignoring the soft dimensions of the natural world by stating that, 'the hard does not last, only the soft lasts'.[15] We must listen and learn to focus on the soft so that we may find a better way of relating to the natural world.[16]

The above insights frame my discussions of recovery in this chapter. I use 'recovery' as an idea that can facilitate 'listening' and the 'entering into a natural contract'. I encourage a broader discussion of how to achieve recovery than how science tells us to engage with the natural world and restore it. This approach is necessary because restoration science is politically reactive. There is now sufficient evidence to suggest that natural recovery will not adequately support restoration goals for ecosystems (such as those to maintain and increase biodiversity).[17] Unfortunately, environmental law around the world is yet to properly engage with the recovery agenda, of which ecological restoration is at times pitched as the stalwart.[18]

What I endeavour to explain in this chapter is that plants and trees have a social life which human beings can do more to understand and explain. While increasing and maintaining biodiversity is essential to ecological restoration activities,[19] I argue that conserving and restoring the social life of plants and trees is even more critical within the broader recovery agenda, which scientific discourses must come to accept and deal with. Wider research suggests that the metaphysical significance of plants and trees has not been adequately recognised.[20] In this chapter I draw

14 Serres (2012), 170–171.
15 Serres (2012) 192.
16 Zournazi (2012) 5.
17 Dobson et al (2009); Hobbs and Harris (2001).
18 See Telesetsky, Cliquet and Akhtar-Khavari (2017); Akhtar-Khavari and Richardson (2017); Richardson (2017).
19 Naeem (2016).
20 Marder (2013).

from the burgeoning literature that considers plant intelligence and how plants receive, store, process and share information. By focusing on recovering the animism inherent in plants and trees, I argue that we can broaden the recovery discourse to avoid focusing on biodiversity just as matter or a commodity.

It is possible that ecologists will find my argument odd. The ecological restoration movement is often said to be historically rooted in plant community ecology, with the occasional suggestion that there is a 'botanical slant' to the discipline.[21] However, recovering plant and tree intelligence and their social life will be significant in terms of 'listening' to the natural world.[22] In section 2.2 I discuss how we must imagine and include the animism and social life of plants in the development of environmental law in the Anthropocene. In section 2.3 I explain that existing scientific approaches in restoration ecology can assist to restore this animism. However, we must restore plants and trees more actively, prior to their demise or degradation, in order to preserve their intelligent life and social organisation. As such, environmental law's recovery agenda must be broadened to include activities that engage us more actively with existing intelligent and socially organised plants and trees. Restoration ecology efforts alone are simply not enough.

## 2.2 The social life of plants and trees

### *2.2.1 Our relationships to plants*

Buhner describes plants as the 'planet's natural healers, stabilisers, and chemists'.[23] Plants are central to recovery, healing, regeneration and restoration efforts. Their role in understanding recovery is increasingly recognised, for example through their ability to capture carbon.[24] Their role in stabilising soil health and providing nutrients is also well known.[25] Yet it is rare to read narratives that consider the benefits of plants without describing their relationship to human beings. However, there are some scholarly exceptions. In his work *Plant-Thinking*, Marder puts plants at the forefront of his 'deconstruction of metaphysics', which he argues can help us avoid turning them into instruments for growth and consumption.[26] Meanwhile, in *The Forest Unseen* Haskell describes, in an intricate and authentic way, the various interactions of life over one year within a one-square-metre section of an old-growth Tennessee forest.[27]

These and other sustained research projects point to the possibility of describing the world of plants and trees through their own interactions with each other and their

21 Vander Zanden et al (2016).
22 See Richardson (2017), who argues that human culture needs to encourage more active 'listening' by thinking about time and space.
23 Buhner (2002).
24 For example, see Toensmeier (2016).
25 For example, see Stirling et al (2016).
26 Marder (2013).
27 Haskell (2013); Haskell (2017).

surroundings. The ability of plants and trees to receive, process, store and distribute information in terms that make them appear intelligent has been undervalued and ignored because of concepts which have been used to differentiate them from animals and human beings. There is a burgeoning body of scholarship exploring plant intelligence that refuses to view plants as cells or commodities, but rather as vibrant, communicating, animated and connected living beings that define and are defined by their relationships, networks and broader environmental conditions. Human language and concepts assist us to engage in conversation, but can also shut us out of seeing and experiencing how plants and trees receive, store, process and communicate information. In the discussion that follows, I describe, perhaps rather briefly, a new direction in terms of how the legal imagination may be enlivened by the animism inherent in plants and trees.

### *2.2.2 Plants' capacity to function in intelligent ways*

The capacity of trees and plants to both act intelligently and feel or perceive has been the subject of substantial recent scholarship. Cahmovitz in 2012,[28] Mancuso and Viola in 2015,[29] and also Wohlleben in 2016[30] have each authored monographs defending the capacity of plants and trees to do intelligent things. Cahmovitz even draws on the word 'knows' to connect plants with their potential to act by exercising volition. Other scholars have used the word 'sentience' to suggest that plants and trees are capable of being sentient because they have volition, exhibit altruism and appreciate living in communities in a similar way to animals.[31] The broader intellectual history and debate surrounding human and animal sentience is far more entrenched and complex than any discussion of potential plant sentience.[32] However, consideration of plant sentience began with philosophers – Democritus, Plato, Fechner and Darwin to name just a few.[33] The 1973 book *The Secret Life of Plants*, written by Tompkins and Bird, argues that plants and animals share and possess similar capacities to sense, and therefore must be as conscious as each other.[34] The authors claim that plants have the capacity to receive and react to human thought and emotion. Upon release, the book failed to achieve any recognition and instead received criticism from well-known biologists. Arthur Glaston rejected the claim that plants exhibit sentience and queried the lack of evidence for claims made by the book.[35]

I will discuss more recent research surrounding plant intelligence below. Such research considers the 'inner workings of trees and forests',[36] so as to reveal their

28 Chamovitz (2012).
29 Mancuso and Viola (2015).
30 Wohlleben (2016).
31 Andrews (2015).
32 See Clark (2000); Matson (1976).
33 Mancuso and Viola (2015), 8.
34 Tompkins and Bird (1973).
35 Galston and Slayman (1979).
36 See comments from Suzanne Simard about the observations of Wohlleben: Wohlleben (2016), 250.

ability to be social. The ability of plants to sense and respond to stimulus allows them to also engage with their surroundings. However, the depth of their interactions requires greater understanding. Plants may not have eyes, a mouth or ears, but research has suggested that plants have a wider range of capacity to perceive, receive, store and react to information than other living creatures with different sensory organs. Scholars argue that plants not only have the senses that are common to human beings, like the ability to see, hear, touch, taste and smell, but plants also have other capacities that influence their physical and social existence.[37] Mancuso and Viola consider the ability of plants to precisely determine available humidity and nutrients in soil, as well as 'sense gravity and electromagnetic fields'.[38] They discuss the ability of plants to detect harmful chemicals metres away from them, and suggest that this is far more sophisticated that what human beings can achieve. Other scholars have not compared plants to humans and the typical five human senses, but instead describe plant behaviour in terms of its general features and characteristics to avoid anthropocentric qualification.[39] Wohlleben lists and discusses over 34 different ways in which trees act socially. For example, trees use the fungal networks that exist around the tips of their roots to facilitate nutrient exchange with trees that are unhealthy.[40] In some cases, a tree may even connect its roots with another to stabilise an unhealthy tree, not necessarily of the same species.[41]

In the discussion that follows, I focus on describing the ability of plants and trees to 'see', particularly as sight is significant for social and cultural development. Human beings have three different kinds of photoreceptors on their retinas that regulate how we react to light and shadows, and also to colours such as red, blue and green. We also have an additional photoreceptor that regulates our circadian rhythms – our internal timing mechanisms as human beings.[42] How we process and store, and what we do with the information that we receive, is so critical for our survival that for many, the fear of blindness is often categorised as second only to cancer.[43]

Plants and trees also have the ability to see, but enjoy a different kind of vision from the one that we experience as humans by taking in light through our eyes and transforming it into an image or picture.[44] Plants and trees do not see in pictures as they do not have a nervous system that turns light absorbed by their leaves into images. Instead, plants and trees can react to light, such as when it is day or night time, or light coming from the left, right or above. They can perceive the ultraviolet and infrared light that burns human beings and heats the planet, respectively.[45] Trees can comprehend when another being or object is competing

37 Chamovitz (2012); Mancuso and Viola (2015).
38 Mancuso and Viola (2015) 78.
39 See Wohlleben (2016).
40 See Ibid, 2 citing Simard et al (1997).
41 See Wohlleben (2016) 2 citing Fraser, Lieffers and Landhäusser (2006).
42 Paul, Saafir and Tosini (2009).
43 Kleege (1999), Ch. 1.
44 See Chamovitz (2012), Ch. 3.
45 Ibid, 9.

for their light and will respond accordingly. Drawing on different bodies of research, Chamovitz argues that 'plant vision is much more complex than human sight at the level of perception'.[46]

Like human beings, plants and trees also have photoreceptors. However, the plant *Arabidopsis*, for instance, has been found to have 11 photoreceptors, compared with the four that humans possess. This gives the plant the ability to know 'when to germinate, … when to bend to the light, … when to flower, and … when its nighttime. Some let the plant know that there's a lot of light hitting it, … that the light is dim, and some help it keep time'.[47] Instead of turning light into an image or a picture, plants 'translate light signals into different cues for growth'.[48] It is the ability of the plant to respond and react to light signals that makes 'sight' important. It is not that the photoreceptors receive the signal through the same chemical processes as animals and human beings, but that plants and trees also respond to light signals, even though they are not developing a picture of their surroundings. However, like animals, plants and trees share a 'circadian clock', which allows plants to respond to changes in daylight and night time.[49]

The significance of 'sight' for plants and trees is that their role and contribution to an ecosystem is dependent on other plants and trees around them. Based on the available research, discussed very briefly above, plants and trees do not just grow indiscriminately, and it should not be presumed that they act mechanistically or that they are not capable of reacting and responding in a variety of ways to their specific environments and conditions. The assumption that plants and trees are intelligent does not necessarily assume consciousness, but as Mancuso and Viola note, it is about the plant being able to calculate 'risk' and 'estimate benefits'.[50] A plant that moves toward the light makes a calculated decision. Mancuso and Viola argue that 'the plant's behaviour shows that it can plan and use resources to bring about future results: in short, this is typical intelligent behaviour'.[51]

The social nature of 'sight' can go much farther than just how an individual plant or tree responds to light. Drawing on unpublished research at the Institute for Environmental Research at Rheinisch-Westfälische Technische Hochschule Aachen, Wohlleben describes how, in an undisturbed beech forest, the trees 'synchronised their performance so that they are all equally successful'.[52] The reason why this is interesting in terms of sight is that each of these trees grows in very different and unique locations, which can vary based on the soil conditions, available nutrients and water. Each tree is likely to photosynthesise and grow at different rates depending on where it is. Wohlleben argues that the variations

46 Ibid, 22.
47 Ibid.
48 Ibid, 24.
49 Ibid, 25.
50 Mancuso and Viola (2015), 48–49.
51 Ibid, 49.
52 Wohlleben (2016), 15.

between the growth conditions for each tree makes the research finding about their performance 'astounding'. He notes that,

> the rate of photosynthesis is the same for all the trees. The trees, it seems, are equalising differences between the strong and the weak. Whether they are thick or thin, all members of the same species are using light to produce the same amount of sugar per leaf. This equalisation is taking place underground through the roots.[53]

This example provides further support for the social nature of trees and the ways in which they respond to each other and their surroundings.

### *2.2.3 Death and memory*

The social life of plants and trees animates their existence and allows for an argument that they are more than just objects that we observe and use. In this section, I discuss an example of how, even through death and dying, plants and trees continue to transmit information in ways that engage other living things. Other academic examples include Sheldrake's consideration of morphic resonance.[54] In the process of dying, a tree is kept metaphorically alive by invigorating and feeding communities of living beings that engage with the dying matter. This is most evident when a tree has died and its rotting trunk becomes an ecosystem or home to a complex array of living matter. Plants and trees animate a forest in different ways when they are dead. For instance, in some forests, hardwood contributes around 10 percent of forest cover and provides forest-dwelling organisms with substrate, foraging sites and a protective dwelling for young ones to be reared.[55] Generally speaking, dead trees have been recognised as significant for the biodiversity of forests.[56] The trunk of a dead tree can take decades to decompose, allowing for larvae to develop and other insects and organisms such as fungi to use the trunk as habitat. One might wonder then whether, after its death, the social life of a tree is more organised in its relations with all kinds of other living things in the forest.

Haskell provides the following explanation for the animism inherent in a trunk and our failure to have a developed means of considering this process properly:

> Our language does a poor job of recognising this afterlife of trees. Rot, decomposition, punk, deadwood: these are slack words for so vital a process. Rot is detonation of possibility. *De*composition is renewed composition by living communities. Duff and punk are smelters for new life. Deadwood is effervescent creativity, regenerating as its 'self' degenerates into the network.[57]

53 Ibid, 15–16.
54 Sheldrake (2009).
55 Bunnell, Wind and Wells (2002).
56 Samuelsson, Gustafsson and Ingelög (1994).
57 Haskell (2017) 9

This observation is significant given that we often limit humans' potential to appreciate the social impact of the death of plants and trees on forests and maintaining biodiversity. Ecologists have argued that, given the significant diversity of habitats in terms of both their history and their organismal composition and structure, it is difficult to plan for what communities will succeed previous generations.[58] The impact and influence of the death of a tree goes beyond its value as a resource. The tree keeps networks alive and brings communities together. Haskell explains further:

> An ecological analog of grief unfolds in the forest: for the other creatures that depend on living trees, death ends the relationship that gave them life. The living tree's partners and foes must all find a new live tree or they will themselves die. Much of the understanding of the forest that dwells embedded in these relationships also passes away. The trees' particular knowledge of the nature of light, water, wind, and living communities, gained through a lifetime of interaction in one location in the forest, dissolves.[59]

Upon its death, a tree can have a significant impact in reorganising the life of the forest. For instance, in 2009, a researcher sprayed a 600-year-old, 17-feet-tall tree in the Bavarian Forest National Park with lethal spray to understand how other species inhabited its upper portions. The spray killed 2,041 animals belonging to 257 different species living within the tree.[60] The importance of research findings like this is not that the tree was housing this incredible number of creatures, but that they had survived together in an ecosystem supported by the living tree. Therefore, this tree was more than just timber but also a host – a hunting and breeding ground for all sorts of species, and a reservoir of moisture and light for smaller creatures that used its height to access food. The tree was a social host living amongst the creatures that it brought together. Accordingly, a forest must react and respond to the change created by the death of a tree.

Death also points us to a new dimension of the social life of plants. It is suggested that, as intergenerational actors, each plant or tree might have the potential to pass memories or information to the next generation of its own species through its seedlings. In 2006, scientists found that *Arabidopsis thaliana* (the cress plant) was able to activate its defences when exposed to stress.[61] The research (from Barbara Hohn's lab at the Friedrich Miescher Institute for Biomedical Research in Basel, Switzerland) found that new generations of *Arabidopsis* inherited the 'homologous recombination' that the parent genes had developed. This is a process where, due to stress, 'one piece of DNA within the genome replaces another fragment of similar sequence'.[62] The importance of this finding is that the genetic

58 Hobbs, Jentsch and Temperton (2007) 150; see also McIntosh (1999).
59 Haskell (2017), 96.
60 See Wohlleben (2016), 132 citing the work of Müller et al (2009).
61 See Molinier et al (2006).
62 Ledford (2006).

make-up of the parent variety of *Arabidopsis* had not changed, but it had managed to pass on to the new generation the behavioural characteristics that it had developed in response to the stress. The offspring of the *Arabidopsis* plant acted as if they had been subjected to the same stresses as their parents.[63] The scientists discovered that 'even if the cells used to create the progeny did not carry the genetic change, the offspring of the stressed plants genes were still more likely to undergo homologous recombinations in their own cells'.[64] Changes to behaviour that do not require mutations to the DNA are known as epigenetics.[65] To have inherited those changes from their parents without direct mutation of the DNA reflects an important finding.

Chamovitz has argued that this kind of behaviour in *Arabidopsis* is a type of procedural memory, that is, a memory of how things are done.[66] This kind of memory might be very basic, but it is part of the simple capacity that all animals and humans have for responding to the way that certain things get done. The fact that *Arabidopsis* can respond to stress by passing down a plant's reactions to the next generation is a significant finding, and important in terms of understanding the social dimensions of a plant's life and death. Intergenerational communication or memory without genetic mutation has important influences not just on the life of the plant *Arabidopsis*, but on other living things with which it comes into contact. The social life of the seedling is influenced by its parent, but in more than just the usual way that we have come to think about plants and trees. The genetic make-up of the plant responsible for creating the seedling is responsible for much of what the plant does. However, its reaction to stimuli in its surrounding environment is also shaped by behaviour, not just its own, but also that of its parent. A memory is shared between the parent plant and its seedling. Although this description is specifically in relation to the *Arabidopsis* plant, epigenetics is now a well-established field of research.[67] More recently, research has even been able to show that plant cells potentially can communicate with one another through receptors that manage calcium ions, which is vital for communication among animals.[68] This research shows the potential of behavioural traits to develop as memory for plants and to be passed onto the next generation.

Therefore, the social life of the plant can also be determined by memory that is not just passed on through DNA or epigenetic change. Death is not sufficient to kill the plant's behavioural response to stress, where stress has not led to mutations in the gene of the parent. This suggests that the complexity of plants and trees is not just as tissue, bark or biodiversity, but also as containing memories of how to deal with stresses in an ecosystem. More importantly, that these traits can be passed on through at least four generations[69] creates challenges in terms of better

63 Chamovitz (2012), 129.
64 Ledford (2006).
65 For example, see Carey (2012); Kovalchuk and Zemp (2010).
66 Chamovitz (2012) 132.
67 Carey (2012); Kovalchuk and Zemp (2010).
68 See Wudick et al (2018).
69 Ledford (2006).

understanding the relationships that emerge between, say, *Arabidopsis* and other living things that exist in relation to it. In other words, it is not just the social life of that plant that is influenced by the behavioural response passed on from its parents, but also those of others within the same ecosystem.

The discussions in this section suggest that human beings have developed a vocabulary and approaches for dealing with and thinking about plants and trees that do not recognise the soft dimensions of their lives. Our ability to enter into a natural contract with the natural world requires a recognition of nature's full potential and capacity to be animated. This observation and critique of modern science is important in the context of discussions about recovery. Once human beings have had to interact with and consider the animated and social life of plants and trees, rather than just seeing them as a contribution to biodiversity, it is likely that we will respond more actively to their injury and harm. Rather than waiting for them to die, to be cleared away or to be overwhelmed by invasive species, human beings will have to think differently and more broadly about what recovery means to plants' animated existence. In the next section I briefly discuss scientific approaches to restoration and explore their role in recovery. I then conclude this chapter with some observations on how environmental law might better embrace Serres' advice to 'listen' to plants and other elements of nature in order to promote more appropriate forms of ecological restoration.

## 2.3 Restoration science and the limits of science

### *2.3.1 Introduction to restoration science*

Restoration ecology aims to better understand processes that drive the natural world to regenerate, reorganise, respond and rebuild resilience to anthropogenic harm and the interactions of planetary systems. Overall, restoration ecology operates with the goal of assisting a system to achieve a less-disturbed environmental state.[70] Currently, a variety of scientific debates exist in relation to how we can achieve the broad goals of restoration ecology.[71] Broadly speaking, ecological restoration and approaches that aim to build novel ecosystems are the main school of scientific thought, with each placing differing values on the input of human beings in the process.[72] Their approaches differ from one another on the basis of how capable human beings are in knowing, understanding and designing for the natural world. Restoration ecology is a broad term for the scientific discipline that explains and drives the recovery and regeneration efforts of the natural world.[73]

70 See Hobbs and Suding (2008).

71 For example, the recent debate amongst scientists surrounding the correct approach to take in defining the work of restoration ecologists was publicly done through Higgs et al (2018); Gann et al (2018).

72 Palmer, Zedler and Falk (2016a); Radeloff et al (2015); also see in this volume Cliquet and Decleer (Chapter 6) and Akhtar-Khavari and Richardson (Chapter 1).

73 Palmer, Zedler and Falk (2016a).

Typically, restoration ecologists use the techniques of ecological restoration to address degraded, damaged or destroyed natural environments and ecosystems.[74] What separates ecological restoration from other restoration efforts is its belief that holistic information about the past is the best way to bring together components, processes and functions of an ecosystem to achieve recovery. A similar approach to recovery is rewilding.[75] Through rewilding, missing keystone species are reintroduced with the intention of restarting historically significant components and processes within a system.[76] Dudley's narrative of working towards 'authenticity' also has appeal in terms of engaging the past.[77] These scientific models and theories have significance for environmental law because they can inform the standards and procedures prescribed by regulators for restoring or managing degraded ecosystems.

### *2.3.2 Ecological restoration*

Ecological restoration is increasingly the most popular and significant approach to ecosystem recovery, and has a strong global following in the scientific community.[78] While intentional human activity substantially assists ecological restoration efforts, the main objective of ecological restoration is to support the recovery of historically valuable functions and structures within ecosystems. The Society for Ecological Restoration in Australia has developed a detailed set of standards and principles to guide restoration activities within Australian ecosystems.[79] Principle 1 refers to the practice of restoring an ecosystem to the local 'indigenous reference ecosystem'. This is significant, as ecologists must first identify the composition, structure and function of an ecosystem's indigenous state, before it can be restored. Therefore, ecologists must identify and analyse this target outcome, against which restoration activities will then be developed over a period of time. Although the concept of a 'reference system' is critical to restoration efforts, it is not imperative that an ecosystem is literally restored to its 'historical conditions as a target for restoration'.[80] A reference ecosystem ensures that holistic recovery efforts do not just prioritise certain dimensions of the environment. Although 'contemporary ecosystems' can be used as reference points, the practice is not recommended as there are often other stresses that must be accounted for.[81]

Furthermore, while the method of restoring an ecosystem to an indigenous reference system may seem straightforward, debate surrounds the application of

74 Ibid
75 Donlan (2005); Monbiot (2014); Fraser (2014); Pereira and Navarro (2015).
76 Fraser (2014).
77 Dudley (2011).
78 The International Society for Ecological Restoration, formed in 1988, is a community of academics and practitioners. Thousands of people attend the Society's bi-annual conferences.
79 See Society for Ecological Restoration Australasia (2017).
80 Palmer, Zedler and Falk (2016b), 16.
81 See Fletcher, Wood and Haberle (2014).

ecological restoration standards to achieve those goals. Scholars argue that restoring to an indigenous reference system is in fact impossible due to the irreversible impact of significant anthropogenic harms such as climate change and the introduction of invasive species.[82] Efforts to restore an ecosystem can take decades and be extremely complicated, depending on the nature of the degradation within that ecosystem.[83] The high costs of ecological recovery efforts may also hinder the whole process.[84] However, a properly recovered ecosystem should be self-sustaining enough to cope with a variety of ongoing stresses, and will not have to be further assisted by humans once it has been restored.[85]

A wide-ranging body of scholarship is critical of ecological restoration, arguing that fundamental changes to the natural world limit what can be done. Elliot argues that it is impossible to replace the value of an ecosystem because there is something that will always be irreplaceable in nature.[86] Katz goes further to argue that ecological restoration is guided by the 'insidious dream of the human domination of nature', which pushes us to think that we can 'recreate' nature.[87] He argues that restored ecosystems are essentially 'artefacts' and human beings cannot recreate natural environments.[88] These critics fail to recognise that ecological restoration does not aim to replicate nature. It draws on scientific techniques and methods that create self-sustaining systems that will continue to function along the historical trajectory of the ecosystem whose functions and processes have been restored. However, ecological restoration scientists agree that even fully restored systems may require continuing intervention due to the severity and frequency of future stresses on certain ecosystems, such as from global climate change.[89]

Restoration ecologists work with the understanding that self-sustaining ecosystems can be restored if the correct historically significant baselines are pre-identified for a particular landscape or region. Components, functions and processes within an ecosystem are important in the trajectory of the ecosystem, and keep scientists focused on restoring the ecosystem in line with the ecological choices that the natural world has made. While this may work in some instances, the capacity of plants and trees to act intelligently and adaptively to changes in environmental conditions suggests that ecosystems are also subject to a dimension of change that traditional scientific literature and restoration critiques have not properly accounted for. If plants and trees are animated and act socially, then simply restoring to a baseline will achieve an important goal, but not necessarily a sufficient one in terms of remediating a breach of our natural contract.

82 For example, see Rohwer and Marris (2016); and Hobbs et al (2013).
83 See Suding et al (2016).
84 Richardson (2016).
85 Society for Ecological Restoration Australasia (2004).
86 Elliot (1997), 91.
87 Katz (1992), 391.
88 Ibid, 392. Also, see Light (2000).
89 Palmer et al (2016b).

### *2.3.3 Novel ecosystems*

Other approaches to recovery in restoration ecology emphasise a different role for human intervention. Novel ecosystems,[90] renewal ecology,[91] and a host of other approaches recognise the role of human beings in designing and shaping an ecosystem or a landscape, or simply responding as needed to different kinds and levels of degradation. Restoration ecologists use the term 'novel ecosystem' to describe situations where human beings imagine, design and frame ways in which an ecosystem is to be restored in order to cope with changes in the environment, including those inflicted by climate change. The concept relies on the assumption that restoring the historically significant components, processes and functions of an ecosystem will be difficult and sometimes impossible to achieve. As such, a different combination of approaches will still restore a landscape, but not necessarily back to the historical trajectory of its components, processes and functions. Many ecologists have criticised the novel ecosystem concept and have critiqued the idea from a diverse range of views and angles.[92]

Bowman and others recently suggested the term 'renewal ecology' to describe recovery efforts employed where a degraded ecosystem or landscape cannot be restored to its historical trajectory.[93] The basis of renewal ecology is in the recognition that the 'rate, scale, and magnitude of the global environmental crisis' requires a different recovery response.[94] Renewal ecologists claim that current ecosystems 'may have no historical analog and harbor a range of non-native species, some of which may be threatened within their historic ranges'.[95] Renewal ecology focuses on managing ecosystems so as to 'maximise both biodiversity and human wellbeing in the face of rapid environmental change'.[96] While renewal ecology may appear effective in its commitment to maximising biodiversity, it does not engage a broader vision for recovery. A commitment to biodiversity alone is the same as creating seed banks in order to recognise the value inherent in the variety of species that exists on Earth. This turns biodiversity into a commodity, and design elements for degraded systems will react to this vision. Renewal ecology in its essence appears to assume, in the same way as novel ecosystems do, that human beings can design ecosystems to withstand the pressures to which they are subjected.

Other activities that aim to achieve certain specific ends are also characterised as restoration initiatives. For instance, the term 'climate restoration' is used to describe the range of activities that potentially contribute to the reduction of carbon dioxide in the atmosphere and reduce the temperature on planet Earth.[97]

90 Hobbs and Harris (2001); Hobbs et al (2013); Zedler (2012).
91 Bowman et al (2017).
92 For example, see Murcia et al (2014); Truitt et al (2015).
93 Bowman et al (2017).
94 Ibid, 674.
95 Ibid, 675.
96 Ibid, 676.
97 For example, see Lempert (2018).

Different cultures draw on restoration ecology to maximise the ecosystem services available from rehabilitated or restored landscapes.[98] In some of these alternative restoration initiatives, an ecosystem or landscape is not being restored, but rather the services provided by an ecosystem are being built up.

Novel ecosystems are pragmatic responses to degradation, and approaches can vary significantly. The scales at which restoration happens, the techniques that are used, and the degree of human intervention involved in the design of what is being done and why can all vary between projects. Whether an ecosystem or landscape that is being restored is going to be natural or not is not really a relevant issue, as restoring to a historical baseline is not necessary. Novel ecosystems can support the recovery of native plants and trees, and create a wide range of responsive and resilient ecosystems. They can also increase biodiversity within an area. As the renewal ecology discussion above suggests, novel ecosystems can help communities to deal with severe and rapid changes in environmental conditions by choosing how to respond to problems while increasing biodiversity. For instance, building coastal wetlands that can cope with sea-level rise may be a better way forward for increasing biodiversity than constructing concrete barriers to stop tidal flows. By not relying on historical baselines, plants and trees have to adapt to function intelligently or socially. Native plants and trees thrive in an area because their various senses are likely to function most acutely. Novel ecosystems simply aim to increase biodiversity or enhance the availability of ecosystem services, such as regulating services in the case of coastal wetlands described above. There is no reason why a restoration project could not focus on rebuilding the conditions necessary for certain native species to thrive and live an animated life. However, plants and trees are just as connected to their surroundings as animals that require certain food groups to survive. Developments that ignore historical trajectories can risk harming the intelligent behaviour that could be habitat-driven, and in the long run the resilience that plants and trees show to their environments.

Novel ecosystems that focus on enhancing services or increasing biodiversity can potentially limit genuine recovery because 'listening' loses its value in some of these projects. The hard and functional use of science simply to create environments with a human value or purpose ignores the importance of native functions and process in supporting information flows through plants and trees. I am not suggesting that increasing biodiversity is not important, but that this alone does not encourage the kind of 'listening' that will ultimately make us more aware of and engaged with the intelligent behaviour of plants and trees. Decreasing carbon emissions by planting trees, especially if they are native to a region, is significant and useful – but not if it comes at the cost of the impact these initiatives can have on existing intelligent behaviour that exists in a semi-natural area. For example, the company Apple recently announced that it had bought a significant mangrove in Columbia to help reforest it with plants and trees so as to increase the potential of the area to capture carbon.[99] While this could be a worthwhile restoration

98 See Allison (2012).
99 See Peters (2018).

initiative, its potential value could be more significantly realised in the long term if investments were made in local communities who could assume ownership over the rehabilitation of the area, as they are more likely to have seen and experienced the intelligence inherent in that ecosystem. Instead, the chance that science will be used to identify the best carbon-capturing plants and trees to plant in that area may create new problems and issues for the region in the long run.

## 2.4 Conclusion and directions for environmental law

This chapter has argued that traditional scientific approaches can narrow or limit the development of good environmental policies and laws. This is because science can be appropriated and used to justify activities that do not ultimately contribute to establishing a natural contract between human beings and the natural environment. Science can signify something as being important and real, when in fact other dimensions of that thing's existence are also valuable, but for a range of other reasons.[100] Biodiversity loss can be measured, but the significance of a tree is sometimes more than what it adds to the diversity of an area, for example because of the signals that its root system is sending to neighbouring trees in that ecosystem.

This argument goes further than characterising an ecosystem service, for instance as having 'cultural value' as well as regulatory and productive significance. This is because science can make some things seem real, when in fact the noises that it silences are also important. What might be real for a tree is its intelligent relationships with insects and other trees and plants in its environment. We may, however, only recognise it for its ability to capture carbon because it is only that part of the tree's reality that seemingly benefits human beings. Environmental law must, as argued by Cliquet and Decleer (Chapter 6 in this volume), deepen its relationship to the science of restoration ecology; it must also broaden what it considers as significant for the recovery agenda in the Anthropocene.

There are ways in which environmental law could facilitate a natural contract that engages human beings with the animism inherent in plants and animals. On the one hand, there are certain kinds of activities that environmental law has rarely regulated or facilitated, that could be important for supporting a broader recovery mindset. For instance, homegardens, which have always been common in Asia and generally in the humid tropics around the world, are recognised by some as 'the epitome of sustainability'.[101] A lot of different structures could potentially be classed as homegardens, but a common definition refers to them as 'intimate, multi-story combinations of various trees and crops, sometimes in association with domestic animals, around the homestead'.[102] Homegardens are both complex and common, but not in Western countries such as the United States and Australia. A lot more could be said about homegardens, but whether they can facilitate deeper

100 For similar critiques of science, see Latour (1987); Latour and Woolgar (1979).
101 Torquebiau (1992), 190.
102 Nair and Kumar (2006), 1.

contact with complex structures of plants and trees will depend on environmental law initiatives. The reason for mentioning the concept of homegardens is simply to identify a potential vehicle, like ecological restoration, which can contribute to the recovery of the animated life of plants and trees by allowing closer connections between them and people who can experience them. Homegardens have been studied scientifically,[103] but they are not commonly legislated for in countries like Australia.[104] The point of homegardens is not that governance arrangements need to be accelerated to permit them, but that they could encourage a kind of interaction with plants and trees that will support the sort of recovery discussed in this chapter.

Another relevant example of environmental husbandry that listens to nature is the land-stewardship practices of Indigenous people, some of who have actively shaped vegetation through use of fire to promote optimal landscape conditions for biodiversity to thrive.[105] In Australia, Indigenous fire-management practices are being revived in some conservation areas to promote their ecological recovery.[106] In the Amazon, researchers have found that certain tree species are unusually abundant because Indigenous tribes cultivated them for eons for their nuts, fruits and utility for building materials.[107] In some places, therefore, achieving a natural contract that respects Serres' advice requires that lawmakers accommodate the ecological wisdom of Indigenous peoples.

Finally, environmental law and governance of recovery efforts, when informed by the animism inherent in plants and trees, can avoid just responding to damage after it has happened. A human being who is ill has to recover and sometimes needs help. Legal systems that can respond more reflexively to damage and harm will have to be more aware and restrained in terms of what activities they allow, and what governance arrangements are put in place to react to harm. The ability of plants and trees to respond intelligently to their environment will mean something more than if they were viewed as a collective system that can be replicated, restored or replaced. Environmental law can encourage and facilitate reflexive engagement with ecosystems, rather than a one-off environmental impact assessment, before major projects are undertaken.

Recovery efforts have to be undertaken proactively in the Anthropocene. This chapter has argued that the more we seek to understand the intelligence of plants and trees, and their social lives, the more likely we are to understand recovery beyond what restoration science can tell us. Environmental law needs to be more deeply infused by a wider recovery ethos, ethics and imagination so that it can creatively foster a different set of relationships between human beings and the natural environment.

103 For example, Peyre et al (2006).
104 Matthews and Akhtar-Khavari (2018).
105 Richardson (2017), 252, 254.
106 Ibid, 254.
107 Levis et al (2017).

## References

Afshin Akhtar-Khavari and Benjamin J. Richardson (2017) 'Ecological Restoration and the Law: Recovering Nature's Past for the Future' 26(2) *Griffith Law Review* 147.

Stuart Allison (2012) *Ecological Restoration and Environmental Change. Renewing Damaged Ecosystems*, Routledge.

Candice Gaukel Andrews (2015) 'Research Shows Plants are Sentient. Will We Act Accordingly?', http://goodnature.nathab.com/research-shows-plants-are-sentient-will-we-act-accordingly/.

David E. Bowman et al (2017) 'Renewal Ecology: Conservation for the Anthropocene' 25 (5) *Restoration Ecology* 674.

Stephen Harrod Buhner (2002) *The Lost Language of Plants: The Ecological Importance of Plant Medicines to Life on Earth*, Chelsea Green Publishing.

Fred L. Bunnell, Elke Wind and Ralph Wells (2002) 'Dying and Dead Hardwoods: Their Implications to Management' in *USDA Forest Service General Technical Report PSW-GTR-181*, 695.

Nessa Carey (2012) *The Epigenetics Revolution*, Icon Books.

Daniel Chamovitz (2012) *What a Plant Knows: A Field Guide to the Senses*, Scientific American.

Austen Clark (2000) *A Theory of Sentience*, Oxford University Press.

Steven Connor (2009) 'Michel Serres: The Hard and the Soft', speech delivered at the Centre for Modern Studies, University of York, 26 November 2009.

Andy Dobson et al (2009) 'The Assembly, Collapse and Restoration of Food Webs' 364 *Philosophical Transactions of the Royal Society B–Biological Sciences* 1803.

Josh Donlan (2005) 'Rewilding North America' 436 *Nature* 913.

Nigel Dudley (2011) *Authenticity in Nature: Making Choices about the Naturalness of Ecosystems*, Routledge.

Robert Elliot (1997) *Faking Nature. The Ethics of Environmental Restoration*, Routledge.

Michael-Shawn Fletcher, Sam W. Wood and Simon G. Haberle (2014) 'A Fire-driven Shift from Forest to Non-forest: Evidence for Alternative Stable States?' 95(9) *Ecology* 2504.

Caroline Fraser (2014) *Rewilding the World: Dispatches from the Conservation Revolution*, Henry Holt and Company.

Erin C. Fraser, Victor J. Lieffers and Simon M. Landhäusser (2006) 'Carbohydrate Transfer Through Root Grafts to Support Shaded Trees' 26 *Tree Physiology* 1019.

Arthur W. Galston and Clifford L. Slayman (1979) 'The Not-So-Secret Life of Plants' 67 (3) *American Scientist* 337.

George D. Gann et al (2018) 'The SER Standards: A Globally Relevant and Inclusive Tool for Improving Restoration Practice – A Reply to Higgs et al' 26(3) *Restoration Ecology* 426.

David George Haskell (2013) *The Forest Unseen: A Year's Watch in Nature*, Penguin.

David George Haskell (2017) *The Songs of Trees: Stories from Nature's Great Connectors*, Black.

Eric Higgs et al (2014) 'The Changing Role of History in Restoration Ecology' 12(9) *Frontiers in Ecology and the Environment* 499.

Eric Higgs et al (2018) 'On Principles and Standards in Ecological Restoration' 26(3) *Restoration Ecology* 399.

Thomas Hobbes (1929 [1651]) *Leviathan*, Clarendon Press.

Richard J. Hobbs and J. A. Harris (2001) 'Restoration Ecology: Repairing the Earth's Ecosystems in the New Millennium' 9(2) *Restoration Ecology* 239.

Richard J. Hobbs, Anke Jentsch and Vicky M. Temperton (2007) 'Restoration as a Process of Assembly and Succession Mediated by Disturbance' in L.R. Walker, J. Walker and R. J. Hobbs (eds) *Linking Restoration and Ecological Succession*, Springer, 150.

Richard Hobbs, Eric Higgs and Carol Hall (2013) 'Defining Novel Ecosystems' in R. Hobbs, E. Higgs and C. Hall (eds) *Novel Ecosystems: Intervening in the New Ecological World Order*, Wiley-Blackwell, Ch. 6.

Robert J. Hobbs and Katharine N. Suding (eds) (2008) *New Models for Ecosystem Dynamics and Restoration*, Island Press.

Eric Katz (1992) 'The Big Lie: Human Restoration of Nature' 12 *Research in Philosophy and Technology* 231.

Georgina Kleege (1999) *Sight Unseen*, Yale University Press.

Igor Kovalchuk and Franz J. Zemp (eds) (2010) *Plant Epigenetics. Methods and Protocols*, Springer.

Bruno Latour (1987) *Science in Action*, Harvard University Press.

Bruno Latour and Steve Woolgar (1979) *Laboratory Life*, Princeton University Press.

Heidi Ledford (2006) 'Stressed-out Plants Warn their Offspring', www.nature.com/news/2006/060731/full/news060731-16.html.

Robert J. Lempert, Giacomo Marangoni and Klau Keller (2018) *Is Climate Restoration an Appropriate Climate Policy Goal?*, Rand Corporation.

C. Levis et al (2017) 'Persistent Effects of Pre-Columbian Plant Domestication on Amazonian Forest Composition' 355(6328) *Science* 925.

Andrew Light (2000) 'Ecological Restoration and the Culture of Nature: A Pragmatic Perspective' in Paul Gobster and Bruce Hull (eds) *Restoring Nature: Perspectives from the Social Sciences and Humanities*, Island Press, 49.

Stefano Mancuso and Alessandra Viola (2015) *Brilliant Green: The Surprising History and Science of Plant Intelligence*, Island Press.

Michael Marder (2013) *Plant-Thinking: A Philosophy of Vegetal Life*, Columbia University Press.

Jamie Matthews and Afshin Akhtar-Khavari (2018) 'Contributing to Nature's Recovery through Urban Agriculture' 5(2) *Journal of Vasyle Stefanyk Precarpathian National University* 45.

Wallace I. Matson (1976) *Sentience*, University of California Press.

Robert P. McIntosh (1999) 'The Succession of Succession: A Lexical Chronology' 80(4) *Bulletin of the Ecological Society of America* 256.

Jean Molinier et al (2006) 'Transgeneration Memory of Stress in Plants' 442 *Nature* 1046.

George Monbiot (2014) *Feral: Rewilding the Land, The Sea, and Human Life*, The University of Chicago Press.

Jörg Müller et al (2009) 'Wie viele Arten leben auf der ältesten Tanne des Bayerischen Waldes? [How Many Species Live on the Oldest Pine in the Bavarian Forest?]' 4 *AFZ-DerWald* 164.

Carolina Murcia et al (2014) 'A Critique of the "Novel Ecosystem" Concept' 10 *Trends in Ecology and Evolution* 548.

Shahid Naeem (2016) 'Biodiversity as a Goal and Driver of Restoration' in M.A. Palmer, J. B. Zedler and D.A. Falk (eds) *Foundations of Restoration Ecology*, Island Press, Ch. 3.

P.K.R. Nair and B.M. Kumar (2006) 'Introduction' in B.M. Kumar and P.K.R. Nair (eds) *Tropical Homegardens. A Time-Tested Example of Sustainable Agroforestry*, Springer.

Margaret A. Palmer, Joy B. Zedler and Donald A. Falk (eds) (2016a) *Foundations of Restoration Ecology*, Island Press.

Margaret A. Palmer, Joy B. Zedler and Donald A. Falk (2016b) 'Ecological Theory and Restoration Ecology' in M.A. Palmer, J.B. Zedler and D.A. Falk (eds) *Foundations of Restoration Ecology*, Island Press, Ch. 1.

Ketema N. Paul, Talib B. Saafir and Gianluca Tosini (2009) 'The Role of Retinal Photoreceptors in the Regulation of Circadian Rhythms' 10(4) *Reviews in Endocrine and Metabolic Disorders* 271.

Henrique M. Pereira and Lautitia M. Navarro (eds) (2015) *Rewilding European Landscapes*, Springer.

Adele Peters (2018) 'Apple is Investing in a Huge Mangrove Forest in Colombia', *Fast Company* 13 September, www.fastcompany.com/90236715/apple-is-investing-in-a-huge-mangrove-forest-in-colombia.

A. Peyre et al (2006) 'Dynamics of Homegarden Structure and Function in Kerala, India' 66(2) *Agroforestry Systems* 101.

Volker C. Radeloff et al (2015) 'The Rise of Novelty Ecosystems' 25(8) *Ecological Applications* 2051.

Benjamin J. Richardson (2016) 'Resourcing Ecological Restoration: The Legal Context for Commercial Initiatives' 24(5) *Restoration Ecology* 686.

Benjamin J. Richardson (2017a) *Time and Environmental Law: Telling Nature's Time*, Cambridge University Press.

Benjamin J. Richardson (2017b) 'Restoring Layered Geographies: Ecology, Society and Time' 26(2) *Griffith Law Review* 154.

Yasha Rohwer and Emma Marris (2016) 'Renaming Restoration; Conceptualizing and Justifying the Activity as a Restoration of Lost Moral Value Rather than a Return to a Previous State' 24(5) *Restoration Ecology* 674.

Jean-Jacques Rousseau (2003 [1762]) *On the Social Contract*, trans. G.D.H. Cole, Dover Publications.

John Samuelsson, Lena Gustafsson and Torleif Ingelög (1994) *Dying and Dead Trees: A Review of Their Importance for Biodiversity*, Swedish Threatened Species Unit.

Michel Serres (1995) *The Natural Contract*, University of Michigan Press.

Michel Serres (2000) *The Birth of Physics*, Clinamen Press.

Michel Serres (2007) *The Parasite*, University of Minnesota Press.

Michel Serres (2011) *Malfeasance: Appropriation Through Pollution?*, Stanford University Press.

Michel Serres (2012) *Biogea*, University of Minnesota Press.

Rupert Sheldrake (2009) *Morphic Resonance: The Nature of Formative Causation*, Park Street Press.

Suzanne W. Simard et al (1997) 'Net Transfer of Carbon Between Ectomycorrhizal Tree Species in the Field' 388 *Nature* 579.

Massimiliano Simons (2017) 'The Parliament of Things and the Anthropocene: How to Listen to "Quasi-Objects"' 21(2/3) *Techné: Research in Philosophy and Technology* 1.

Society for Ecological Restoration Australasia (2004) *The SER International Primer on Ecological Restoration*, www.ser.org/page/SERDocuments.

Society for Ecological Restoration Australasia (2017) *National Standards for the Practice of Ecological Restoration in Australia*, http://seraustralasia.com/standards/contents.html.

Graham Stirling et al (2016) *Soil Health, Soil Biology, Soilborne Diseases and Sustainable Agriculture*, CSIRO Publishing.

Katharine Suding et al (2016) 'Ecological Dynamics and Ecological Restoration' in M.A. Palmer, J.B. Zedler and D.A. Falk (eds) *Foundations of Restoration Ecology*, Island Press, 27.

Anastasia Telesetsky, An Cliquet and Afshin Akhtar-Khavari (2017) *Ecological Restoration in International Environmental Law*, Routledge.

Eric Toensmeier (2016) *The Carbon Farming Solution*, Chelsea Green Publishing.

Peter Tompkins and Christopher Bird (1973) *The Secret Life of Plants*, Harper & Row.

Emmanuel Torquebiau (1992) 'Are Tropical Agroforestry Home Gardens Sustainable?' 41 (2) *Agriculture, Ecosystems & Environment* 189.

Amy M. Truitt et al (2015) 'What is Novel about Novel Ecosystems: Managing Change in an Ever-Changing World' 55 *Environmental Management* 1217.

M. Jake Vander Zanden et al (2016) 'Food Web Theory and Ecological Restoration' in M. A. Palmer, J.B. Zedler and D.A. Falk (eds) *Foundations of Restoration Ecology*, Island Press, 301.

Peter Wohlleben (2016) *The Hidden Life of Trees: What They Feel, How They Communicate – Discoveries from a Secret World*, Black.

Michael M. Wudick et al (2018) 'CORNICHON Sorting and Regulation of GLR Channels Underlie Pollen Tube $Ca^{2+}$ Homeostasis' 360(6388) *Science* 533.

Joy B. Zedler, James M. Doherty and Nicholas A. Miller (2012) 'Shifting Restoration Policy to Address Landscape Change, Novel Ecosystems and Monitoring' 17(4) *Ecology and Society* 36.

Mary Zournazi (2012) 'Cosmocracy: A Hymn for the World? Reflections on Michel Serres and the Natural World' 9(2) *Journal of Multidisciplinary International Studies* 1.

Mary Zournazi and Michel Serres (2003) 'The Art of Living' in M. Zournazi (ed.) *Hope: New Philosophies for Change*, Routledge, 192.

# 3 Timescapes of ecological restoration

*Benjamin J. Richardson*

## 3.1 Nature's timescapes

This chapter evaluates eco-restoration governance through the intertwined lenses of time and space (aka 'timescapes')[1] because they offer novel and important insights into restoration decision-making that analyses focusing on actors, tools and other commonly evaluated dimensions do not reveal so clearly. Drawing on some examples and ideas from my recent book *Time and Environmental Law*,[2] but extending it considerably with fresh insights on the spatial context, the chapter has two aims: first, to demonstrate how time and space can be used to critique current thinking about eco-restoration, especially in light of the Anthropocene, which is re-shaping human perceptions of time and space; and second, to illuminate potential pathways for governance reform that can improve eco-restoration practices.

Ecological recovery, especially at landscape or ecosystem scales, alters the temporal and spatial dimensions of natural systems and human relationships with them. Most obviously, restoration shifts ecologies to a different time, both to their former condition (usually in part) while changing their future trajectory, and re-organises their spatial context and material elements as well, such as by re-connecting fragmented wildlife habitat or re-establishing extirpated fauna and flora. Culturally, eco-restoration can either enrich or impair the 'sense of place' communities may have with their environs, and governance regimes themselves can embody new temporal and spatial relationships. But not all restoration has these characteristics or consequences; rehabilitation of former mines or contaminated industrial zones is far less likely to alter the timescapes of restored environments than landscape 'rewilding'.

In focusing on restoration that reflects the latter approach, this chapter grapples with a variety of governance challenges through the portals of time and space. Humankind's prospective orientation and desire to master its future can make it difficult culturally to engage with the past[3] to address the deep history of anthropogenic environmental

1 The concept of 'timescapes' in the environmental context was first coined by Adam (1998).
2 Richardson (2017).
3 Niedzviecki (2015).

change. Human beings are not disinterested in their ancestry, and some Aboriginal cultures' and other local communities' identities are strongly wedded to their environmental history.[4] The 'cultural landscapes' that embody these intertwined biological and cultural histories must be reckoned with in many restoration projects if they are to be socially acceptable.[5] Furthermore, the large scale of some restoration ventures faces impediments where diverse property tenures and governmental jurisdictions that have legally fragmented space must be transcended. In other words, the disconnections between physical and legal spaces have the potential to skew restoration projects. Another governance difficulty for eco-restoration festers in the 'empty' spaces – the realms beyond the formal jurisdiction of any state, such as detailed in Richard Ellis' book the *Empty Ocean*, where international environmental law is often merely notional.[6] This chapter thus identifies and maps these and other ways in which time and space affect ecologies and their restoration through law.

In some guises, time and space should already be familiar concepts to seasoned scholars of environmental law. The former is seemingly fundamental to the philosophy of 'sustainable development' or 'sustainability' in its emphasis on foresight and long-term decision-making.[7] The sustainability discourse encompasses the associated concepts of precaution and intergenerational equity, as found in some environmental legislation, which dovetail with this prospective outlook, as well as the theory of adaptive management and resilience that is percolating into some environmental regulation to boost decision-makers' responsiveness to changing and uncertain circumstances.[8] Furthermore, time can be conceptualised as a structural or operational feature of environmental law (and legal systems in general): illustratively, time defines the duration of laws (eg, period of a pollution licence) and triggers environmental compliance requirements (eg, timing of actors' environmental reporting duties), among numerous examples.

Likewise, space is a fundamental concept that both influences and is influenced by environmental law.[9] While physics distinguishes between 'space' and 'matter', in that matter is an object with mass that occupies space, in this chapter space and matter are generally conflated as constitutive of places or realms such as forests or oceans, as well as conceptual spaces such as property tenure created by the law. Environmentalists' motto to 'Think Global, Act Local' evokes the centrality of space to their agenda in responding to the Anthropocene. But not all space is alike to policy-makers: they usually pay most attention to problems having the most overt, physical presence (eg, urban smog) rather than invisible or distant impacts (eg, marine plastic debris). Environmental regulators' jurisdiction also has a spatial quality because of the dominant territorial basis to the reach of the law. Similarly here, property tenure shapes legal entitlements to forestry, farming and other

4 Taylor, St Clair and Mitchell (2015).
5 Richardson (2016b).
6 Ellis (2003.
7 See generally Richardson and Wood (2006); Schrijver (2008).
8 Angelo (2009).
9 Space has also shaped entire fields of legal scholarship: eg, Holder and Harrison (2003).

resource harvesting that occupy specific physical realms. Carbon and biodiversity 'credits' and 'offsets' flexibly extend property mechanisms to new spatial contexts.[10] Conceptually, environmental law also functions via several spatial categories including 'commons' (eg, the high seas), and 'world heritage' (places of universal, outstanding natural or cultural heritage).[11] Space also figures prominently in some proposals to reinvigorate environmental law, such as that relating to global environmental constitutionalism and trusteeship.[12]

Promoting ecological recovery requires conceptualising time and space in environmental law in more complex guises than the foregoing overview suggests. Unlike the temporal linearity of sustainability, restoration also engages with the past by looking at the environmental history of localities targeted for recovery. Restoration also challenges the scourge of shifting baselines by which imperceptible, cumulative environmental degradation persists and amplifies with time. A restoration project responds to nature's 'counter memories', the traces of bygone times that linger in the landscape, such as once-abundant but now remnant vegetation and fauna being targeted for recovery. On space, restoration can strengthen a community's sense of place in the recovering environments by helping to reconnect people to their natural surroundings; but sometimes the opposite effect can arise where a community has become accustomed to a degraded or heavily modified landscape to which they attach cultural significance. Restoration also can fashion novel governance spaces that combine state and non-state actors collaborating in new legal geographies.

Existing laws worldwide for environmental restoration have quite narrow temporal and spatial parameters, often serving just to remediate a pollution spill or former mining site, rather than addressing holistically ecological changes at landscape scales over long timeframes. Among many examples, Canada's Environmental Protection Act 1999 illustrates the practice of many states of having environmental damage liability and clean-up provisions that target contaminated lands and emergency disasters.[13] More ambitious models of ecosystem-scale restoration include, in domestic law, the Everglades Forever Act 1994[14] and the Trinity River Basin Fish and Wildlife Restoration Act 1990,[15] both from the United States, and in international law, the European Landscape Convention[16] and the European Union's Habitats Directive.[17] Community groups and non-governmental organisations are also leading some innovative ecological recovery projects that address these larger temporal and spatial scales, as evident in the

10 Lazarus (2004), 56–60.
11 Buck (1998); Noyes (2012).
12 Kotze (2016); Bosselmann (2015).
13 Eg, s. 235 (environmental protection compliance orders).
14 Fla. Stat. § 373.4592 (1994).
15 Public Law No. 98–541.
16 European Treaty Series No. 176, 2000, article 1 ('Landscape planning', a term used in the Convention, is defined as 'strong forward-looking action to enhance, restore or create landscape').
17 Council Directive 92/43/EEC on the Conservation of Natural Habitats and of Wild Fauna and Flora, OJ L 206/7 of 22 July 1992, articles 1(a), 2–4, annex III, A(c) and B(b).

Initiative 2020 in Latin America and the Caribbean,[18] and A2A (Algonquin to Adirondacks) in North America.[19]

The next section of this chapter unpacks the nature of time and space, so that we are more familiar with key terminology. Moving from this general orientation, section 3.3 focuses on the roles of time and space in natural systems and their shifts in the Anthropocene. It enables us to understand more clearly how ecosystems function and the impact of anthropogenic changes. Providing the core of this chapter, sections 3.4 and 3.5 discuss space and time, respectively, as dimensions of eco-restoration governance. Section 3.6 concludes the chapter with advice for a better governance agenda for eco-restoration that is more sensitive to these timescapes.

## 3.2 Unpacking time and space

Although time pervades our daily lives, people struggle to define it. Apart from physicists' models of an objective time in the universe (an enquiry outside the scope of this chapter), time serves several practical functions in societies. In social life, time is a metric of efficiency, as epitomised in the aphorism 'time is money', as well as a way to sequence events from the past to the present and future, and to measure the intervals between them. The experience of time can be intensely personal, being influenced by emotions, occupation and lifestyle, among other factors,[20] which has implications for environmental governance because decision-making is very much influenced by human agency. Behavioural economists' analysis of time as a factor in human economic decisions, as explained through concepts such as 'time preferences' and 'time discounting', is one field of research that also has implications for human environmental behaviour.[21]

More relevant to understanding time in environmental law is the literature that examines how humans' experience of time is culturally and institutionally mediated. This 'social psychology' of time has been investigated by a range of scholars, such as Joseph McGrath and Janice Kelly who pioneered some early research in this field.[22] This perspective highlights how institutions such as schools, religious organisations and corporations inculcate specific temporalities into human activities and relationships. Cross-cultural differences in time perception, such as that between Indigenous and non-Indigenous peoples, are another strand of enquiry.[23] Other research has examined time through broader economic, political and legal systems. New scales of economic production through Taylorism and Fordism in the early twentieth century helped to 'accelerate' and 'commodify' time in aid of industrial capitalism.[24] The political system, and specifically electoral politics, has

18 www.wri.org/our-work/project/initiative-20x20.
19 www.a2acollaborative.org.
20 Hammond (2012); Zimbardo and Boyd (2010).
21 Frederick, Loewenstein and O'Donoghue (2002).
22 McGrath and Kelly (1986).
23 Allen (1986).
24 Kern (2003)

been correlated with the inculcation of relatively short-term time horizons in government policy-making.[25] Time also permeates the very structure of legal systems[26] through precedent and tradition that confer legitimacy for legal norms and institutions, access to legal entitlements (eg, voting rights for adults), imposition of duties (eg, age of criminal responsibility), and periods of legal accountability (eg, statute of limitations), for instance.[27] Moreover, the law shapes how society uses time, such as with regard to use of natural resources and consideration of long-term environmental impacts.[28]

Superficially, space (including matter, namely the objects occupying space) is more discernible for lay people because of its physical essence, such as the geographical distance between cities, as well as its association with three-dimensional objects or places such as a building or topographical feature. Many other understandings of space are conceivable, as Kirsten Simonsen identifies: the social spatiality that inheres in a variety of social processes; and differences in the material and social qualities of localities.[29] Another way to think about space, explain geographers Yi-Fu Tuan and Edward Relph, is the distinction between a space and a place.[30] A place differs from abstract space because of the valuable meanings people attach to places due to their history and cultural context. Places are thus realms for social relationships, from a community neighbourhood to a global online network, and have the capacity to focus and structure human activities.

Space also has controversial connotations because power shapes the construction and control of places.[31] Cities are among the most socially constructed spaces, shaped by their architecture, land-use zoning and other elements, in which diverse subcultures and communities are spatially arranged. The social construction of place includes its institutionalised 'regulatory spaces'. Environmental governance affects the social construction and physical uses of places in myriad ways, regulating their occupation, economic activities and conservation. Environmental law also seeks to 'connect' spaces, such as by bringing parcels of wildlife habitat together through protecting habitat corridors, or managing a watershed holistically through catchment plans, or connecting land-use activities through biodiversity and carbon offsets. Environmental law itself is a 'regulatory space', namely a realm in which actors and instruments from the state and non-state sectors interact to create decision-making processes and generate norms.[32] Such regulatory spaces are increasingly scaled up to the global level under the aegis of transnational social movements addressing causes such as ecological restoration or corporate social responsibility. We can also differentiate regulatory spaces by their degree of formality and

25 MacKenzie (2016), 25–28.
26 Eg, Postema (2004); French (2001).
27 Greenhouse (1988–89).
28 Adam (1998)
29 Simonsen (1996).
30 Relph (1976); Tuan (1977).
31 Harvey (1993), 5.
32 Hancher and Moran (1989); Scott (2001).

institutionalisation, with some such spaces operating quite informally and even clandestinely where lawmakers are subject to 'regulatory capture' by their clients.

Space thus has physical and social dimensions, which together influence the environmental practices of people as well as being impacted by those practices. Space provides a key lens through which to insightfully understand the relationships between different places from local contexts to larger national and global scales. Time provides a way to perceive and experience *changes* in space: the lives of individual creatures or the evolution of ecological systems, as well as shifts in human social and economic systems. Time is embedded in cognitive and cultural contexts that shape its meanings in multiple ways, including the direction of time (cyclical versus linear), disciplining and regularising social order ('clock time'), and commodification of time as a measurement of economic value. Space and time thus come together as vital axes to critically evaluate eco-restoration governance.

## 3.3 Natural timescapes

### *3.3.1 The Anthropocene and ecological upheaval*

Traces of time permeate our natural surroundings – the daily ebb and flow of the ocean tides, the seasonal fruiting and flowering of vegetation, and the recurrent migrations of peripatetic birds and animals.[33] These life-sustaining cycles follow specific temporal cues embedded in climatic, evolutionary and ecological processes, and some, such as those relating to species evolution, unfold over periods indiscernible within individual human lifespans. Environmental scientists have documented countless ways by which ecological and biological systems are in nearly constant flux,[34] whose permutations become meaningful in specific places, for time itself just measures change rather than constituting change itself. Nature's fauna, forests, geology, climate and other spatial elements are never static, but shift and fluctuate through interwoven biological and ecological processes, including human influences.

Unrepaired, cumulative environmental damage, the hallmark of the Anthropocene, is altering the spatial and temporal context of nature. The Earth's vegetation patterns have already shifted dramatically, and many forests have been felled to make way for agricultural landscapes that now cover huge expanses. These timescapes give rise to 'novel ecosystems' that replace the majority of species that perish or diminish.[35] These and other upheavals have become the staples of best-sellers such as Elizabeth Kolbert's *The Sixth Extinction*[36] and Gaia Vince's *Adventures in the Anthropocene.*[37] While the Anthropocene signifies the dominant influence of humankind, especially industrial capitalism, in shaping Earth, it also

33 Murphy and Doherty (2000).
34 Matthews (2013).
35 Low (2017).
36 Kolbert (2014).
37 Vince (2015).

paradoxically represents our increasing loss of control over the upheavals humankind has unleashed. The Anthropocene is thus equally a story about the *agency of nature* in defining new timescapes that defy regulatory control, and thus undermining the law's governance of space and time. We see the signs in oceans filling with non-degradable plastic debris; or species on the move in response to changing conditions. Space is also being homogenised by these global forces – biodiversity is diminishing while the few ubiquitous invasive species, such as European wasps and South American cane toads, zealously colonise new terrain to create increasing biological monotony and uniformity.

The 'shock of the Anthropocene', as Christophe Bonneuil and Jean-Baptiste Fressoz describe its impact,[38] demands that people re-imagine their lives in unfamiliar temporal and spatial contexts. Timothy Morton coined the term 'hyperobject' to describe some of these contexts: hyperobjects, from Styrofoam to plutonium, are 'so massively distributed in time, space and dimensionality' that we struggle to perceive and comprehend their impacts.[39] The epochal shift is clearly not a singular event in any specific location and instance, but rather unfolds across vast stretches of space and time. Temporally, we must imagine ourselves as members not just of a single human generation, but connected to the deep time of Earth history and its indefinite and uncertain future.[40] Spatially, we must see ourselves connected physically to areas once seemingly remote from our daily lives because of the interwoven matrices of culpability and vulnerability that link us to the rest of the biosphere. Re-imagining our lives in nature's changing timescapes is also a prerequisite for serious restoration; the dominant parsimonious approach that focuses on repairing 'niche' areas once conscripted for mining or industry leaves untouched an ocean of cumulative degradation in the intervening spaces.

### 3.3.2 Temporal and spatial ecology

Environmental law, including restoration governance, must be attuned to the spatial and temporal dimensions of the Earth's systems in order to improve its forecasting and thus the efficacy of policy decisions. Climatology, conservation biology, botany and related disciplines that inform restoration science rely heavily on time and space in their concepts, models and applications. Time is a primary denominator of numerous ecological properties such as population trends, predator–prey dynamics, ecological succession and the hydrologic cycle, as well as their responses to anthropogenic influences.[41] Space likewise exerts a seminal influence in helping scientists to understand biotic systems, with the field of 'spatial ecology' relying on spaces to decipher species' patterns of diversity, co-existence, conflict and distribution.[42]

38 Bonneuil and Fressoz (2017).
39 Morton (2013).
40 Glikson and Groves (2016).
41 Bissonette and Storch (2007).
42 Tilman and Kareiva (1997).

Time and space provide similar and contrasting ways to understand natural systems. A commonality is that larger spatial and temporal scales enable patterns to emerge: climate change trends, for instance, become more apparent as the vista is enlarged to average out seasonal fluctuations. Over larger spatial and temporal scales, greater perceived risks and uncertainties about changes in natural systems can arise. Conversely, the major distinguishing feature between time and space is directionality: a location in space can be influenced by other spatial locales while a moment in time can be influenced only by its antecedents.[43] Because time flows, history can shape subsequent activities, but such activities cannot shape history except in its interpretation. Another difference is that while humankind can alter space by changing a locality's physical properties, by building dykes or clearing forests, for example, we cannot alter absolute time except in a relative sense, such as by the chosen timing of pollution clean-ups or post-mining rehabilitation.

Eco-restoration must understand and predict the spatial and temporal dynamics of natural systems.[44] Models built on assumptions about static conditions such as biodiversity habitat or climatic conditions are becoming unreliable in the Anthropocene as frameworks for restoration science, as Phillipa McCormack explores in detail in her chapter in this book. Many concepts and theories in ecology assume stationarity,[45] perhaps a reasonable assumption when variables change slowly but are increasingly untenable as global warming alters both the extent and pace of change in many environmental systems. The spread of invasive species, for instance, can intensify, leading to flow-on shifts in biodiversity that tip systems to different ecological trajectories, or 'novel ecosystems'.[46] The temporal incidence of biotic and abiotic phenomena, including fire, drought, frosts and so on, are changing in a warming planet. Spatial decisions, such as the best locations to restore habitats and manage connectivity corridors, will increasingly need to be revisited in light of anthropogenic changes: global warming-induced shifts in the suitable habitat of a species may necessitate identifying new abodes outside existing conservation reserves. Climate change also affects temporal scaling, the scale to observe events over time – studying a short timeframe can result in important long-term cycles being overlooked. The consequences of using faulty temporal scales can be damaging: for instance, assumptions about rainfall for a wetland being restored can lead to planting inappropriate vegetation dependent on precipitation volumes that might be declining due to climate change. Similarly, for space, landscape ecology may need to change its relevant spatial scale for analysis in order to properly understand the ecological processes of interest, such as species' movements or emergence of novel ecosystems. The incomplete theoretical models thus exacerbate the difficulty of forecasting for many temporal and spatial dimensions of ecology that eco-restoration managers must utilise.

43 Wolkovich et al (2014), 1366.
44 This discussion draws on Wolkovich et al, op. cit.
45 Betancourt (2012), 307.
46 Hobbs, Higgs and Hall (2013).

A further challenge with any theoretical model of nature's timescapes is that our knowledge of them is always anthropomorphically framed by human cognition and culture, and thus potentially an imperfect approximation of the real world on which to base eco-restoration decisions.[47] This difficulty is not unique to eco-restoration, however, as it pervades all environmental regulation that must make assumptions about the behaviour of natural systems. Science provides our most valuable resource in interpreting nature's biorhythms.[48] Of course, science has cognitive and cultural biases;[49] consider the industry-paid scientists who once defended the safety of agrichemicals in the face of Rachel Carson's polemical *Silent Spring*,[50] and recent disputes among 'experts' over their climate change predictions. Informed public participation in environmental governance is potentially a check against abuse of scientific knowledge, as well as to help solve ethical dilemmas that transcend scientific 'expertise', such as how to ensure social justice in eco-restoration decisions.[51]

In sum, although we cannot avoid the anthropomorphic framing of time and space in environmental governance, the risk of harmful biases or gaps in understanding nature's timescapes can be mitigated through decision-making systems that provide for checks and balances and make reasoned, evidence-based choices. The imprimatur of science in environmental regulation is mediated through numerous processes, from environmental impact assessments to public inquiries. In acknowledging science's limitations, the precautionary principle behoves society to act cautiously to avoid grave environmental damage.[52] The law may also concede space for other types of erudition about nature's timescapes, such as Indigenous people's ecological wisdom. Transparent, consultative and evidence-based decision-making are our best tools for melding scientific knowledge and community participation.[53] Furthermore, technical data alone will hardly engender social change in our environmental behaviour, given the need to engage people emotionally and creatively in their attitudes and thinking as well.[54] Eco-restoration in particular depends on community participation because of its labour-intensive needs.

The next two sections of this chapter shift the focus to eco-restoration specifically, first its spatial context and then its temporal, using space and time as lenses to generate critical new insights into this field of environmental governance.

## 3.4 Space in ecological restoration

Restoration projects are inherently about space because they target interventions in specific, physical areas. The size of some landscape-scale restorations, such as in

47 On social constructions of time and space, see Massey (1994).
48 Chapman and Driver (1997).
49 Jasanoff (2004); Turner (2014).
50 Carson (1962).
51 Rozzi et al (2013).
52 Trouwborst (2002).
53 Eg, Duane (1997).
54 Buell (1995).

Australia's Gondwana Link and North America's Yukon-to-Yellowstone, stretch over thousands of square kilometres in attempting to reconnect fragmented eco-systems.[55] The relationships between space and eco-restoration law reveal several distinct issues.

First, the law draws on a scientific understanding of space as a physical system. Spatial ecology provides knowledge to guide what to restore and how; regulators may incorporate spatial ecology into decisions about threatened species recovery programs, habitat restoration management, adaptive planning for climate change impacts, and so on. But spatial ecology does not recognise the cultural dimension of space, a dimension made clear by the distinction between space and *place*. As explained earlier, geographers Yi-Fu Tuan and Edward Relph showed that a place differs from a space because of the meanings people attach to it, meanings we commonly connect with history and culture that make places socially or personally valuable, rather than mere abstract spaces we encounter.[56] Spaces are thus also realms for social relationships. This space versus place distinction matters because landscapes are embedded with cultural meanings, a relationship that can be so strong that Yi-Fu Tuan invented the word 'topophilia' – meaning the instinctive affinities between people and cultural landscapes. Thus, if eco-restoration focuses just on spatial ecology criteria it may jeopardise that sense of place, that topophilia, which is necessary for community engagement in restoration governance. This sense of place relates to the notion of 'cultural landscapes' in some environmental law. The World Heritage Convention's criteria for listing cultural landscapes, adopted in 1992, acknowledge that a 'long and intimate relationship between peoples and the natural environment' may exist, requiring legal protection.[57]

The sheer size of some restoration projects creates another spatial issue. The large areas being restored can often span numerous public and private property tenures and many human communities, with resulting risks that some affected landholders will be unwilling to cooperate. The law defines ownership of space, which influences who has control over or responsibility for eco-restoration activities. Regulations may oblige property owners to remove pests and weeds, or require authorities to obtain the consent of landowners to undertake restoration on their properties. Property tenure, which delineates the location and ambit of entitlements to land and other resources,[58] usually bears no relation to ecological patterns, resulting in restoration interventions potentially working against the grain of the landscape. On the other hand, the absence of property tenure carries risks. Garret Hardin's tragedy of the commons parable, a seminal influence in the development of environmental law, suggests that legally 'empty' space invites its reckless exploitation.[59] 'Empty' spaces, without owners, may fail to be restored because of the absence of any legally

55 See www.gondwanalink.org; Yellowstone to Yukon Conservation Initiative (2017).
56 Relph (1976); Tuan (1977).
57 World Heritage Centre, UNESCO, 'Cultural Landscapes', http://whc.unesco.org/en/culturallandscape. See Convention Concerning the Protection of World Cultural and Natural Heritage, 1972 11 ILM 1358.
58 Blomley (2014).
59 Hardin (1968).

responsible actor: eco-restoration of marine environments outside national jurisdiction is neglected for this reason. And lack of property rights can also lead to no person having standing to seek judicial remedies for protection of such spaces: courts in many jurisdictions have ruled that a mere intellectual concern about harm to nature, without any legally recognised spatial connection such as via a property right, is insufficient to have standing to seek a remedy.[60] Some jurisdictions are also notable for countering this spatial bias, such as in New South Wales, Australia, which offers open standing for any person to enforce provisions in its lodestar Environmental Planning and Assessment Act 1979.[61]

Globalisation adds a further complication to how the law defines our relationships with space. In the accounts of Anthony Giddens and David Harvey, globalisation is associated with the shrinkage of space in which events in formerly distant locations increasingly have ramifications in our local communities or environments.[62] While Giddens and Harvey are interested mainly in political and economic events that shrink space globally, such as international finance capitalism, environmentalists also talk about the contraction and warping of space in the context of the Anthropocene and global ecological upheavals such as climate change.[63] The shrinkage of space affects eco-restoration specifically – a seemingly successful restoration project in New Zealand or the Netherlands might eventually unravel because the Chinese and Americans are burning too many greenhouse gases and the resulting climate warming undermines the viability of the restored environment. The global dissemination of plastic debris through the oceans is another example one could give here.

Unlike Giddens and Harvey, who depict space in the abstract as simply a frontier or distance being overcome by globalisation, a cluster of scholarship has examined how local cultures and places affect globalisation as much as it affects them. Roland Robertson coined the term 'glocalisation' to emphasise how local cultures can affect globalisation as much as it affects them.[64] Some cities are incubators of policy innovation in addressing eco-restoration and other environmental problems; one outstanding example in New Zealand is 'Zealandia', a sanctuary of 225 hectares whose flora and fauna are being restored as closely as possible to the site's original condition 500 years ago, located near the heart of Wellington, the busy capital city of New Zealand.[65] Globalisation should not mean that eco-restoration governance is swept up into homogenous transnational agenda, as suggested by the Bonn Challenge of 2011[66] or the Aichi Targets of the Convention on Biological Diversity,[67] but rather that decision-makers should still

60 Eg, *Sierra Club v Morton* (1972) 405 US 727; *Australia Conservation Foundation v Commonwealth* (1980) 146 CLR 493.
61 Section 9.45.
62 Giddens (1990); Harvey (1990)
63 Philippopoulos-Mihalopoulos (2012).
64 Robertson (1994).
65 See www.visitzealandia.com.
66 See www.bonnchallenge.org.
67 (1992) 31 ILM 818.

be very attentive to local places and conditions. A successful eco-restoration in one local space thus might not necessarily be a role model for other places. New Zealand eco-restoration projects are dominated by pest-removal actions that target invasive species such as stoats and ferrets that decimate native wildlife, and the New Zealanders have pioneered world-class methods to control invasive species; but Dutch restoration ecologists would not regard invasive species as their principal challenge. In other words, local spaces and contexts still matter despite globalisation.

Third, the law structures the relationships between spaces, facilitating their connection or separation in ways that can help or hinder eco-restoration. Spatial connections can be made vertically and horizontally. The development of multiscale governance has grown with globalisation, with environmental decisions now increasingly made within nested Russian dolls: action on biodiversity restoration and climate change adaptation spans international treaties to local government initiatives. Horizontally, the law defines spatial connections through diverse instruments: land-use zoning has been a powerful tool to codify representations of space and to inscribe them into the physical uses of land and water.[68] Environmental regulations may also allow for 'offsets' to manage land-use impacts.[69] Offsets aim to compensate for the adverse environmental impacts of an activity that the user cannot adequately mitigate on the affected site – they have become increasingly applied in climate change mitigation and biodiversity management. Environmental offsets emerged in governance from at least the 1980s, when US President George H.W. Bush introduced a policy of 'no net loss' of wetlands, allowing for some wetlands to be cleared or impaired so long as the impacts were offset by creating or restoring other wetlands.[70] Offsets have a variety of potential limitations, including a possible disparity between the environmental spaces impaired and improved; and the delay, often considerable, between the initial loss (eg, forest clearance) and the future environmental gain on the offset restored site.[71] Through property tenure, the law can also fragment and separate ecological space to the detriment of its restoration. Likewise, the law also manipulates spatial connections through liability rules for environmental damage that determine whether a source (spatial location) of pollution is responsible for damage in another location; too often, there is no legal way of attributing responsibility for undertaking eco-restoration when the sources of the damage arose from 'distant' spaces, as with the case with marine plastic pollution or climate change.

Finally, environmental law not only governs space but also creates its own space. These 'regulatory spaces' may be formal governance regimes anchored in specific legal instruments, such as a treaty or legislation, coupled with an administering agency and associated mechanisms for engaging stakeholders; or might be

68 For an excellent case study of this phenomenon in the US, see Hirt (2014).
69 Eg, Department of Sustainability, Environment, Water, Population and Communities (2012).
70 Deland (1992).
71 Maron et al (2012).

informal regulatory spaces such as that developing in the community sector for landscape restoration.[72] Globalisation is altering these regulatory spaces, as governance is often coupled with decisions made in international forums. The weakening of nations' authority has created space for civil society and business actors to assert a greater role for themselves in governance; the emergence of polycentric governance through transnational networks is evident in a variety of environmental contexts, and specifically in eco-restoration through the work of the international Society for Ecological Restoration – a global network of restoration ecologists who have developed soft law codes and guidance to facilitate eco-restoration worldwide.[73] Rewilding Europe is another such transnational organisation, contributing to restoration governance spaces through its role in facilitating the design and implementation of projects such as restoring European native carnivore populations.[74] In other words, just as globalisation is closing the spatial distances between environmental impacts and sources, so too it is bringing environmentalists closer to forge more effective collaborations to address such problems.

## 3.5 Time and ecological restoration

Just as space informs the architecture of eco-restoration law, so does time. Three particular patterns are evident.

First, time is often invoked to set a historical baseline for restoration projects. Although restoration ecologists debate considerable difficulties in choosing a viable historical baseline, given possible lack of knowledge of past environmental conditions, the impossibility of recovering extinct species, or just the difficulty of returning a 'novel ecosystem' to its former condition, some restoration projects use very specific historical benchmarks. In Western Australia, conservation authorities are restoring Dirk Hartog Island to its condition in year 1616, the date when European settlers first visited it and brought substantial ecological changes, including the introduction of invasive species.[75] Another example is the restoration of Scotland's Carrifran Wildwood: the project led by a nongovernmental organisation is rewilding this denuded valley of 1,600 acres to its condition of 6,000 years ago.[76] The land was purchased by the conservation organisation with the aim of recreating the vegetation and ecosystem of its historical condition, although some extirpated fauna, such as wolves and bears, probably cannot be returned in the foreseeable future. In a variety of environmental remediation regulations, such as for post-mining rehabilitation and clean-up of pollution spills and contaminated land, specific time baselines are set to return the impacted sites to their condition immediately prior to the environmental degradation.

72 On the theory of 'regulatory spaces', see Hancher and Moran (1989).
73 Society for Ecological Restoration, www.ser.org.
74 Rewilding Europe, www.rewildingeurope.com.
75 Department of Parks and Wildlife, 'Dirk Hartog Island: Return to 1616', www.sharkbay.org/restoration/dirk-hartog-island-return-1616.
76 See further www.carrifran.org.uk.

More often than not, however, it is impossible or unrealistic to restore nature to specific time points, especially in the distant past. A major obstacle to some eco-restoration is the temporal inertia embedded in the legal landscape. Environmental law does not work from a blank slate, but rather functions within a layered history of existing uses and societal expectations of stability and certainty of legal rules. Many environmental users are 'grandfathered', meaning that their impacting activities are shielded indefinitely or temporarily from transitions to stricter legal standards.[77] Although grandfathering avoids the seeming injustice of coercing a person to incur costs in adjusting to a new regulation after having made investments to conform to a prior standard, it cements temporal inertia in environmental governance.[78] Grandfathering can be a problem for eco-restoration when existing uses and developments cannot be moved or closed to enable restoration to occur. Massive environmental transformation through cities, farms, economic infrastructure and so on for many centuries, nearly all legally grandfathered, creates an enormous barrier to some types of restoration aimed at reconstituting ecosystems to their pre-modified conditions. In these cases, the best strategy might be 'ecological reconciliation', a form of restoration intended for human-dominated ecosystems that aims to improve biodiversity and enrich people's engagement with nature.[79] Occasional exceptions to this pattern should be noted: one is the Australian state of Victoria's intervention in the 1980s to compulsorily acquire and close the entire town of Summerland in order to safeguard and restore a threatened population of penguins that could not co-exist with the human settlement.[80]

Temporal proximity is a second important facet of time in environmental law of relevance to eco-restoration, as it can demonstrate the salience of legally relevant relationships.[81] 'Proximity' as a judgement of lawmakers of course can be rather unlike the natural time scales of proximity, such as influences on the climate or species evolution. Environmental law often reflects the human preoccupation with the most visible and dramatic environmental damage, such as events we label 'disasters' like the Deepwater Horizon oil spill in 2010 that led to a major environmental clean-up and rehabilitation program in the Gulf of Mexico.[82] Conversely, the law can be insouciant about the slow, incremental poisoning of the environment. Rachel Carson's *Silent Spring* provoked international awareness about slow damage – unveiling the protracted and festering accumulation of dichlorodiphenyltrichloroethane (DDT) and other toxic chemicals in the food chain.[83] Carson's warnings spurred the development of modern environmental law, but Rob Nixon's recent critique of 'slow violence' suggests that gradual, cumulative environmental damage continues to gestate beyond the reach of the

77 Richardson (2017), 161–63.
78 Nash and Revesz (2007).
79 Rosenzweig (2003).
80 Pepper (2018).
81 Richardson (2017), 114.
82 Fargle and Esmail (2012).
83 Carson (1962).

law.[84] Gradual environmental decay is pervasive – the warming climate, acidifying oceans, and plastic debris filling the seas – and it may lead to shifting environmental baselines that undermine societal awareness of the need to restore past damage.

Another problem with temporal proximity arises in addressing deep historical grievances. The difficulty many Indigenous peoples suffer in proving ownership of their traditional territories based on historical use is associated with the lack of temporal proximity, as such claims may hinge on unverified evidence thousands of years old, and there may be a lack of continuity in the claimants' connections to the land because of colonial dispossession.[85] With increasing recognition of the important role of Indigenous peoples in eco-restoration in landscapes socialised by Indigenous husbandry over thousands of years, the denial of Aboriginal land and water rights has ramifications for restoring the cultural history to landscapes as well as improving environmental stewardship.[86] If we wish to overcome the destructive nature-versus-human dichotomy, there must be a place for humankind in restored landscapes, rather than aspiring to restore pristine wilderness. Some initiatives in Australia that promote Aboriginal 'caring for country', such as via Indigenous Law Use Agreements or Indigenous Protected Areas, provide innovative governance regimes that synthesise cultural and ecological issues in environmental stewardship.[87]

The third key temporal dimension of eco-restoration governance relates to its future, namely how to govern projects that may have a very long gestation and indefinite time horizon. At what point in time can we judge that a restoration project is successful? Does a project ever in fact have a completion date, given the dynamic quality of ecosystems? These questions create a governance problem, because there may be a need for legal accountability for those obliged to undertake restoration and to determine a point in time when a legal responsibility to restore is fulfilled. Such legal accountability is less of an obstacle for projects that rehabilitate former mining sites or pollution incidents where the spatial and temporal parameters are relatively discrete, but more so at a landscape or ecosystem scale. The long-term, indefinite future of many eco-restoration projects also can hinder community support and participation, as the benefits of tree planting, pest removal and other restorative actions can be so far into the future that community interest will wane. And without community participation, some restoration projects will fail. One approach to this dilemma is the Trees for Life project in Scotland, which since 1989 has been planting up to 1.5 million trees over 230,000 hectares in order to regenerate Scotland's ancient Caledonian forest. Trees for Life's success owes apparently to combining a grand, long-term vision (over at least half a century) with shorter milestones (annually or within a decade), which

84 Nixon (2011), 2.
85 See generally Richardson, Imai and McNeil (2008).
86 Mitchell and Buggey (2000).
87 Altman (2012).

helps keep the voluntary participants motivated, and helps attract funding and measure progress.[88]

## 3.6 Conclusions: an agenda for better eco-restoration law

Aligning eco-restoration governance with nature's timescapes is a multi-dimensional task. This chapter concludes with some advice for law reform, and suggests the key criteria for better eco-restoration law are: (1) biological feasibility based on temporal and spatial ecology; (2) community engagement that nurtures a sense of place in ecological space; (3) inclusion of multi-level governance regimes; and (4) removal of grandfathered uses. The approach to each of these issues has a variety of legal implications, as well as political and economic ramifications.

Nature's recovery begins with policies and laws that are aligned to the temporal and spatial characteristics of ecological and biological systems. While restoration requires decision-makers to respect the social context, the threshold consideration should be a scientific judgement about the biological feasibility of restoration.[89] Otherwise, scarce financial and social resources are wasted on irreparably damaged environments, or decisions are made to manipulate biological elements or ecological systems at odds with their temporal and spatial characteristics. A science-based approach cannot be achieved through ad hoc restoration projects that assess biological feasibility in isolation from the larger forces of environmental change in the Anthropocene. Therefore, governments should invest heavily in national and global environmental research to understand the larger temporal and spatial context of changing natural systems and ensure that such information feeds into specific restoration schemes.

The foregoing has a variety of implications for the law. Legal frameworks for restoration must be supported by clear terminology, robust standards and adequate information.[90] The law must codify scientific criteria into policy standards to which regulators are publicly held accountable. Current legal mandates and procedures for restoration frequently lack reference to such criteria, with governing agencies having broad discretionary powers to make judgements on goals and methods. While administrative discretion offers case-by-case flexibility, it can damage the integrity of legal governance. Restoration standards should also be linked to a clear statutory definition of 'restoration' and enunciation of its difference from the narrower approaches of remediation or reclamation, to ensure that executive agencies act within legally acceptable parameters. The collection and evaluation of environmental information must also be mandated by legislation. Without knowing the condition of natural systems and how they have changed historically and may do in the future, decision-makers risk not recognising any shifting baselines. Environmental monitoring that is scientifically robust and

88 Manning, Lindenmayer and Fischer (2006); Richardson (2017), 243–45.
89 Richardson and Lefroy (2016).
90 Richardson (2017), 366.

involves the community, such as through citizen science projects, is also essential for the ongoing implementation of restoration projects over many decades.

The second key criterion for a new agenda of eco-restoration law relates to public participation, specifically participation that can help communities to develop a sense of place in recovering ecologies. As noted earlier in this chapter, the 'sense of place' is a vital socio-spatial concept that must thrive in our lives if nature is to be more than just a repository of material resources and ecosystem services, bereft of spiritual or cultural significance. All the scientific wisdom about natural systems will not alone suffice to connect human beings emotionally and aesthetically to their environments. Furthermore, scientific expertise is not value-free and thus restoration or any other type of environmental decision-making cannot rest solely on scientific judgements.[91] Furthermore, communities provide a valuable resource for many environmental actions; restoration ecologists bring technical skills to design projects, but they do not supply the personnel needed for planting trees, fencing reserves, weeding and so on. With community involvement, eco-restoration can improve individuals' emotional affinity with nature and enhance their ecological literacy, as reconciliation ecology demonstrates.[92] Governance must thus improve processes for engaging with affected communities through disclosure, education, consultation and participation.

Some of the best legal models for the foregoing tasks are the regimes for managing cultural landscapes recognised by the World Heritage Convention.[93] The cultural landscape framework also may provide a way to address the related methodological and philosophical challenge, noted earlier in this chapter, about how to accommodate the human history layered in some landscapes targeted for restoration. Another legal strategy used with Aboriginal communities is to negotiate new forms of Indigenous co-management. To illustrate possible solutions, in Australia governments negotiate Indigenous Land Use Agreements that allow the Aboriginal stakeholders to negotiate resource management and restoration in their traditional territories.[94] One was negotiated in 2009 between the federal and Queensland governments and the Kuuku Ya'u people to recognise the latter's rights and interests in environmental management of nearly 2000 $km^2$ of the Great Barrier Reef.[95] These examples illustrate the intertwined and often inseparable ecological and social dimensions of restoration.

A third piece of the agenda for future eco-restoration governance relates to scale. As restoration moves from small niches to large landscapes, its scale spatially and temporally increases dramatically, with concomitant governance challenges.[96]

91 Jasanoff (2004)
92 Rosenzweig (2003).
93 World Heritage Centre, UNESCO, 'Cultural Landscapes', http://whc.unesco.org/en/culturallandscape. See Convention Concerning the Protection of World Cultural and Natural Heritage, 1972 11 ILM 1358.
94 Richardson (2017), 259.
95 Kuuku Ya'u People's Marine Park Indigenous Land Use Agreement, available at www.atns.net.au/agreement.asp?EntityID=5007
96 Cohen (2014); Gomez-Baggethun et al (2013); Hobbs et al (2014); Richardson and Lefroy (2016).

Restoring ecosystems such as an entire watershed or seascape dramatically enlarges the scale to govern beyond that normally encountered in environmental legislation. As restoration scales up, it likely requires involvement of multiple tiers of government, as well as participation by nongovernment stakeholders. In some cases this extends to global-level governance that might come under the auspices of an international treaty or supranational organisation.[97] This engenders challenges evaluated by multi-level governance theorists regarding introduction of rival policy goals, more complex negotiations, and difficulties in coordination of numerous actors.[98] The temporal scale also enlarges: because reclaimed ecosystems may need stewardship over decades or longer, they require multi-generational efforts.

To address these challenges requires a variety of strategies. First, better coordination mechanisms are needed across different property tenure and jurisdictional regimes. The introduction of bio-regional planning that is able to transcend the jurisdictional limitations of existing land-use management regimes is particularly important, as recommended by the Australian Panel of Experts on Environmental Law in its 2017 proposals for a new generation of environmental law.[99] Second, intergovernmental environmental agreements for local and national governments are necessary to facilitate exchange of information and coordination of regulatory approvals, and to achieve economies of scale in funding restoration projects. Third, larger scales should ensure that local grass-roots actors such as local governments have adequate discretion to apply overall restoration plans; landscapes often comprise a patchwork of ecosystem types in variable conditions, each requiring customised management as well as integrated approaches that link local interventions to a broader restoration plan.

Finally, an essential part of the timescapes of eco-restoration governance is to deal with existing uses. As previously explained in this chapter, environmental law does not work from a blank slate, but generally governs within an array of existing land uses and economic activities, most of which tend to be accommodated for political reasons. These, however, create considerable temporal inertia that limits the scope for responding to a changing future and undoing past mistakes in environmental decisions. Resolving the dead hand of grandfathered uses is ultimately a political and financial matter rather than a legal conundrum. The costs of removing existing legal rights through financial compensation can be high, but the political case for doing so will be stronger if policy-makers can demonstrate the long- and short-term benefits.

The initial benefits of restoration projects include direct economic gain from job opportunities and value-added activities. Data on the restoration economy suggest considerable short-term benefits alone.[100] Whatever the benefits of healthy ecosystems and more economically productive lands, someone must still pay the upfront costs. Environmental liability regimes are designed to recover the costs of environmental degradation from actors with spatial and temporal proximity to the problem,

97 Telesetsky, Cliquet and Akhtar-Khavari (2017).
98 Winter (2006).
99 Australian Panel of Experts on Environmental Law (2017), 25.
100 BenDor et al (2015).

such as a mining company obliged to rehabilitate its former mine. But the deep historical dimensions of much environmental decline defy such liability regimes as there are no longer any identifiable responsible parties, and much environmental degradation is the result of cumulative impacts from general social and economic activity rather than specific malfeasance. The solution therefore may reside in new environmental taxes imposed on existing and future uses that can create large eco-restoration funds to pay for the removal of grandfathered activities that impede ecological recovery. Encouraging more philanthropic support for eco-restoration, such as via better tax incentives for charitable investments, is also worthwhile.[101]

The foregoing points are clearly not the only issues one should consider in revitalising eco-restoration law, but they are particularly important in regard to the temporal and spatial dimensions of restoration on which this chapter focuses. Other chapters in this book canvass a variety of other issues for a new agenda of eco-restoration law. Time and space provide both a powerful way to illuminate the strengths and weaknesses of current approaches to eco-restoration law, and guidance on how to improve it. As time and space are embedded in our psychology and culture, rather than providing a fixed lens for experiencing the world, time and space are continually being redefined. Changes in institutions, including the law, are central to the dynamic of timescapes, and thus eco-restoration practices should evolve as these timescapes change.

## References

### *Primary sources (law materials)*

*Australia Conservation Foundation v Commonwealth* (1980) 146 CLR 493.
Convention Concerning the Protection of World Cultural and Natural Heritage, 1972 11 ILM 1358.
Convention on Biological Diversity (1992), 31 ILM 818.
Environment Act 1995 (England and Wales).
Everglades Forever Act, Fla. Stat. § 373. 4592 (1994).
European Landscape Convention, European Treaty Series No. 176, 2000.
European Union's Habitats Directive, Council Directive 92/43/EEC on the Conservation of Natural Habitats and of Wild Fauna and Flora, OJ L 206/7, 22 July 1992.
*Sierra Club v Morton* (1972) 405 US 727.
Trinity River Basin Fish and Wildlife Restoration Act 1990, Public Law No. 98–541

### *Secondary sources*

Barbara Adam (1998) *Timescapes of Modernity: The Environment and Invisible Hazards*, Routledge.
Jon Altman (2012) *People on Country: Vital Landscapes, Indigenous Futures*, Federation Press.

101 Richardson (2016a).

Paula Gunn Allen (1986) 'The Ceremonial Motion of Indian Time: Long Ago, So Far' in P.G. Allen (ed.) *The Sacred Hoop: Recovering the Feminine in American Indian Traditions*, Beacon Press, 147.

Mary Angelo (2009) 'Stumbling Toward Success: A Story of Adaptive Management Law and Ecological Resilience' 87 *Nebraska Law Review* 951.

Australian Panel of Experts on Environmental Law (2017) *Blueprint for the Next Generation of Australian Environmental Law*, APEEL.

Todd BenDor et al (2015) 'Estimating the Size and Impact of the Ecological Restoration Economy' 10(6) *PLoS ONE*, doi.org/10.1371/journal.pone.0128339.

Julio Betancourt (2012) 'Reflections on the Relevance of History in a Nonstationary World' in J.A. Wiens et al (eds) *Historical Environmental Variation in Conservation and Natural Resource Management*, Wiley-Blackwell.

John A. Bissonette and Ilse Storch (eds) (2007) *Temporal Dimensions of Landscape Ecology*, Springer.

Nicholas Blomley (2014) 'Property, Law and Space' 3(3) *Property Law Review* 229.

Christophe Bonneuil and Jean-Baptiste Fressoz (2017) *The Shock of the Anthropocene: The Earth, History and Us*, Verso.

Klaus Bosselmann (2015) *Earth Governance: Trusteeship of the Global Commons*, Edward Elgar Publishing.

Susan Buck (1998) *The Global Commons*, Island Press.

Lawrence Buell (1995) *The Environmental Imagination: Thoreau, Nature Writing and the Formation of American Culture*, Harvard University Press.

Rachel Carson (1962) *Silent Spring*, Houghton Mifflin.

Graham Chapman and Thackwray Driver (eds) (1997) *Time-Scales and Environmental Change*, Routledge.

Alice Cohen (2014) 'Rescaling Environmental Governance: The Case of Watersheds as Scale-sensitive Governance?' in F. Padt et al (eds) *Scale-sensitive Governance of the Environment*, Wiley-Blackwell.

Michael Deland (1992) 'No Net Loss of Wetlands: A Comprehensive Approach' 7(1) *Natural Resources and Environment* 3.

Department of Sustainability, Environment, Water, Population and Communities (2012) *EPBC Act Environmental Offsets Policy*, Commonwealth of Australia.

Timothy Duane (1997) 'Community Participation in Environmental Management' 24 *Ecology Law Quarterly* 771.

Richard Ellis (2003) *The Empty Ocean*, Island Press.

Lisa A. Fargle and Ashraf Esmail (eds) (2012) *Black Beaches and Bayous: The BP Deepwater Horizon Oil Spill Disaster*, University Press of America.

Shane Frederick, George Loewenstein and Ted O'Donoghue (2002), 'Time Discounting and Time Preference: A Critical Review' 49 *Journal of Economic Literature* 351.

Rebecca French (2001) 'Time in the Law' 72 *University of Colorado Law Review* 663.

Anthony Giddens (1990) *The Consequences of Modernity*, Cambridge University Press.

Andrew Glikson and Colin Groves (2016) *Climate, Fire and Human Evolution: The Deep Time Dimensions of the Anthropocene*, Springer International.

Erik Gomez-Baggethun et al (2013) 'Scale Misfit in Ecosystem Service Governance as a Source of Environmental Conflict' 26 *Society and Natural Resources* 1202.

Joseph McGrath and Janice Kelly (1986) *Time and Human Interaction: Toward a Social Psychology of Time*, Guilford Press.

Carol Greenhouse (1988–89) 'Just in Time: Temporality and the Cultural Legitimation of Law' 98 *Yale Law Journal* 1631.

Claudia Hammond (2012) *Time Warped*, Canongate.

Leigh Hancher and Michel Moran (1989) 'Organizing Regulatory Space' in L. Hancher and M. Moran (eds) *Capitalism, Culture and Regulation*, Clarendon Press, 271.

Garrett Hardin (1968) 'The Tragedy of the Commons' 162 *Science* 1243.

David Harvey (1990) *The Condition of Postmodernity*, Blackwell Publishing.

David Harvey (1993) 'From Space to Place and Back Again' in J. Bird et al (eds) *Mapping the Futures: Local Cultures, Global Change*, Routledge.

David Harvey (2009) *Cosmopolitanism and the Geographies of Freedom*, Columbia University Press.

Sonia A. Hirt (2014) *Zoned in the USA: Origins and Implication of American Land-use Regulation*, Cornell University Press.

Richard Hobbs, Eric Higgs and Carl Hall (eds) (2013) *Novel Ecosystems: Intervening in the New Ecological World Order*, Wiley-Blackwell.

Richard Hobbs et al (2014) 'Managing the Whole Landscape: Historical, Hybrid and Novel Ecosystems' 12 *Frontiers in Ecology and the Environment* 557.

Jane Holder and Carolyn Harrison (eds) (2003) *Law and Geography: Current Legal Issues*, Oxford University Press.

IUCN Inter-Commission Task Force on Indigenous Peoples (1997) *Indigenous Peoples and Sustainability: Cases and Actions*, IUCN.

Sheila Jasanoff (2004) *States of Knowledge: The Co-production of Science and the Social Order*, Routledge.

Stephen Kern (2003) *The Culture of Time and Space, 1880–1918*, Harvard University Press.

Elizabeth Kolbert (2014) *The Sixth Extinction*, Henry Holt and Co.

Louis Kotzé (2016) *Global Environmental Constitutionalism in the Anthropocene*, Hart Publishing.

Richard J. Lazarus (2004) *The Making of Environmental Law*, The University of Chicago Press.

Tim Low (2017) *The New Nature*, Penguin.

Michael MacKenzie (2016) 'Institutional Design and Sources of Short-termism' in I. González-Ricoy and A. Gosseries (eds) *Institutions for Future Generations*, Oxford University Press.

Adrian Manning, David Lindenmayer and Joem Fischer (2006) 'Stretch Goals and Backcasting: Approaches for Overcoming Barriers to Large-scale Ecological Restoration' 14(4) *Restoration Ecology* 487.

Doreen Massey (1994) *Space, Place and Gender*, University of Minnesota Press.

Martine Maron et al (2012) 'Faustian Bargains? Restoration Realities in the Context of Biodiversity Offset Policies' 155 *Biological Conservation* 141.

John Matthews (2013) *Encyclopedia of Environmental Change*, Sage.

Nora Mitchell and Susan Buggey (2000) 'Protected Landscapes and Cultural Landscapes: Taking Advantage of Diverse Approaches' 17 *George Wright Forum* 1.

Timothy Morton (2013) *Hyperobjects: Philosophy and Ecology After the End of the World*, University of Minnesota Press.

Pat Murphy and Paul Doherty (2000) *Traces of Time: The Beauty of Change in Nature*, Chronicle Books.

John Nash and Richard L. Revesz (2007) 'Grandfathering and Environmental Regulation: The Law and Economics of New Source Review' 101 *Northwestern University Law Review* 1677.

Hal Niedzviecki (2015) *Trees on Mars: Our Obsession with the Future*, Seven Stories Press.

Robert Nixon (2011) *Slow Violence and the Environmentalism of the Poor*, Harvard University Press.

John E. Noyes (2012) 'The Common Heritage of Mankind: Past, Present, and Future' 40 *Denver Journal of International Law and Policy* 447.

Fiona Pepper (2018), 'When Phillip Island Penguins Won, and Land Owners Lost', *ABC Radio National*, 8 April, www.abc.net.au/news/2018-04-08.

Andreas Philippopoulos-Mihalopoulos (2012) *Spatial Justice: Body, Lawscape, Atmosphere*, Routledge.

Gerald Postema (2004) 'Melody and Law's Mindfulness of Time' 17 *Ratio Juris* 203.

Edward Relph (1976) *Place and Placelessness*, Sage.

Benjamin J. Richardson and Stepan Wood (eds) (2006) *Environmental Law for Sustainability*, Hart Publishing.

Benjamin J. Richardson, Shin Imai, and Kent McNeil (eds) (2008) *Indigenous Peoples and the Law: Comparative and Critical Perspectives*, Hart Publishing.

Benjamin J. Richardson (2016a) 'Resourcing Ecological Restoration: The Legal Context for Commercial Initiatives' 24(5) *Restoration Ecology* 686.

Benjamin J. Richardson (2016b) 'Restoring Layered Geographies: Ecology, Society and Time' 26(2) *Griffith Law Review* 1.

Benjamin J. Richardson (2017) *Time and Environmental Law: Telling Nature's Time*, Cambridge University Press.

Benjamin J. Richardson and Ted Lefroy (2016) 'Restoration Dialogues: Improving the Governance of Ecological Restoration' 24(5) *Restoration Ecology* 668.

Roland Robertson (1994) 'Globalisation or Glocalisation?' 1(1) *Journal of International Communication* 33.

Michael Rosenzweig (2003) *Win–Win Ecology*, Oxford University Press.

Ricardo Rozzi et al (2013) *Linking Ecology and Ethics for a Changing World*, Springer.

Nico Schrijver (2008) *The Evolution of Sustainable Development in International Law*, Hague Academy of International Law.

Colin Scott (2001) 'Analysing Regulatory Space: Fragmented Resources and institutional Design' (summer) *Public Law* 329.

Kirsten Simonsen (1996) 'What Kind of Space in What Kind of Social Theory?' 20(4) *Progress in Human Geography* 494.

Katharine Suding et al (2015) 'Conservation: Committing to Ecological Restoration' 348 *Science* 638.

Ken Taylor, Archer St Clair and Nora J. Mitchell (2015) *Conserving Cultural Landscapes*, Routledge.

Anastasia Telesetsky, An Cliquet and Afshin Akhtar-Khavari (2017) *Ecological Restoration in International Environmental Law*, Routledge.

David Tilman and Peter Kareiva (eds) (1997) *Spatial Ecology*, Princeton University Press.

Arie Trouwborst (2002) *Evolution and Status of the Precautionary Principle in International Law*, Springer.

Stephen Turner (2014) *The Politics of Expertise*, Routledge.

Yi-Fu Tuan (1977) *Space and Place*, University of Minnesota Press.

Gaia Vince (2015) *Adventures in the Anthropocene: A Journey to the Heart of the Planet We Made*, Milkweed.

Gerd Winter (ed.) (2006) *Multilevel Governance of Global Environmental Change: Perspectives from Science, Sociology and Law*, Cambridge University Press.

Elizabeth Wolkovich et al (2014) 'Temporal Ecology in the Anthropocene' 17 *Ecology Letters* 1365.

Yellowstone to Yukon Conservation Initiative (2017) *2017 Annual Report*, Y2Y.

Philip Zimbardo and John Boyd (2010) *The Time Paradox: Using the New Psychology of Time to Your Advantage*, Rider.

# 4 The story of stewardship and ecological restoration

*Emily Barritt*

## 4.1 Introduction

Stories have tremendous power. Through stories we can challenge and reimagine assumptions about our privilege and place in the world. And in this reimagining, stories help us to cultivate empathy and compassion, thus allowing us to move closer to ideals of justice and fairness.[1] J.M. Coetzee's novel *Waiting for the Barbarians* depicts the growing consciousness of the unnamed Magistrate as he begins to recognise the part that he has played in the dominance and devastation of his landscape.[2] Whilst primarily a critique of the exercise of colonial power, through images of ruined crops, disintegrating bodily functions and trapped wolves the book also has something to say about our relationship with nature.

The allusion to the wrong of colonisation as an illustration of the imbalance that also exists in our relationship with the natural world is one that has been explored by scholars across disciplines, and Coetzee's novel is a subtle yet formidable critique of colonial assumptions. What follows is not intended as a full-scale comparison of the parallels between our colonisation of fellow human beings and our colonisation of the natural world. Rather, my intention in opening with this colonial narrative is to draw attention to the difficulty of reordering imbalances of power. Decolonisation can never hope to restore the world to its pre-invasion state, the displacement of peoples is historically entrenched, sovereign resources have been exploited beyond reparation and the collective memory of wrongdoing is indelible. The best that can be hoped for is that institutions of colonial power are dismantled, past injustices remedied, and relationships of equality between nations and people established.

A similar pattern of arrogant exploitation can be identified in the colonisers' relationship with the natural world – ecosystems are devastated, biological resources depleted and wildernesses tamed. All of which is possible because the inherent value and agency of nature has been ignored. Restoration ecology therefore has identifiable parallels with the broader project of decolonisation and, through the return of stolen lands to Indigenous communities, can also be part of the project

1 Nussbaum (1995), xvii.
2 Coetzee (1980).

of decolonisation. And whilst the task of recreating past landscapes is appealing,[3] there are many ways in which such ambition is impossible, some damage is irreversible and exploitative ways of life are severely entrenched. Pinpointing 'pristine' points in time can be artificial, and our very judgments about what in nature is authentic are inherently anthropocentric.[4]

Instead of simply prioritising the restoration of particular landscapes or ecosystems as a matter of natural fact, a central aim of the ecological restoration project should be on imbuing the human/nature relationship – across the different scales at which these relationships occur – with characteristics that will allow each to flourish. In this contribution I suggest that this necessary reconciliation can be facilitated, maintained and built upon if the human/nature relationship is rooted in the idea of stewardship values. From individual landowners to states, from small communities to the global one, stewardship can help 'restore the disrupted balance between humanity and nature'.[5] Thus, rather than characterising the restoration project as a meticulous recreation of historical landscapes, restoration can have both a backward-looking dimension, but also a forward-looking one in order to maintain restored landscapes.[6] In short, restoration is as much about achieving a *restored relationship* to our environment, as it is about the state of the environment itself.

As claimed in my opening, the power of storytelling is in its ability to prick the imagination, to open up avenues of critical thought and to leave the reader changed. The story I wish to tell is the story of stewardship and its role in reframing and, ultimately, restoring our relationship with the natural environment. Stewardship is a deeply value-laden concept that has the potential to reshape how we understand our appropriate place in the world – to *reconcile* human beings to their natural environment.[7] There is emotive, even spiritual, power in the concept of stewardship that has the potential to deeply reorder the way that human beings treat and interact with their natural environments.

Equally, stewardship, as a concept, itself is not immune from imagery of dominance and dominion – this is a historical reality about which it is necessary to be frank if stewardship is to have any role in reshaping human/nature relationships. However, stewardship in its purest sense embodies ideas about an ethics of care or a land ethic, that can imbue our understanding of our proper place within the world order. Its persuasive power is compelling and I suggest that the story of stewardship, properly articulated, has the potential to reshape human/nature relationships. Stewardship moves the focus of environmental law and eco-restoration away from technical targets, which can alienate the non-specialist, to a shared idea with cross-cultural significance and a deep normative core. Stewardship is therefore the ethical companion to the scientific practice of restoration.

3 Jordan (2012), 13.
4 Graham and Bartel (2017), 222.
5 Telesetsky (2013), 175.
6 Hayes (2002).
7 Graham and Bartel (2017), 222.

Introducing this normative element could risk reopening political debates about how to protect nature, which the faithful replication of ecosystems can seemingly displace. However, even within the technical movement, there is disagreement about the contours of restoration, for example whether it is creative or non-creative, whether it can occur on miniature scales as well as grand ones, and whether it is a luxury or an essential.[8] These questions cannot be resolved without reference to ethical values and neither can they be resolved by scientists alone. Stewardship helps to reveal these hidden political questions about restoration as well to resolve them. Through its explicitly ethical complexion and through its communitarian emphasis on participation, stewardship can guide restoration choices and focus restoration towards justice.

As an ethical value, stewardship supports connective justice in particular.[9] Connective justice pushes the idea of eco-restoration beyond the particular act of restoring landscapes, to a broader idea that restoration should also be about a restored relationship with our environs. The act of restoring is a recognition of the fact that something has gone wrong in the world order, that there is a need for reconciliation between human beings and the natural world.[10] William R. Jordan describes it as an act of obedience and a 'stilling of the will' that helps to repair our relationships with the natural world.[11] Implicit in this relationship reordering is a recognition of the inherent value of nature, that is not merely instrumental, but that it has agency which must be taken seriously in our choices about how we manage and interact with the natural word.

By reframing eco-restoration as a relationship-focused activity, I hope to address the concerns of scholars such as Robert Elliot, who are critical of some forms of environmental restoration because, as often conceived, it amounts to an artificial meddling that fails to properly cherish nature, or recognise its agency, and can never hope to replace the lost ecological value caused by human destruction.[12] If we focus on reconciling human/nature relationships, then it will also be necessary to determine 'the intrinsic moral value exemplified by nature'.[13] Placing the eco-restoration project on a foundation of stewardship stands to both properly recognise the moral worth of nature and embed the enterprise with a necessary normative core of connective justice.

My contribution to this book is not intended as rejection of the restoration project as currently conceived, rather a reframing of its goals. In what follows, I elaborate my vision of eco-restoration as relationship restoration and show what a restored, stewardship-rooted relationship should (and should not) look like, closing with some suggestions as to how environmental law at different scales can embody this sense of stewardship more generally. The chapter does not focus on any specific jurisdiction but rather draws illustrations from around the world.

8 Jordan (2012), 25.
9 Telesetsky, Cliquet and Akhtar-Khavari (2017), 288.
10 Graham and Bartel (2017), 222.
11 Jordan (2012), 24.
12 Elliot (1997), 75.
13 Ibid, 121.

## 4.2 Reframing restoration – restored relationships

Ecological restoration emerged as a counterpoint to the future focus of the environmental movement on sustainable development[14] and its limited vision of conservation and preservation,[15] which was an inadequate response to environmental degradation.[16] Ecologists began to recognise that there was also a need to undo the damage of our past interactions with the natural world – to restore. The early restoration movement was in the truest sense grassroots as it involved interested individuals engaged in private restoration projects: replanting prairie grasses, rehydrating marshlands and returning lost species to their original habitats.[17] There was an organic quality to the movement, which reflected the ethos of its early practitioners. However, ecological restoration as the pet project of individual, ecologically astute landowners was fragmented and geographically limited.[18] In recent years, this scientists' movement has caught the attention of lawyers and lawmakers, who have offered their 'intelligent tinkering' to promote, motivate and strengthen restoration efforts globally.[19] Restoration practices have become more widespread, coalescing around the restoration of former industrial sites and mines, and reversing the damage of environmental disasters such as oil spills. And through this powerful coalition of lawyers and scientists, restoration concepts have been embedded in environmental governance regimes, both international and national.[20]

One challenge for the conversion of narrow, environmental restoration from a primarily scientific practice to landscape-scale restoration that is informed by intelligible legal frameworks is the issue of definition. Lawyers are eminently concerned with definition because if law is to play a useful role in driving action, definitions are needed to set standards, develop procedures, guide behaviour, assess success and challenge failures.[21] The difficulty with technical definitions in the context of restoration is that their rigidity does not necessarily allow for the plurality of approaches that might be necessary for particular landscapes. For example, in one instance a restored landscape might be one that perfectly matches a historical snapshot, whilst in another it might be one that recognises irreparable damage but finds alternative species to repair ecological integrity. There is an artistry to restoration[22] that defies legalistic definition.[23]

14 Richardson (2017), 159; Richardson (2016); Duncan (2003).
15 Leopold (1949), viii.
16 Telesetsky (2017), 7.
17 Court (2012), 1–17.
18 Although, as Telesetsky points out, even with the intervention of lawyers and policy makers there remain issues of fragmentation in terms of how ecological restoration projects are administered Telesetsky (2013), 176.
19 Telesetsky, Cliquet and Akhtar-Khavari (2017), 6.
20 Ibid, 59–195.
21 Ibid, 6.
22 Turner (1987), 48.
23 For a more detailed discussion on the complex interaction of legal definition and artistic standards see Richardson, Barritt and Bowman (2019); and for a more general discussion on the difficulty of defining art see Perry (2014).

Restoration as a technical practice, with rules of conduct and prescriptive targets, therefore risks becoming an artificial exercise in replicating past landscapes that fails to adequately respect its inherent value.[24] As Elliot laments, 'in intervening we are dominating nature, holding nature captive, taking nature prisoner, incarcerating nature or some such, and probably for our own end'.[25] Compounding the potential arrogance of some forms of restoration is the risk that it is vulnerable to being used as a tool for the commodification of ecological resources as a way to offset environmental damage caused elsewhere, or to bank biological integrity to facilitate future damaging activities.[26] It is conceivable that enterprising ecologists could restore and then sell off parcels of land as a means to offset damage by corporate actors elsewhere in the world. This practice of 'offsetting' would be particularly problematic if areas in the Global North were restored and sold as a replacement for the destruction of ecosystems in the Global South, as it would exacerbate existing ecological imbalances and environmental justice challenges. A technical approach to restoration therefore risks a rupture between the letter of restoration and its spirit.

Another problem with the technical focus of restoration is that it can be exclusionary.[27] Restoration is a privileged practice, a form of 'boutique conservationism', available only to those who have some pre-existing (often legal and proprietorial) relationship to the natural environment, or who have a sufficient degree of ecological training for their contribution to meet the high technical bar necessary for restoration.[28] This is made apparent through Anastasia Telesetsky's typology of restoration actors: those who have engaged in large-scale restoration projects, such as the restoration of US/Mexico borderlands or the mangroves of Rookery Bay in Florida, have generally been those with the financial capital ('eco barons' and corporate entities) or the technical know-how (Aldo Leopold being chief amongst the expert restorationists) needed to embark on restoration projects.[29] The exclusivity of restoration is especially apparent in leading restorationist William R. Jordan's intentionally restrictive approach to restoration, which seeks to distinguish restoration from other forms of land management such as reconciliation, rehabilitation and reclamation.[30] Restoration in this technical sense excludes and thus disenfranchises the city dweller or the suburbanite, and although movements such as reconciliation ecology have emerged to include city dwellers in restoration-type activities, restoration in its purest sense remains a practice from which most people are alienated.[31] As Eric Freyfogle acknowledges: 'preserving the wilderness does little to make our cities more liveable; it doesn't bring our life practices into natural balance'.[32] I can cram my little green square of

24 Elliot (1997), 5–11.
25 Ibid, 135.
26 Holligan (2018), 60–62.
27 Holligan (2018), 66.
28 Jordan (2012), 17.
29 Telesetsky (2017), 251–259.
30 Jordan (2012), 22.
31 For more on reconciliation ecology see: Rosenzweig (2003), in particular Ch. 1.
32 Freyfogle (1993), 14.

London with flowers that attract bees and fill my birdfeeders to keep the resident robins and goldfinches content, but my landlady would no doubt be very distressed if I attempted to return my garden to a manifestation of ancient Battersea gravel lands.[33]

A less prescriptive way to characterise ecological restoration would therefore be in terms of relationship restoration. That is, rather than the focus of ecological restoration being the physical return of natural environments to points in time that are deemed to be pristine or ecologically worthy, the focus of restoration should be the repair of the relationship between the human community and the rest of the natural world. In this way, a restoration ethos can imbue all our interactions with our environment, so that an office worker in Bangkok can be as much engaged with the concept of restoration as a wealthy landowner in Wyoming – for example, by engaging city dwellers in restoration projects that are geographically remote, or establishing pockets of nature in urban centres. Thus, the aim of restoration should be the building of a 'permanent and mutually beneficial relationship between civilised men [sic] and civilised nature',[34] a harmonious relationship built on the recognition of the intrinsic value of nature and 'more healthy images of our planetary role'.[35] As Tony Arnold warns, 'relationships between humans and the natural environment … pervasively threaten ecological health and integrity'.[36] Reimagining the relationship between human beings and the natural world is therefore an urgent priority for those concerned with environmental protection and restoration.[37]

My suggestion that ecological restoration should be conceived as a form of relationship restoration, rather than simply an ecologists' practice of returning an ecosystem to a past state, is not a novel one. From Aldo Leopold in his seminal *Sand County Almanac* in 1949, to Jordan in his more recent work *The Sunflower Forest*, this relational view of restoration is apparent. Nicole Graham and Robyn Bartel describe this approach as ecological reconciliation,[38] and Telesetsky and her co-authors advocate connective justice as the foundation of restoration ecology in order to reconnect humans with their natural environs.[39] In the specific context of Indigenous peoples, M. Kat Anderson describes restoration as both the 'return of landscapes to historical conditions' and the restoration of 'the place of humans in their continuing management'.[40] In sum, this view of restoration regards it as

33 Although there might be fewer objections if I attempted to reinstate the historical lavender fields of the area.

34 Greenwood (2017), 684 quoting Aldo Leopold from his 20 September 1934 address to the Nakoma District Welfare League, 'What is the Arboretum?'. Although I am not able to consult Leopold on his meaning here, I would suggest that his use of 'civilised nature' was intended as a literary flourish rather than a statement about the need to civilise what he would no doubt would have thought ought to be wild.

35 Freyfogle (1993), xv.

36 Arnold (2004), 5.

37 Davies (2017), 3.

38 Graham and Bartel (2017), 222.

39 Telesetsky, Cliquet and Akhtar-Khavari (2017), 288–289.

40 Anderson (2005), 335.

defining 'the terms of our relationship with natural landscapes'.[41] Recast in terms of human/nature relationships, ecological restoration can be closer to 'intelligent tinkering' than the nefarious manipulation that Elliot finds so repulsive.

There is, however, a deep tension that sits within this relationship-focused brand of ecological restoration, namely the proper place of human beings within an ecosystem. Is it that a truly restored landscape is one where human beings withdraw entirely, or is a restored relationship one of harmony and symbiosis between all the Earth's inhabitants? The 'withdrawal' approach is particularly emblematic of conceptions of wilderness, which are spaces that Freyfogle has characterised as free 'from all signs and symbols of human occupancy', including imperceptible trails, discrete signage and even the aeroplanes that fly above wilderness spaces.[42] Thus a fully restored relationship with nature might be one where humans learn to keep a respectful distance and graciously vacate wilderness spaces.

The wilderness, 'museum-piece' view of restoration is, however, a limited and hyperbolic vision of restoration[43] to which few rewilders would subscribe.[44] This is particularly so because it exacerbates the fragmentation of landscapes that an overly technical approach to restoration can lead to. That is, certain landscapes will be categorised as wilderness preserves, immune from human influence, whilst the rest of the world is for unchallenged cohabitation. Such an approach does little to address our limited imagination about how humans ought to interact with their habitats. Instead, a restored relationship is one where there is a 'harmonious re-entry into nature'.[45]

One reason why Indigenous land relationships feature so strongly in restoration scholarship is that they seem to typify the respectful stewardship of the natural environment that restorationists are aiming for. Indeed, the Society for Ecological Restoration acknowledges that Indigenous practices are integral to the restoring and preserving natural environments and at its 1995 annual meeting, the Indigenous People's Restoration Network was formed in part to embed Indigenous practices into emerging ecosystem management schemes.[46] Therefore, if we are to restore our relationship with nature, it will be necessary to dissolve the bicameral attitude towards humanity and nature that pervades of much of Western society to adopt the monist approach of Indigenous communities.[47]

A particular advantage of relationship-focused eco-restoration that is implicit in restoration scholarship, but rarely articulated, is the need for restoration of the whole community.[48] Restoration is therefore not only about individuals repairing

41 Jordan (2012), 15.
42 Freyfogle (1993), 101.
43 Graham and Bartel (2017), 228.
44 Nash (1982); Scott (2001); Foreman (2004).
45 Elliot (1997), 148.
46 Although the Network is no longer in operation, the SER continues to recognise and support indigenous restoration through the Full Circle Award, designed to support restoration projects that significantly incorporate indigenous knowledge and techniques. See Anderson (2005), 337.
47 Ibid, 358.
48 Arnold (2004), 5.

their relationships with their natural environment, although this is of course a necessary aspect of restoration; it is about a collective restoration relationship. Arnold characterises restoration as the building of relationships and connections, not just between particular humans and their local environment, but also across the whole human community.[49] In his account of the restoration of Mono Lake in Northern California, it was the restorers' efforts to engage and involve the geographically proximate, but mentally remote, city communities that helped generate the attention and support necessary to ensure that the project was a success. The building of stronger relations between expert restorers and members of the local community typifies the interdisciplinary approach to restoration that is advocated by An Cliquet and Kris Decleer.[50] Thus, the end goal of restoration should not simply be the restoration of particular landscapes that remain fragmented from each other, but a reimagined earthly community.

Framing environmental restoration in terms of the need to establish stronger human communities, with a collective imagination for protecting the environment, is already an integral aspect of environmental law. The focus on participation, ensuring wide access to environmental information and participation in environmental decision-making, is in part a way of strengthening the connections between humans and their natural environments in order to strengthen an ethos of environmental democracy.[51] And a profound commitment to environmental restoration requires a further leap of imagination, to find ways to 'consult'[52] or 'listen' to nature, thus recognising its agency.[53] Ambitious and imaginative environmental democracy can therefore help to establish a more respectful relationship between humans and their natural environment.

How then do we characterise this newly restored relationship? It is my contention that a restored relationship is a stewardship relationship.

## 4.3 The story of stewardship

Stewardship is a value-laden idea, richly suggestive of how to properly maintain precious resources. The good steward is someone who exercises virtues of consideration, planning, care and even love. In the environmental context, stewardship brings to mind images of Benedictine monks worshiping the divine with their hands in the earth, Californian Indians cultivating bracken fern to weave beautiful baskets, or the yogi practising sun salutations in the early morning dew. If ecological restoration, and environmental law more generally, is to address the innumerable pressing environmental crises, it is necessary 'to develop a vision of the natural order that helps us to learn how to live, day by day without abusing our

49 Ibid.
50 Cliquet and Decleer (2017), 179.
51 Although public participation in restoration opens up challenges about public distaste for the uglier aspects of natural processes, see for example Prior and Brady (2017), 42–45.
52 Jordan et al (1987), 13.
53 Dryzek (2000), 149.

native surroundings'.[54] The critical question then is how to develop such a vision of the natural order. As Martha Nussbaum claims, law too infrequently attempts to engage our literary imagination.[55] The same may be said of the technical language employed by restorationists. Thus, reframing our relationship with the natural world in terms of stewardship accesses a rich language, imagery and narrative which can, as a persuasive exercise, help to facilitate a restoration of our relationship with the natural environment. What is more, the conceptual dynamism[56] of stewardship can move us beyond an expanded imagination to realise restored human/nature relationships.[57]

However, the story of stewardship is not an uncomplicated one, and there are those who are understandably sceptical of the concept of stewardship because of its potential to reinforce, rather than confront our exploitative, anthropocentric view of the world.[58] Yet stewardship is an idea, with numerous layers and facets of meaning, capable of containing this paradox. There is a cross-cultural power to the concept of stewardship that, fully embraced, has the power to transform human/nature relationships. In addition to its cross-cultural heritage and its relevant normative content, stewardship is an idea with a considerable cosmopolitan lineage. For example, post-Platonic philosophers in the Roman Empire, such as Iamblichus, took as their theme Plato's claim in *Phaedrus* that 'it is everywhere the responsibility of the animate to look after the inanimate';[59] and the Confucian model of governance was one of 'political stewardship' that linked authority with virtue.[60]

Stewardship, in its various guises, also manifests in each of the major world religions. From the Qu'ran's repeated calls to care for the environment, to the Hindu concept of *ahimsa* (respect for all life); from the Jewish concept of justice – *Tzedek* – that can encompass environmental values, to the Buddhist reverence for life, stewardship ideals pervade. A stewardship ethic also pervades the human/nature relations of many Indigenous communities. For example, the Māori have an obligation of guardianship and protection – *kaitiakitanga* – that characterises their interactions with the natural environment, and for the Saami people of Northern Europe, their cultural heritage is wrapped up in their biological heritage and so protecting the environment safeguards important cultural values. This is also true of many of the traditional African religions; for example, in the Yoroba tradition humans are viewed as co-tenants of God's Earth alongside all other members of the natural world.[61] There is therefore a universality to the concept of stewardship, such that it serves as a rule of recognition that can guide our interactions with the natural world.

54 Freyfogle (1993), 10.
55 Nussbaum (1995), xv.
56 Davies (2017), 47.
57 Cooper (2014), 35–39.
58 Graham and Bartel (2017), 222.
59 Passmore (1974), 28.
60 Greer and Piow Lim (1998), 87.
61 Babalola (2009), 127.

Elsewhere I have elaborated the concept of stewardship in terms of its different elements – duty object, beneficiary and duty bearer – and the different types of stewardship relationship.[62] The four stewardship relationships I identify – custodial, managerial, proprietorial and spiritual – are not intended to be exhaustive or even discrete. Rather, they are illustrative of the different ways that stewardship duties can manifest to demonstrate the layered potential of the concept of stewardship where one aspect may come to the fore in a given setting whilst the others recede to distal relevance. Understanding stewardship in this dynamic way recognises 'conceptualisation ... as a process, and as responsive', as Margaret Davies explains, which allows concepts, particularly legal ones, to 'engage and respond'.[63] Thus the different relationships, actors and scales of application allow the concept of stewardship to adapt and grow as our understanding of our connectively just place in the world is refined.

A custodial idea of stewardship represents a minimum content for a stewardship duty, and can be characterised as 'the socially and ecologically responsible custody of land'.[64] The managerial form of stewardship is a more active form of stewardship, it is a cooperation with nature that is typified by many Indigenous land relationships. For example, M. Kat Anderson describes the careful ways in which Californian Indians manage, cultivate and tend their environs. Stewardship as a proprietorial relationship either maps environmental duties onto ownership such that stewardship becomes 'a requirement of landownership',[65] or transforms the legal idea of property into one of stewardship.[66]

The final form of a stewardship relationship that I articulate is a spiritual one. As I have already indicated, most faith systems encompass some idea of ethical environmental responsibility, and some even incorporate a mystic and reverential attitude towards the natural world that adds spiritual bite to the ethical obligations. There is, then, a rich backstory to the concept of stewardship that is deeply expressive of humanity's relationship with our natural habitat. And it is the dynamic and transcendent nature of the concept of stewardship that provides its poetic potential as a descriptor of our restored relationship with the natural environment.

In appealing to the poeticism of stewardship, I am attempting to avoid what Arnold describes as the 'the legal centralism of environmental law scholarship'; to consider how lawyers can bring to bear their rhetorical gifts to help to foster better human/nature relationships.[67] Carefully considered linguistic choices alongside storytelling have the potential to nurture our imagination and therefore our ability to develop a 'just way of living on Earth'.[68] Nussbaum builds a powerful argument for a more poetic approach to our legal reasoning and interactions in public life, to

62 Barritt (2014).
63 Davies (2017), 47.
64 Caldwell (1986), 322 and 327.
65 Caldwell (1986), 330.
66 Lucy and Mitchell (1996), 586.
67 Arnold (2004), 11.
68 Freyfogle (1993), xix.

help to build a more authentic and nuanced understanding of our interactions with each other. The emotionally neutered intellect that is characteristic of public life (particularly in the West) causes us to be 'value-blind'[69] and thus less able to 'imagine one another with empathy and compassion', and this imaginative failure needs to be overcome in order to stitch stewardship values into our interactions with the natural world.[70]

This imaginative failure can also be recognised in relation to our interactions with our natural environment. Environmental legislation is often deeply technical, uninspiring and impenetrable to the non-expert. Restoration, as broadly conceived, and focused as it is on a backward-looking end, can also be characterised in terms of particular modes of conduct or ecological targets that can fail to engage the imagination of the non-specialist.[71] There is then a need for language and imagery that can move and delight in the context of environmental law. To this end, I extend Nussbaum's argument to include our interactions with the natural environment as well as each other. This is not a presumption about the ability of the natural environment to participate in or enjoy our poetic endeavours; rather, adopting Nussbaum's vision is intended to reframe our common cultural behaviours beyond the law so as to incentivise more respectful human/nature relationships. As Jane Bennett explains, 'my hunch is that the image of dead or thoroughly instrumentalized matter feeds human hubris and our earth-destroying fantasies of conquest and consumption'.[72] It is my hope, therefore, that the poetic potential in the idea of stewardship[73] can engage our imaginations, challenge our assumptions about the agency of nature and, in so doing, move us towards better, more just interactions with our natural environment.[74]

An illustration of the problem of our imaginative myopia is recounted by Isabella Tree in her book about of the restoration of Knepp farm in West Sussex.[75] Tree had engaged Ted Green, a leading authority on ancient woodlands, to assess the health of a 550-year-old oak that was beginning to separate in its middle. During the Second World War, Canadian soldiers had attempted to remedy the situation by strapping the tree up, which had provided a temporary solution. When Green inspected the great oak, he noticed that its low-hanging branches had also been removed in an attempt to save the tree. In removing these branches,

69 Nussbaum (1995), 68.
70 Ibid, vxii.
71 For example, the Estuary Restoration Act as examined in Telesetsky, Cliquet and Akhtar-Khavari (2017), 177–179; see also Jordan's argument that restoration is necessarily non-creative and requires obedience to the existing patterns of nature: Jordan (2012), 24.
72 Bennett (2010), ix.
73 In advocating for the concept of stewardship as the best characterisation of human/nature relationships, I am not wedded to the nomenclature, rather the layers of meaning that are embodied in the idea of stewardship. Thus, ideas like trusteeship, guardianship, tending, custodianship, *kaitiakitanga, khalifa* all represent the spirit I am intending to capture.
74 Bennett (2010), xi; Cooper (2014), 35.
75 Tree (2018), 13–15.

however, the soldiers had removed the buttresses that the tree had grown in order to support the vast weight that would soon need stabilising. Similarly, in relation to younger oaks on the estate, frequent ploughing in their vicinity was causing them distress, as Green lamented during his consultation, 'we never think of what's going on below ground'.[76] Previous custodians of the Knepp farm forests had failed to understand the complex mechanisms and far-reaching ecological conditions necessary for the health and longevity of these ancient trees because they had applied a human-centred understanding of structure and growth to their interactions with the forest.

Despite the emotive power of stewardship, it would be remiss not to acknowledge the complex lineage of the idea. At its worst it has the potential to perpetuate anthropocentric ideas of entitled destruction,[77] and more passively it 'continues to privilege the role of the human', viewing humans as 'separate from nature rather than part of it'.[78] These are significant challenges to the literary potential of stewardship. However, the potential for corruption of an idea does not make the idea in and of itself wrong. Rather, I would suggest that just because a love of nature, which can characterise the stewardship relationship, can manifest as a greedy overconsumption of our natural resources, it does not automatically mean that stewardship as a motivating idea is in and of itself problematic.

Of course, there can be legitimate disagreement as to whether stewardship does represent the zenith of our restored relationship with land, just as there is legitimate disagreement about whether restored landscapes are ones that are remote from all human interactions. However, I would hold firm to the idea that the antecedence and universality of stewardship make it a desirable candidate for describing and defining the restored human/nature relationship. Thus, it is, in part, the imagery of the careful custody of children or precious artefacts, and partly the transcendent history of the concept, that enables stewardship to hold this potential.

By way of an example, the Country Stewardship Scheme promoted by Natural England relies on the persuasive power of the concept of stewardship to encourage farm managers to care for the natural environment.[79] The scheme offers financial incentives to engage farmers, woodland owners, foresters and land managers in making environmental improvements, for example restoring wildlife habitats and reducing agricultural pollution. The linguistic tenderness but normative strength of stewardship is intended to empower as well as to nudge towards better ecological practices.

This soft power is important if good ecological stewardship is to be practised by more than just the environmentally woke. For example, although the accounts of three ecologically astute Australian farmers, detailed by Graham and Bartel, provide a hopeful picture of how farming practices can be reconciled to the natural

76 Ibid, 16.
77 White (1967), 1206; Palmer (2006), 75; Attfield (2006), 79–82; Black (2006), 92–93.
78 Graham and Bartel (2017), 228.
79 Natural England (2015).

world, the farmers are already environmentally conscious. Their accounts do not provide much advice on how to engage the farming community more generally. If membership of Telesetsky's typology of restoration actors is to be expanded, then there is a need to engage others in restorative activities to change the current pattern of fragmented restoration. Schemes such as the one administered by Natural England harness the persuasive potential of the concept of stewardship in an effort to engage a wider range of stewardship actors. Stewardship in its most rudimentary sense is about care,[80] and it is this focus on care that teaches us to 'live in partnership with the Earth'.[81] As R.J. Berry argues 'relationship is at the root of stewardship'.[82]

## 4.4 The different guises of stewardship

In order to illustrate the full scope of potential stewardship relationships, I will examine a number of examples of the embodied stewardship relationship, starting with the recent legislative innovations in New Zealand designed to embed stewardship values and practices in legal frameworks, and finishing with geoengineering, which promotes human intervention in ecological process in an attempt to reverse environmental harm. I move from the exemplary to the aberrant to demonstrate what a restored stewardship relationship should (and should not) look like.

The full spiritual sense of stewardship can be exemplified by two legislative interventions in New Zealand: The Te Urewera Act 2014 and the Te Awa Tupua (Whanganui River Claims Settlement) Act 2017.[83] These statutes grant legal personality to Te Urewera (a former national park) and the Whanganui River, sacred Māori landscapes, and prescribe governance structures to ensure the stewardship of these natural resources. They also represent a settlement between the Māori and the Crown, to redress colonial wrongs, restoring (at least partially) the Māori relationship with their natural resources.[84] Katherine Sanders describes the results of these acts of parliament as 'a symbolic reframing of relationships between people and the environment' as well as providing 'new frameworks for relationships between people', in particular between Indigenous and settler peoples.[85]

The use of legal personality in both these cases was intended to provide a legal mechanism to recognise Māori obligation of *kaitiakitanga* and thus restore them to their right of stewardship with the natural environment.[86] Similarly, in a 2018

80 Berry (2006), 1.
81 Lovelock (2000), 111.
82 Berry (2006), 1.
83 Rodgers (2017)
84 Sanders (2017), 3.
85 Ibid, 3, 5.
86 Although, as Sanders (2017) explains, the focus on a stewardship relationship rather than the proprietary relationship in these cases was a way of sidestepping politically contentious claims to ownership by the Māori, the concept of stewardship that is reflected in these statutes is at least a more accurate reflection of the relationship the Māori have with their natural environment, even if it does raise challenges about the failure of the Crown to cede full legal control.

decision of the Supreme Court of Colombia, the recognition of legal personality of the Colombian Amazon was a way to instil stewardship values and recognise the entitlement of the Amazon to 'protection, conservation, maintenance and *restoration*'.[87] Critical to both of these legal innovations is the recognition of the intrinsic value of nature, as well as of its agency.

The second example of what a restored stewardship relationship might look like is a communitarian form of stewardship, exemplified by restoration projects that actively engage with local communities. One challenge for scientifically authentic restoration projects is that they can be exclusionary; thus, although they might achieve an authentically restored ecosystem, the impact of the project is geographically contained and its impact potentially limited. Reframing restoration in terms of reconciliation therefore embeds an impulse to include as many participants as possible: to restore the community's relationship with the natural environment, not just the specially qualified restorers.

A particularly striking instance of this community-based stewardship is provided by the restoration of Mono Lake, a rare lake ecosystem in Northern California, fed by the snowmelts of Sierra Nevada.[88] Efforts to restore the lake began with a small group of scientists and environmentalists in the 1970s, led by David Gaines, a biologist, ornithologist and ecologist educated at Stanford University (the embodiment of the expert restorer). As part of its restoration efforts, the Mono Lake Committee engaged in the usual practice of environmental litigation to fight for legal recognition by the Los Angeles Department of Water and Power of the need to protect the lake. But, critically, they also embarked on a complementary programme of education and community engagement. Indeed, the inclusion in restoration efforts of the Southern Californian public, who were recipients of the diverted tributaries of Mono Lake, resulted in modified water usage. Thus, the efforts of the expert restorationists to include the wider public generated a 'rise of the environmental ethic and the force with which that ethic would be bought to bear'.[89]

The third example of the manifestation of the stewardship relationship is conservation covenants. These are a voluntary (usually) and private conservation agreement whereby a landowner can come to an agreement with a conservation body to ensure the conservation of natural or cultural features of land and water.[90] Critically, conservation agreements remain in place even after the land changes hands, thus they are a mechanism by which landholders can lock conservation standards into their property in perpetuity. Conservation covenants are already available in Scotland, the USA, Canada, Australia and New Zealand, and the Law Commission has produced a report on the desirability of adopting the practice in England and Wales.[91]

87 *Demanda Generaciones Futuras v Minambiente* (2018) (emphasis added).
88 Arnold (2004), 2, 12.
89 Koehler (1995), 564.
90 Law Commission (2014), 1.
91 Ibid, 35–45.

Conservation covenants, however, are vulnerable to being treated as a formalistic version of the stewardship relationship, that does not necessarily grapple with what it means to have a truly restored relationship. As conservation covenants are parasitic on institutions of private ownership, they can perpetuate a relationship of human dominance, as well as excluding the public from conservation commitments. As Bonnie Holligan suggests, conservation covenants are criticised as 'typifying an unaccountable exploitative relation in which ecological interests are subordinated to those of capital'.[92] A chilling example of this is highlighted in research by Vanessa Adams and Katie Moon, whereby land in Australia protected by conservation covenants can be exploited for minerals, because the surface rights are treated as separate from the mineral rights.[93] Further, the privatisation of conservation control lacks democratic legitimacy and administrative accountability: rights guaranteeing access to information about the landscape, or rights to participate in decisions about appropriate conservation measures, are not embedded within the structure of conservation covenants.[94] Thus, conservation covenants can be vulnerable to becoming a form of a stewardship relationship that does not engage unequivocally with the substance and spirit of what a restored human/nature relationship should look like.

My final example is a perversion of stewardship – geoengineering. The Royal Society has described geoengineering as 'the deliberate large-scale manipulation of the planetary environment to counteract anthropogenic climate change'. Geoengineering can manifest in one of two ways: through carbon removal, commonly conducted by fertilising the oceans with iron filings to stimulate the growth of microscopic phytoplankton that will absorb dissolved carbon and then drop to the seabed, thus reducing both climate change and ocean acidification; and solar radiation management, which attempts to reduce global temperatures by reflecting incoming solar radiation, for example through stratospheric aerosol injection which mimics the cooling effect of large volcanic eruptions by introducing into the stratosphere a layer of particles that will reflect solar radiation.[95]

Admittedly, both techniques have appeal for their potential to alleviate the looming climate crisis; however, they represent a perversion of the stewardship relationship at two levels and thus exemplify a failure to adequately reconcile human beings to their natural environment. In the first place, geoengineering emerges out of our failure to steward our climate system adequately. A failure to readjust behaviour and restore the human/nature relationship at an early stage, in response to the troubling findings of climate science, has resulted in a situation of global urgency that sparks anxious scientific responses. Second, geoengineering is intended to remedy that failure through manipulation of complex climate systems in ways that have the potential to cause profound damage. For example, a practice of stratospheric aerosol injection might temporarily cause a decrease in global

92 Holligan (2018), 55.
93 Adams and Moon (2013), 117.
94 Holligan (2018), 73.
95 Brent et al (2015), 1–2.

temperatures, but a failure to continue a state of stratospheric injection could result in a sharp increase in global temperatures that would immediately reverse the dip in temperatures initially precipitated.[96] Thus, whilst it might be possible to characterise this as a managerial form of stewardship, I would reject such labelling. Stewardship requires explicit engagement with environmental values, and geoengineering represents a failure to engage with those values and to recognise the implicit value choices that are being made, even within the context of a scientific practice.[97]

In each of these examples, the different guises of stewardship have different consequences for how humans interact with their environment, and thus how the specific legal and regulatory regimes can support, or antagonise, restored human/nature relationship. In closing, I make two broad suggestions as to how law and legal institutions more generally might accommodate my vision of restoration, through the emergence of a principle of environmental stewardship in international environmental law and a reimagining of proprietary land relationships.

## 4.5 Law and restoration

A serious commitment to a restored relationship will require a transformation of not just our conceptual relationship to the land, but also our legal relationships as well as legal instruments themselves. Law has already adapted imaginatively to this challenge in a number of ways. At one level, all environmental law is ostensibly directed towards good environmental stewardship, and thus reconciling the human/nature relationship. There are, however, some specific instances of environmental law that are directed particularly at restoring our relationship with our natural world. For example, by granting legal personality to specific eco-systems, and participatory environmental rights to non-landowners, legal systems have found creative ways to respond to Christopher Stone's challenge of 1972: 'should trees have standing?'[98] Thus, citizen stewards are mobilised with legal tools to respond actively and robustly to their environmental responsibilities.

Further, the parameters of good stewardship have been set through established principles of precaution and environmental impact assessment, and through severely curtailing damaging practices such as ocean fertilisation or the production of ozone-depleting substances. Governments have also found ways to instil stewardship values into the proprietary relationship, through Countryside Stewardship Agreements that incentivise good behaviour, and by limiting landownership through conservation covenants and special habitat designation. Similarly, on an international level, good stewardship is being promoted in the form of a new implementing agreement to the UN Law of the Sea Convention.[99] Environmental law therefore already recognises the central importance of restoring humanity's

96 Ibid, 22.
97 Mason (1999), 50.
98 Stone (1972).
99 General Assembly Resolution on oceans and the law of the sea (2013).

relationship with the natural world. However, if a meaningful reconciliation between nature and human beings is to be achieved, two further steps should be taken. The first – the emergence of a principle of environmental stewardship in international environmental law – is already on the horizon; the second – a transformation of property rights in land and natural resources – is significantly more ambitious, but not beyond the realms of legal imagination.

The recognition of environmental stewardship as a principle of international environmental law is already emerging as part of the international legal order, for example through initiatives such as the Earth Charter[100] and the Ecological Law and Governance Association.[101] International relations scholars Robert Falkner and Barry Buzan trace the 'deep normative development' of environmental stewardship in global society, arguing that it has the potential to become a 'durable institution' of the international order.[102] The normative appeal of stewardship is particularly evident in relation to the governance of the oceans in areas beyond national jurisdiction, whereby stewardship bridges the conceptual conflict arising from the legal division of different parts of the oceans into the 'common heritage of [hu]mankind' and a free-for-all.[103] It is therefore possible to identify hints of a principle of stewardship in the international legal order.

However, further developments will be necessary for environmental stewardship to be fully recognised as a principle of international law.[104] As Eloise Scotford explains of environmental principles more generally, what is necessary for them to crystallise as a general principle of international environmental law is for it to be given recognition within discrete legal systems, and although there are isolated instances of this happening with environmental stewardship, the scale of its recognition does not match that of other principles of environmental law.[105]

One way in which the emergence of a principle of environmental stewardship will be valuable is that it will be able to shape existing legal frameworks and approaches. Much of environmental law is focused on the procedural, through participatory rights and access to information; a principle of stewardship would therefore complement this procedural focus, imbuing the exercise of participatory rights with a spirit of stewardship. And although stewardship is essentially a duty of conduct, there are implicit indications about the quality of that conduct embedded in the idea of stewardship. In the context of restoration, it would help to unify the disparate and often ambiguous restoration provisions in environmental law. Thus, obligations of restoration will be flavoured with a sense of a particular outcome.

My second broad suggestion for how legal infrastructure can enable restored stewardship relationships is through the transformation of the institution of

100 Earth Charter (2000).
101 Ecological Law and Governance Association (2016) *Oslo Manifesto for Ecological Law and Governance*.
102 Falkner and Buzan (2018).
103 Ridings (2018), 483, see also Turnipseed et al (2013); Henriksen (2016).
104 Cf. the difficult progress of the precautionary principle/approach in international law.
105 Scotford (2017), 11.

property, as discussed in more detail in Chapter 5 in this volume from Robyn Bartel and Nichole Graham. That is, rather than characterising legal land relationships as a trinity of rights (exclusivity, enforceability and transferability) in the same way as ownership can be claimed of an inanimate, discrete object, legal land relationships should be framed in stewardship terms.[106] This is not to ignore the reality that humans will inevitably exercise some degree of control over the physical spaces that they occupy, but it would hold them to a higher standard as to how they interact with their particular habitats. This vision for stewardship is admittedly wildly ambitious. But a truly restored relationship will have to be one that dispenses with legal institutions that encourage the pursuit of economic exploitation and obscure the imperatives of responsibility and care.[107] Just as it was impossible to even begin to unravel the profound wrong of colonisation whilst human beings remained the property of others, it is also difficult to reframe our relationship with the natural world when our primary legal interaction with it is one of dominance and subjugation.

## 4.6 Conclusion

What enabled Coetzee's Magistrate to close his ears to the sounds of human suffering, indeed what made enslavement and colonisation possible, was a failure to acknowledge the intrinsic value and agency of other human beings.[108] This wilful blindness is also evident in our interactions with the natural world. Restoration projects have the potential to illuminate the intrinsic value of our surroundings, to recognise the agency of nature and to instil a stewardship ethic.

Of course, this story could be dismissed as mere fairy tale, a fantasy in the face of the immovable fortress of global capitalism, neo-liberalism and new, destructive technologies. These structural challenges are profound. But no good story is without its moments of jeopardy and obstacles to overcome. Restored ecological relationships will require restored human relationships, stronger communities and better articulated ethical values, able to resist the intoxication of unbridled consumption that the Capitalocene engenders.[109]

The ethical character of stewardship and its commitment to community participation are intended to provide a firm foundation for this resistance, to embed restoration practices across political, social and legal scales as well as geographical ones. In so doing, the ostensibly scientific practice of restoration can begin to rebalance the disorder of the Capitalocene/Anthropocene era. This rebalancing is already evident in the evolution of the character of environmental law, shifting its approach from command-and-control to 'rules of engagement'.[110] It is also evident in the increasing jurisprudential and legislative innovation that recognises the

106 Lucy and Mitchell (1996), 586.
107 Ibid.
108 Arendt (1958), 50.
109 Patel and Moore (2017), 2–3.
110 Telesetsky (2017), 6.

rights of nature and emphasises the importance of humans as stewards of their natural environment.

In Arnold's account of the restoration of Mono Lake, the success of the project was in its ability not just to bring into the fold those with a pre-existing interest in the restoration of the lake, but to engage entire communities in the efforts. They also found ways to work within, but ultimately challenge, corporate and government structures resistant to the project. Mono Lake offers encouragement that the story of stewardship can help cultivate our relationship with nature, not just on an individual level, but also within a geographically wide community, in turn allowing for a transformative response to the environment. We must therefore find ways to reimagine our relationships with the natural world if restoration in a deep and enduring sense is to be achieved.

## References

### *Primary sources (law materials)*

Convention on Access to Information, Public Participation in Decision-making and Access to Justice in Environmental Matters (UNECE), 25 June 1998, 2161 UNTS 447.

*Demanda Generaciones Futuras v Minambiente* 11001-22-03-000-2018-319–301 (Colom. Sup. Ct. Apr. 5, 2018).

Earth Charter (2002) *The Earth Charter Initiative.*

Estuary Restoration Act, Public Law 106–457 (United States)

General Assembly, Resolution of 9 December 2013 on oceans and the law of the sea. UN Doc. A/RES/68/70 of 27 February 2014.

Law Commission (2014) *Conservation Covenants*, Law Com No. 349.

Te Urewera Act 2014 (New Zealand)

Te Awa Tupua (Whanganui River Claims Settlement) Act 2017 (New Zealand).

### *Secondary sources*

Randall S. Abate (2016) 'Ocean Iron Fertilization and Indigenous Peoples' Right to Food: Leveraging International and Domestic Law Protections to Enhance Access to Salmon in the Pacific Northwest' 20 *UCLA Journal of International Law and Foreign Affairs* 45.

Vanessa Adams and Katie Moon (2013) 'Security and Equity of Conservation Covenants: Contradictions of Private Protected Area Policies in Australia' 20 *Land Use Policy* 114.

M. Kat Anderson (2005) *Tending the Wild*, University of California Press.

Craig Anthony Arnold (2004) 'Working Out an Environmental Ethic: Anniversary Lessons from Mono Lake' 4 *Wyoming Law Review* 1.

Hannah Arendt (1958) *The Human Condition*, Chicago University Press.

Fola D. Babalola (2009) 'Roles of and Threats to Yoruba Traditional Beliefs in Wilderness Conservation in Southwest Nigeria' *Science and Stewardship to Protect and Sustain Wilderness Values: Ninth World Wilderness Symposium* 125.

Emily Barritt (2014) 'Conceptualising Stewardship in Environmental Law' 26 *Journal of Environmental Law* 1.

Jane Bennett (2010) *Vibrant Matter*, Duke University Press.

R.J. Berry (2006) 'Stewardship: A Default Position?' in R.J. Berry (ed.) *Environmental Stewardship: Critical Perspectives – Past and Present*, T&T Clark International, 5.
John Black (2006) 'The Dominion of Man' in R.J. Berry (ed.) *Environmental Stewardship*, T&T Clark International.
Kerryn Brent et al (2015) 'The Governance of Geoengineering: An Emerging Challenge for International and Domestic Legal Systems' 24 *Journal of Law, Information and Science* 1.
Lynton Caldwell (1986) 'Land and the Law: Problems in Legal Philosophy' *University of Illinois Law Review* 319.
An Cliquet and Kris Decleer (2017) 'Halting and Restoring Species Loss: Incorporating the Concepts of Extinction Debt, Ecological Trap and Dark Diversity into Conservation and Restoration Law' 26 *Griffith Law Review* 178.
John M. Coetzee (1980) *Waiting for the Barbarians*, Vintage.
Davina Cooper (2014) *Everyday Utopias*, Duke University Press.
Franklin E. Court (2012) *Pioneers of Ecological Restoration*, University of Wisconsin Press.
Margaret Davies (2017) *Law Unlimited*, Routledge.
John S. Dryzek (2000) *Deliberative Democracy and Beyond, Contestations*, Oxford University Press.
Richard Duncan (2003) 'Ecosystem Restoration: The New New Thing' 37 *Minnesota Law Review* 1209.
Ecological Law and Governance Association (2016) *Oslo Manifesto for Ecological Law and Governance*, ELGA.
Robert Elliot (1997) *Faking Nature: The Ethics of Environmental Restoration*, Routledge.
Robert Falkner and Barry Buzan (2018) 'The Emergence of Environmental Stewardship as a Primary Institution of Global International Society' *European Journal of International Relations*, https://doi.org/10.1177/1354066117741948.
Dave Foreman (2004) *Rewilding North America*, Island Press.
Eric T. Freyfogle (1993) *Justice and the Earth*, The Free Press.
Nicole Graham and Robyn Bartel (2017) 'Farmscapes: Property, Ecological Restoration and the Reconciliation of Human and Nature in Australian Agriculture' 26 *Griffith Law Review* 221.
David Greenwood (2017) 'Making Restoration History: Reconsidering Aldo Leopold's Arboretum Dedication Speeches' 25(5) *Restoration Ecology* 681.
Steven Greer and Tiong Piow Lim (1998) 'Confucianism: Natural Law Chinese Style?' 11 *Ratio Juris* 80.
David J. Hayes (2002) 'Land Conservation and Restoration: Moving to the Landscape Level' 21 *Virginia Environmental Law Journal* 115.
Tore Henriksen (2016) 'The Arctic Ocean, Environmental Stewardship, and the Law of the Sea' 16 *UC Irvine Law Review* 61.
Bonnie Holligan (2018) 'Narratives of Capital Versus Narratives of Community: Conservation Covenants and the Private Regulation of Land Use' 30 *Journal of Environmental Law* 55.
William R. Jordan III et al (1987) 'Restoration Ecology: Ecological Restoration as a Technique for Basic Research', in W.R. Jordan III et al (eds) *Restoration Ecology: A Synthetic Approach to Ecological Research*, Cambridge University Press.
William R. Jordan III (2012) *The Sunflower Forest*, University of California Press.
Cynthia Koehler (1995) 'Water Rights and the Public Tryst Doctrine Resolution of the Mono Lake Controversy' 22 *Ecology Law Quarterly* 541.
Aldo Leopold (1949) *A Sand County Almanac*, Oxford University Press.

James Lovelock (2000) *Gaia: A New Look at Life on Earth*, Oxford University Press.
William Lucy and Catherine Mitchell (1996) 'Replacing Private Property: The Case for Stewardship' 55 *Cambridge Law Journal* 566.
Michael Mason (1999) *Environmental Democracy*, Earthscan.
Roderick Nash (1982) *Wilderness and the American Mind*, Yale University Press.
Natural England (2015) 'Countryside Stewardship', www.gov.uk/government/collections/countryside-stewardship-get-paid-for-environmental-land-management.
Martha C. Nussbaum (1995) *Poetic Justice*, Beacon Press.
Clare Palmer (2006) 'Stewardship: A Case Study in Environmental Ethics' in R.J. Berry (ed.) *Environmental Stewardship*, T&T Clark International.
John Passmore (1974) *Man's Responsibility for Nature*, Gerald Duckworth & Co Ltd.
Raj Patel and Jason W. Moore (2017) *The History of the World in Seven Cheap Things*, University of California Press.
Grayson Perry (2014) *Playing to the Gallery*, Penguin UK.
Jonathan Prior and Emily Brady (2017) Environmental Aesthetics and Rewilding' 26(1) *Environmental Values* 31.
Benjamin J. Richardson (2016) 'The Emerging Age of Ecological Restoration Law' 25 *Review of European Community and International Environmental Law* 277.
Benjamin J. Richardson (2017) 'Restoring Layered Geographies: Ecology, Society and Time' 26 *Griffith Law Review* 154.
Benjamin J. Richardson, Emily Barritt and Megan Bowman (2019) 'Beauty: A Lingua Franca for Environmental Law? 8(1) *Transnational Environmental Law*, doi.org/10.1017/S2047102518000195.
Penelope Ridings (2018) 'Redefining Environmental Stewardship to Deliver Governance Frameworks for Marine Biodiversity Beyond National Jurisdiction' 75 *ICES Journal of Marine Science* 435.
Christopher Rodgers (2017) 'A New Approach to Protecting Ecosystems: The Te Awa Tupua (Whanganui River Claims) Settlement) Act 2017' 19 *Environmental Law Review* 266.
Michael Rosenzweig (2003) *Win–Win Ecology*, Oxford University Press.
Katherine Sanders (2017) '"Beyond Human Ownership?" Property, Power and Legal Personality for Nature in Aotearoa New Zealand' 30(2) *Journal of Environmental Law* 207.
Eloise Scotford (2017) *Environmental Principles and the Evolution of Environmental Law*, Hart Publishing.
Douglas W. Scott (2001) *Wilderness-Forever Future*, Pew Wilderness Centre.
Christopher D. Stone (1972) 'Should Trees Have Standing? Towards Legal Rights for Natural Objects' 45 *Southern California Law Review* 450.
Anastasia Telesetsky (2013) 'An Emerging Legal Principle to Restore Large Scale Ecoscapes' in C. Voigt (ed.) *Rule of Law for Nature*, Cambridge University Press, 175.
Anastasia Telesetsky (2017) 'Eco-restoration, Private Landowners and Overcoming the Status Quo Bias' 26 *Griffith Law Review* 248.
Anastasia Telesetsky, An Cliquet and Afshin Akhtar-Khavari (2017) *Ecological Restoration in International Environmental Law*, Routledge.
Isabella Tree (2018) *Wilding: The Return of Nature to a British Farm*, Picador.
Fredrick Turner (1987) 'The Self-effacing Art: Restoration as Imitation of Nature' in W.R. Jordan III et al (eds) *Restoration Ecology: A Synthetic Approach to Ecological Research*, Cambridge University Press.
Mary Turnipseed et al (2013) 'Using the Public Trust to Achieve Ocean Stewardship' in C. Voigt (ed.) *Rule of Law for Nature*, Cambridge University Press.
Lynn White (1967) 'The Historic Roots of our Ecological Crisis' 155(3767) *Science* 1203.

# 5 Ecological reconciliation on private agricultural land

## Moving beyond the human–nature binary in property-environment contests

*Robyn Bartel and Nicole Graham*

### 5.1 Introduction

Resolution of the troubled relationship between private property and the environment appears to have reached an impasse. Two debates in particular have ossified in unproductive and polarising ways. The first are the arguments concerning property, an institution that underpins dominant human–nature relationships within capitalist economies. While the privatisation of natural resources is still promoted as a means to ensure protection from over-use, the commoditisation agenda has been undermined by the failure of ownership, including of real property, to avert exploitation and environmental harm. The second debate concerns the positions taken with respect to managing our environment into the future, which at one extreme advocate for half the Earth[1] to be preserved in a pristine, 'restored' and almost human-free state; and on the other advance human coexistence as not only ubiquitous, but necessary.[2]

Our analysis is based on the documented experiences of leaders in regenerative agriculture, who have attempted to go beyond ecological restoration through creating 'farmscapes'[3] that reconcile agricultural practices with the Australian environment. Informed by a growing body of work in the areas of new materialism, new environmental governance and new 'localism', we consider how private land ownership, and in particular intensive uses such as agriculture, can support rather than undermine environmental ends through recognising that humans are always and already situated within an agential 'nature'. We argue that place-based approaches that reconcile humans with nature, and the particular potentials and emergent conditions of living landscapes, rather than restoring a 'nature' that has been 'lost' or is external to humans, can address the underlying anthropocentrism of Anglo-Australian property. Reconciliation approaches de-centre the human, and recognise that the environment is dynamic, reflexive and a relational–material co-becoming,[4] rather than an end point. We submit that the reconciliation of human and nature can resolve the tensions between private property and the environment.

1 Wilson (2016).
2 Purdy (2015).
3 Smukler et al (2010), 81. See also Imhoff (2003).
4 Bawaka County et al (2013); Bawaka County et al (2016).

## 5.2 Private property, agriculture and environmental law: problems and possibilities

Australian legal theorist Margaret Davies contends that '[p]roperty is a particularly expansive practice of dominance and deserves to be singled out for both its centrality in present forms of global ordering and its potential to be constructively rethought'.[5] Private property has been identified by Australian environmental law scholar Jan McDonald as a key legal institution that is resistant to change, or changes too slowly, and thereby becomes a 'barrier to adaptation'.[6] For example, private property was designed for productivity and profit rather than facilitating positive environmental outcomes. This connection derives from eighteenth-century English property law, founded on a moral rationale of 'improving' nature by 'developing' land and on an economic theory of infinite growth.[7] Its geographical expansion as a legal regime attended the colonisation of Antipodean, American, African and Asian nations, whose sovereign status was not recognised precisely on the basis of the apparent absence of 'improved' lands constituting private property. The environmental consequences of this expansion are diverse and often adverse, particularly with regard to vegetation and biodiversity.[8] European land-use practices have been recognised as maladapted to non-European geographical contexts, and yet the logic of the private property laws that facilitated and sometimes required these practices remains in place centuries later.[9]

Private property has been unable to avert, and indeed may have perversely guaranteed, the environmental conditions that define the Anthropocene. These encompass anthropogenic climate change, land and water degradation, habitat destruction and modification, and the sixth great extinction.[10] Conservation objectives are largely pursued through environmental laws, which, until recently, have focused on the reserve estate. Such a demarcation reflects the fundamental separation of human and nature, and between public and private goods, law and property, notwithstanding that environmental issues are impervious to tenure.[11] Off-reserve conservation, including through the regulation of habitat destruction, water use and invasive species (with notable exceptions for agricultural production, companion animals and ornamental plants), has grown in recent years. However, after several decades of expansive development, environmental law is being eroded through resistance[12] and regression.[13]

Off-reserve conservation requires the 'tacking on' of obligations to a pre-existing legal landscape which may be averse to an environmental agenda.[14] American environmental law scholar Joseph Sax has observed that the priority of private

5 Davies (2016), 38.
6 McDonald (2010), 8. See also Sax (2008), 10.
7 Graham (2011, 2014).
8 Williams (2006); Graham (2011); Muir (2014)
9 Graham (2011); Muir (2014).
10 Ceballos et al (2015).
11 Hobbs (2013) v; Norton and Reid (2013), 2; Godden and Cowell (2016).
12 Bartel and Barclay (2011); Barclay and Bartel (2015).
13 Prieur (2012); Bartel and Graham (2016).
14 Sax (2008); Adams and Moon (2013); Bartel et al (2014).

property both legally and culturally 'handicaps' environmental law 'in doing the job it needs to do' with regard to environmental protection.[15] Property law 'promotes and encourages' activities that environmental law endeavours to limit: '[d]espite a plethora of environmental laws and regulations' there has been little or no reform of the 'basic structural laws' that 'essentially drive behaviour'.[16] Sax identifies property law as the most important category of law in this regard and contends that the relationship between environmental law and property law is 'an unequal battle'.[17]

Property is regarded as the foundation stone of Anglo-Australian law, and environmental objectives may seem inconsistent with its purpose, particularly according to the discourse of rights and entitlement.[18] Most legislation that applies to private land is regulatory, and more lucrative land uses may subordinate or marginalise environmental aims. Restrictions on land clearing are one example: clearing was classified as a land improvement historically, and the practice continues in order to facilitate agricultural, residential, industrial, infrastructure and resource-extraction uses. It is a major threat to biodiversity and its legacy goes beyond environmental damage; productivity and wellbeing have been undermined.[19] However, land clearing and similar regulations may be considered unwarranted.[20] Farmer lobby group submissions to government inquiries emphasise the disempowerment felt in response to environmental regulation, the economic hardship perceived to be imposed by such laws, and expressions of desire for greater freedoms and entitlements with regard to land use.[21] Indeed, the legitimacy of environmental law has become subject to regular interrogation not only from lobbyists, but also from the (then) Commonwealth Attorney-General in his description of the 'encroachment' by public environmental law on 'fundamental rights and freedoms' in Australian private law.[22]

The introduction of largely voluntary[23] ecological restoration programs on private land emerged as a soft strategy to improve off-reserve conservation. Conservation covenants[24] allow governments to meet their obligations under international environmental law[25] by the removal of part or all of a privately held

15 Sax (2008), 10.
16 Ibid.
17 Ibid.
18 Australian Law Reform Commission (2015); Productivity Commission (2016). See also Sax (2008), 10.
19 Norton and Reid (2013); Massy (2017).
20 Bartel and Barclay (2011); Graham (2011); Barclay and Bartel (2015); Bartel and Graham (2016).
21 See submissions to Australian Law Reform Commission (2015) and Productivity Commission (2016).
22 Australian Law Reform Commission (2015); Graham (2016).
23 Most covenants are entered into voluntarily. However, some conservation covenants are required under offset agreements in development approvals. Fitzsimons and Carr (2014), 611.
24 Richardson (2015; 2017); Godden and Cowell (2016); Lindsay (2016); Hardy et al (2017).
25 Fitzsimons and Carr (2014), 606; Lindsay (2016), 698.

property from productive land-use practices[26] that commonly (but not necessarily) involve biodiversity loss. In practice, covenants involve the absence of agricultural land-use practice on a small number of isolated[27] parcels (or parts of a parcel) of private land. Positive obligations, such as for ecological restoration, can be legally unenforceable without special statutory provisions.[28] Effectiveness of conservation covenants is further hampered by poor monitoring[29] and government-issued mining licences.[30]

The dominant focus of conservation thus remains public land, for both the legal system and environmental scientists. Eminent biologist E.O. Wilson advocates for the preservation of half the Earth, accessible primarily to scientists for research purposes.[31] Without discounting the importance of reserves, nor how their objectives are undermined,[32] 'expansionist' arguments are not uncontested. Beyond the human equity – and more pragmatic feasibility – questions, pursuing a 'fortress' nature agenda may reinforce rather than remedy human/nature divides that are the root cause of problems that this approach is therefore unlikely to be able to solve.[33]

There are alternatives to drawing the boundaries between human and nature more widely and deeply still. Coexistence embraces the empathetic[34] idea of 'cooperation or working together'[35] and views humans as part of nature rather than apart, as interdependent rather than in control, thus refuting essentialist and exceptionalist claims.[36] While the term coexistence has gained some prominence in apex predator conservation, it is also used to refer to human coexistence with nature more broadly, including the co-creation and co-evolution of humans and nature.[37] Coexistence recognises non-human agency and describes humans as participating in a dynamic multi-agential process. Rather than aiming at human-defined ends, it focuses on reflexivity and responsiveness to natural dynamism, emergence, variation and variability. Coexistence may be critiqued[38] as conservation 'lite'; however, the concept has particular potential for private property and agriculture, which is significant considering the extent and consequences of these practices.[39] Coexistence may be a more open-minded and forward-looking frame

26 For example, grazing, vegetation clearing, pesticide use. Fitzsimons and Carr (2014), 609.
27 The program lacks scientific selection criteria: ibid, 613.
28 Australian covenants 'run with the land' or bind successors in title only where they create negative obligations. Edgeworth (2017, 617).
29 Hardy et al (2017), 227; Fitzsimons and Carr (2014), 606, 610–611.
30 Adams and Moon (2013), 115; Hardy et al (2017), 223.
31 Wilson (2016).
32 Lovejoy (2016).
33 Siurua (2006); Bartel et al (2014).
34 Bekoff (2014).
35 Heikkurinen (2017), *ix*.
36 Bartel et al (2014).
37 Bonnedahl (2017); Carter and Linnell (2016).
38 Woods (2017).
39 See Muir (2014), 7.

than those that inform ecological restoration, which may maintain problematic classifications of both human and nature.

## 5.3 Ecological restoration and reconciliation: promises, practices, pitfalls and potentials

Ecological restoration (also environmental and habitat restoration)[40] is defined as the process of assisting the recovery of an ecosystem that has been degraded, damaged or destroyed.[41] Full recovery is 'the condition of the ecosystem as it would be had it not been degraded'.[42] This objective, derived from restoration ecology and conservation biology,[43] is based on a fundamental normative distinction whereby nature is idealised as human-free, and human interference is viewed as almost inherently malign.[44] The restoration ideal has been criticised for its exclusivity, particularly for which types of nature and human activity may be excluded, including First Nations Peoples, as well as introduced species and novel ecosystems.[45] The 'nativist' species position is dependent on historical accounts of landscapes that classify species as indigenous or exotic, invasive and alien.[46] These categories are often value-laden and arbitrary rather than evidence-based.[47] An alternative position is that the world is and has always been altered (or co-created) by humans.[48]

The international standards for ecological restoration recognise cultural ecosystems, although they appear to be conflated with native ecosystems, at least where these 'were modified prior to industrialisation and so exhibit states very similar to those occurring in unmodified areas'. However where 'dissimilar states' with 'substantially different species composition' have been produced, these 'may not be appropriate reference models for ecological restoration but may still be 'valued' as 'semi-natural/cultural ecosystems'.[49]

Idealising nature as human-free is highly problematic in the Australian context, with an imagined historical baseline of 1788 serving to discount 60,000 years of ongoing human co-creation of landscapes, including agriculture.[50] History is central to the restoration project, but how far back is far enough?[51] We must be wary of 'shifting baseline syndrome', in which depauperate landscapes are viewed by

40 Miller and Hobbs (2007); Hobbs and Cramer (2008). For terminology see Telesetsky et al (2017).
41 Society for Ecological Restoration International (2004).
42 McDonald et al (2016a, 2016b).
43 Meffe et al (2006).
44 Berkes (2003); Miller et al (2011). See also Monbiot (2013). And save for restoration activities themselves, Elliot (1985); Katz (1992).
45 Hobbs et al (2006); Atchison (2015); Orion (2015); Pearce (2015).
46 Keulartz and van der Weele (2009). Also Frawley and McCalman (2014).
47 Warren (2007); Head (2012); Head et al (2015).
48 Marris (2011).
49 McDonald et al (2016b), 12.
50 Gerritsen (2008); Gammage (2011); Pascoe (2014).
51 Corlett (2016), 453. Also Hall (2010); Monbiot (2013); Akhtar-Khavari and Telesetsky (2016).

contemporaries as 'normal'.[52] But should the megafauna be resurrected? Rainforests expanded? More practically, historical data are often unavailable, incomplete, inaccurate and/or unreliable, and additional pragmatic factors affecting feasibility include degradation of seed banks, soil, water, and the loss of many species and adaptation of others to new conditions.[53] It may be preferable to adopt a 'recombinant', dynamic approach, looking forward rather than back.[54] The international standards allow that the

> use of reference models in ecological restoration does not signify in any way an attempt to immobilize an ecological community at some point in time, but rather to optimize potential for local species and communities to recover and continue to reassemble, adapt, and evolve.[55]

While the concept of restoration may be broad enough to encompass new ideas and widen its definition, there remains the question of what happens to land that exists beyond the bounds of those areas that have been (to whatever standard and however successfully) 'restored', even aside from the perverse consequence of intensifying land degradation in non-restored areas.[56]

More fundamentally, restoration may reinforce human responsibility and capacity for restoration, without sufficiently acknowledging the arrogance and therefore paradox that lies within this assertion. For, if we have caused environmental harms, there may well be something inherently problematic in our worldview that warrants reconsideration, prior to instituting attempts to resolve the issues using the very same tools that were responsible for their creation.[57] For example, even if exploitation is recrafted as stewardship,[58] and even if this is benevolent, environments are often still conceived as passive and humans are its controlling agents.[59] The ontological binaries thus reinforced by certain restoration approaches may perpetuate the problems that are sought to be resolved, and therefore it is important to explore the possibilities of a de-centred approach when remodelling the human–nature and property–environment relationships.

It is time to question whether restoration, as practised by mainstream restoration ecologists and environmental managers, may have outlived its conceptual utility, particularly where blunt categorisations of species and ecosystems are adopted, nature is framed as other, and human interventions (other than restoration) are ascribed as often egregious and always exceptional, and especially where projects are narrowly site-focused and colonial timelines and

52 MacKinnon (2013), 16; Massy (2017), 291.
53 Mabey (2010).
54 Hobbs et al (2006); McDonald et al (2016a, 2016b); Low (2017). Also Keulartz and van der Weele (2009); Wuerthner et al (2014).
55 McDonald et al (2016b).
56 Massy (2017), 407.
57 See Bartel et al (2014). Also Goodman (1999); Howitt and Suchet-Pearson (2006).
58 See Winter (1991).
59 See Goodman (1999).

anthropocentric objectives are deployed.[60] As legal geographer Irus Braverman has argued in relation to the under-interrogated disjunctures between creating and saving nature within wildlife conservation, 'our definition of nature urgently needs rethinking'.[61]

We propose an alternative approach of *ecological reconciliation*, by which we mean reconciling human and nature, and land-use practices on private (both leasehold and freehold) property with the environment of particular Australian places. Reconciliation is suggestive of a different ontological relation between humans and nature; it moves beyond the limits of the subject/object dichotomy by acknowledging their connectivity and relationality.[62] Some reconciliation advocates reinforce rather than repudiate anthropocentrism. James Miller defines reconciliation as 'the purposeful design of human land use to meet the needs of human species';[63] and Michael Rosenzweig defines it as seeking 'environmentally sound ways for us to continue to use the land for our benefit'.[64] Rosenzweig views it as one option to consider in addition to, rather than in replacement of, reservation and restoration approaches, and primarily applicable to highly modified and human-dominated environments.[65] The agricultural lands explored here are precisely such environments. However, the farmers' accounts in our chapter reveal very different ways of relating to land. These narratives convey relational and material connections between land-use practices and the landscape. This reframing has placed them at odds with conventional farming and governance practices that prescribe and produce essentialist, dichotomous and hierarchical human/nature relations. Drawing on these farmers' accounts, we contend that ecological reconciliation requires place-based practices that support the diverse and changing interrelationships of humans in nature. In this way, reconciliation can be distinguished from restoration by departing from the dualism and anthropocentrism inherent in restoring the 'other' and with 'pre'-human baselines for naturalness.

## 5.4 Farmscapes: an analysis of the environmental narratives of four Australian farmers

### *5.4.1 Farmer narratives*

We focus on 'local environmental narratives'[66] because we regard the observations and experiences of the possibilities of reconciliation in Australian agriculture as similar to the value of patient histories in medical practice and oral testimony in

60 See for example Katz (1992); Higgs (1997); Hobbs et al (2006); Siurua (2006); Keulartz and van der Weele (2009); Marris (2011); Head (2012); Ryan and Hartter (2012); Frawley and McCalman (2014); Orion (2015); Pearce (2015); Purdy (2015).
61 Braverman (2012), 9.
62 Rozensweig (1995); Miller (2006); Quinn (2009), 4; Davies (2016), 38.
63 Miller (2006), 357.
64 Rosenzweig (2003), 1.
65 See also Miller (2006) and the treatment of rehabilitation in McDonald et al (2016a, 2016b).
66 Robertson et al (2000), 119.

law.[67] The accounts provide 'a rich data base for assessing environmental conditions and change'.[68] Robertson and her colleagues argue that such narratives may form a statement of 'cultural and ecological reconciliation', which Sinatra and Murphy describe as 'a landscape ethos which is sensitive to place'.[69] We selected these four narratives, derived from different Australian jurisdictions and geographies,[70] as examples of exceptional farming practices that are achieving recognised environmental ends beyond compliance[71] (and sometimes are non-compliant) with cultural and legal standards. They have intentionally departed from orthodox and maladapted farming techniques notwithstanding legal, financial and social obligations and pressures.[72] The farmers lament the dissonance of European approaches to Australian landscapes[73] and the environmental (and social) disaster of modern industrial agriculture.[74]

Their stories are about learning from the land itself and from Indigenous Australians' land-use practices where possible. They are not aiming to achieve an ecology devoid of humans, but an ecology that aligns human land-use practices to functional landscapes – farmscapes. The term 'farmscape' describes land-use practices that 'enhance biodiversity and improve specific ecosystem functions' on private (farm) land.[75] Their stories reveal relationships with landscapes that unsettle the subject–object relation of property in cultural and legal discourse, and subvert the ontological ideals of ecological restoration. The farmers do not see themselves and their land-use practices as separate from the environment. Their objectives have been to reconcile themselves and their farming practices with the particularities of specific places, rather than to restore their properties to an imagined pristine state.

### *5.4.2 New materialism and relationality: interacting with nature and respecting non-human agency*

New materialism refers to the recent re-engagement of scholars with 'things', particularly via plural and relational ontological lenses that pursue de-binarised

67 Ibid, 122; see also Massy (2017), 500.
68 Showers and Malahleha (1992) quoted in Robertson et al (2000), 122.
69 Sinatra and Murphy (1999), 8.
70 We do not refer to these narratives as 'case studies' because we employed no empirical methods to obtain them. They derive entirely from the accounts published in the farmers' own names. Although not scholarly, we regard these sources as part of the literature that forms the data of this article.
71 Gunningham (2002).
72 We note that recently some farmers have grouped together to form the Environmental Farmers' Network and their work can be found on their website www.environmentalfarmersnetwork.net.au/ accessed 30 April 2018. However, the farmers referred to in this article are not connected to each other (to our knowledge) and have published their accounts under their own names with separate publishers.
73 Purvis (1999), 74; Andrews (2008), 17; Fenton (2010), 96; Massy (2017).
74 Massy (2017).
75 Smukler et al (2010), 81. See also Imhoff (2003).

explorations of humans and nature, and foreground non-human agency.[76] All four farmers articulate an empathy with landscapes that are recognised as being dynamic, living things, and that land-use practices must be sympathetic and responsive to material conditions.[77]

Bob Purvis developed his philosophy 'by trial and error over 25 years'.[78] He inherited his property 'Woodgreen' in arid central Australia in 1960, after his father, Robert Henry Purvis, first 'settled' the station in the early 1920s. Earlier grazing practices had 'left the country in a state of almost permanent drought'.[79] His father had overgrazed 'ten-fold', destroying the vegetation and consequently '[c]attle died … every year, all year'.[80] After destocking failed to improve soil fertility and vegetation cover he concluded that: 'time alone would not repair the damage … Where overgrazing had eliminated the palatable perennials, it was essential to introduce a quick responding, palatable, and if necessary exotic, perennial grass to take their place'.[81] He had to increase water retention across the property and stem the severe soil erosion caused by run-off during rain events. After years 'of practical experience and numerous failures',[82] he developed a system of 'ponding banks'. These operate by dissipating water energy, increasing silt deposition and reducing gully erosion. The resultant fertility increase led to significant revegetation. Purvis has built 600 ponding banks since 1960 and is adamant that they are 'the only way you can reclaim land'.[83]

Using a similar trial-and-error technique, Peter Andrews developed his Natural Sequence Farming approach, focused on soil fertility and hydrological processes, over several decades on properties in New South Wales (NSW) and South Australia. He insists that plants should be treated with the respect they deserve as agents of fertility, 'yet we treat plants as if they were of minimal importance to us, as if they were completely expendable'.[84] Biodiversity, he says, is 'one of the pillars of a sustainable agricultural system', and he urges farmers to 'allow the Australian landscape to regain the cover of vegetation that it once had'.[85] Andrews also identifies plants as critically important for productivity and raising quality horses.

Andrews is characterised by scientific colleagues[86] as capable of 'reading' the land, a skill which can be related to empathy, in this case with non-human life. He concluded that the intensive use of chemicals, and loss of vegetation and trees,

76 Goodman (1999); Trentmann (2009); Bennett (2010); Connolly (2013); Bawaka Country et al (2016); Davies (2017); Bartel (2017; 2018).
77 For example Massy (2017), 298.
78 Purvis (1986), 111.
79 Ibid.
80 Purvis (1999), 73.
81 Purvis (1986), 113.
82 Ibid, 115.
83 Purvis (1999), 78.
84 Andrews (2008), 4.
85 Ibid, 2.
86 Several scientists are acknowledged in, and contribute prefaces to, both of Andrews's books. See also Goldney cited in ABC TV (2005) and Williams and Mitchell cited in ABC TV (2015).

were the key reasons for the current crisis in agriculture and, more broadly, the Australian environment.[87]

When John Fenton inherited his property 'Lanark' in south-western Victoria in 1956, it was severely degraded, a 'bare expanse – a moonscape almost'.[88] It had lost most of its vegetation and 80 percent of its 200 bird species. Like Andrews, Fenton is adamant that 'trees create fertility by raising minerals to the surface and by simply dropping their leaves and branches'. Trees shelter 'stock from the worst of the winter weather' and 'crops and pasture from the wind, thereby reducing evaporation from the soil in summer and generally encouraging growth'.[89] It helps too, he says, that trees 'raise the capital value of any farm'.[90] He planted 100,000 trees on his property over several decades. Fenton contends that clearing trees was the performance of a European ideology in foreign soil, based on a view that trees are 'obstacles to progress' and productivity.[91] Fenton's own early farming practices included heavy reliance on fertilisers, pesticides and high stocking rates.[92] However, he came to see that the land was severely overgrazed and the soils degraded, and realised that his farming business could not continue indefinitely under these conditions.[93]

Charles Massy subscribes explicitly to an 'Earth empathic'[94] view and explains how this may be developed through gaining an understanding of landscape functions, through humbly listening to the land,[95] recognising non-human agency and de-centring the human. He describes a fellow regenerative farmer thus: '[t]he empathic feel for nature and the love of its beauty that this helped engender in him was part of what led to his farm regeneration'.[96]

Massy observes that the impetus for change from conventional to regenerative agriculture may be gradual, or crisis-led.[97] Massy's own 'ponderous',[98] 'delayed and bumbling'[99] learning curve from an arrogant 'young industrial farmer'[100] arose from several incidents, including when a freshly ploughed field was washed away in a rainstorm: 'I drove up the next morning to find a scene of destruction … My heart was stricken and I felt like a vandal as I comprehended the enormity of what I had done'.[101] Massy faced debt, as well as the material reality of the landscape: '[o]nly … after decades of block-headedly trying to "fight"

87 Andrews (2008), 5 and 18.
88 Fenton (2010), 60.
89 Ibid, 10.
90 Ibid.
91 Ibid, 3.
92 Ibid, 6, 85.
93 Ibid, 2.
94 Massy (2017), 216 and 333.
95 Massy (2017), 303.
96 Massy (2017), 298.
97 Ibid, 295, 408, 500.
98 Ibid, 184.
99 Ibid, 490.
100 Ibid, 183.
101 Ibid, 184.

perceived droughts, did I finally come to terms with … Antipodean reality'.[102] Massy previously viewed the land as 'an inanimate resource',[103] 'an inert substrate',[104] instead of appreciating it as a self-organising system.[105] He had believed that he was 'in control of this inert agricultural environment'.[106] For the past 40 years he has practised regenerative agriculture on his Merino sheep stud, Severn Park, near Cooma on the Monaro plains of NSW, using a blend of Yeoman's keyline (essentially a property-wide system of water retention) and holistic management approaches.[107] Massy focuses on five landscape functions: solar energy (accessed through plant photosynthesis); the water and soil-mineral cycles; dynamic ecosystems; and the human–social aspects.[108]

Massy attributes his earlier failings to having 'ignored vital warning signs revealed by the landscape'.[109] He identifies 'ecological literacy' as the solution, defined as 'the ability to read a landscape'[110] which requires 'a great empathy with, and understanding of, nature and how it functions'.[111] He believes it is essential to 'work with and not against nature and its systems'[112] and that it would be better if 'farmers get out of the way' and let nature self-organise.[113] Massy identifies past practices as culpable, although he differentiates between regenerative European practices, like biodynamic farming,[114] that can be applied in Australia, and those, such as chemical-intensive industrial agriculture, which suit neither Europe nor Australia, nor anywhere else. And in an echo of Rachel Carson's arguments in *Silent Spring*[115] he observes how poor practices have been 'encouraged and supported by government and private extension and research bodies'[116] and lays blame with the chemical industry, reductionist science and governments eager to promote technical over holistic solutions.

### *5.4.3 New environmental governance: acting beyond compliance and convention*

New environmental governance refers to the distribution of power beyond the state, and draws particular attention to the interactions between government and non-government actors.[117] Australian examples include volunteer activity to

102 Ibid, 291, 293.
103 Ibid, 62.
104 Ibid, 183.
105 Ibid, 64.
106 Ibid, 183.
107 Ibid, 155.
108 Ibid, 56.
109 Ibid, 183.
110 Ibid, 64.
111 Ibid, 65.
112 Ibid, 118.
113 Ibid, 138, 303, 378.
114 Ibid, 226ff.
115 Bartel (2018).
116 Massy (2017), 294.
117 Gunningham (2009); Bartel and Barclay (2011); Etemire and Muzan (2017).

restore biodiversity, ranging from individual landholders to nationwide programmes. Many of these are publicly funded and underpinned by state and federal policies.

The farmers' stories demonstrate occasional benefit from government support, but more often their actions have been performed despite, and sometimes in contravention of, the law, as well as being non-compliant with social convention. They all express frustration and anger with the persistence of maladapted farming practices, and with the laws and policies that support them. Fenton maintained wetlands on his property despite government financial incentives, and peer pressure, to drain them.[118] Against environmental law and policy, Andrews champions weeds for restoring degraded soils. The second author of the present chapter once attended a meeting between Andrews and NSW State government officers, the purpose of which was to discuss at length his violation of several regulations pertaining to his failure to remove 'weeds' and 'invasive species' from the property and the financial penalties for this inaction. Andrews is emphatic that he does not prefer weeds 'for their own sake' and does not advocate 'spreading weeds over the entire landscape'.[119] It is the function that plants, classified by law as weeds, perform that Andrews says is crucial to revitalise degraded soils. Informed by the materiality of specific places, rather than by law that abstracts place, he regards weeds as useful plants that grow rapidly and repair the environment by accumulating minerals, and thus fertility.[120]

Purvis also disregarded the native/exotic dichotomy, instead introducing buffel grass, a 'quick responding, palatable perennial grass' from Africa. Once this had taken off, other native grasses returned and today the result is a more biodiverse landscape than he inherited. Purvis realised from the outset that he needed to 'reclaim my degraded country' by developing a 'philosophy' of land use that departed from the 'conventional wisdom' that he thought was part of the problem: '[f]or many years I have found myself quite at odds with my pastoralist colleagues and many government bureaucrats who are charged with overseeing land management'.[121] The legislation at the time 'condoned, even endorsed' overgrazing.[122] The land Purvis inherited was leasehold to which covenants were attached, including minimum stocking rates. Purvis wished to reduce the stocking rate applied to his father, of 3,000 head of cattle, to address soil erosion due to vegetation loss. Purvis reduced the rate to 300, placing him in default of his lease.[123] He asked for an exemption: '[y]ou've got to convince the bureaucrat that

118 Fenton (2010), 91.
119 Andrews (2006), 126.
120 Ibid, 128–9.
121 Purvis (1986), 110.
122 Ibid, 116.
123 Purvis (1999), 74, 83, 116. The pre-1970 stock figure was 3,000 and the 1995 rate was 900. The figure of 3,000 head of cattle was the minimum stocking rate stated in the lease agreement. The Pastoral Land Act 1992 (NT) replaced maximum stocking rates with mandatory rangeland monitoring programs that include assessment of carrying capacity of property.

the stocking rate is unrealistic. It took me twenty-odd years to get them to change, and it only changed for me'.[124]

Purvis developed his herd towards quality rather than quantity.[125] He became one of the most successful graziers in the district through sustaining profitable, high-quality beef cattle production.[126] The success of his farming business, like the health of his farm, depended on convincing government officials to accept a substantial variation in the terms of his lease regarding its most important feature – the stocking rate. Purvis's decision to challenge its terms is antithetical to one of the most iconic species of Australian land law. The purpose of a pastoral lease is pastoral activity.[127] Purvis became frustrated not only with government policy and a one-size-fits-all approach to leasehold covenants, but also with farmers who failed to change their practices.

Other farmers have found offence, madness and foolishness in the views, concerns and practices of these 'dissident' farmers.[128] Andrews has been decried as a 'crazy' man and '(f)or years his ideas have been ignored and ridiculed'.[129] Professor John Williams, former Chief of the CSIRO's Land and Water division, and a founding member of the Wentworth Group of Scientists, observes: 'Peter has faced much opposition from many different quarters as he has sought to put into practice activity on the ground that runs against conventional wisdom, but he has continued and prevailed.'[130] The changes Fenton made to his property isolated him from his community, which regarded him 'as eccentric or worse'.[131] He says: '[f]armers who heard I had given up using artificial fertilisers like superphosphate viewed me as a kind of heretic'.[132] Fenton was also pressured to conform to conventional farming practices by government officials and bank managers. He accepts his ostracism as 'understandable':

> For years, various government authorities had been encouraging and financing farmers to clear their land of trees and to drain swamps. Yet here I was doing exactly the opposite – planting tens of thousands of trees and restoring wetlands – and I was doing it all at my own expense.[133]

The experiences of these farmers indicate a substantial lack of support and encouragement for farmers who question conventional approaches. Massy

124 Purvis (1999), 77.
125 Some of the most restrictive conditions and covenants attached to pastoral leases in the Northern Territory were removed in a major reform in 1992. See Pastoral Land Act 1992 (NT).
126 Walsh (2009), 3–4.
127 Productivity Commission (2002).
128 We are not proposing this as a category, but see also Howden et al (1998); Cross and Ampt (2016).
129 Gerry Harvey, in Andrews (2006), xii.
130 John Williams, in ibid, *x*.
131 Fenton (2010), 118.
132 Ibid, 85.
133 Ibid, 118 and 120.

describes the adoption of regenerative agriculture as 'a path that leaves scars', due to being shunned.[134] He describes how there is a 'dearth of Earth-empathic thinking … [in] the economic rationalist world of Australian farming'.[135] However, none of the farmers were anti-production. All needed to maintain farm productivity; however, they saw that this could not be sustained through conventional (i.e. modern industrial) practices, which they all perceived were disconnected from the specific geographical conditions of the places in which they lived.

### *5.4.4 New localism: vernacular knowledge, place attachment and decolonising the placelessness of property*

We use 'new localism' to refer to the 'spatial turn' and scholarship that reflects a renewed emphasis on the explanatory power of geography.[136] Notwithstanding the geographical differences between the landscapes in which these farmers lived and worked, and indeed because of them, all demonstrated the development of unique place-based knowledge that situates human land-use practices within, and adapted to, the specific conditions and limitations of particular landscapes.

Fenton highlights the centrality of place-based knowledge not only as valuable in itself, but as critical to successful farming practice. He laments that there was no information, education or advice available to him and that his lessons were hard learned over a long period of time and at great personal and financial expense. Nevertheless, he is optimistic about the outcome of his decades of 'radical' farming practices. According to Fenton's analysis, conventional practices don't work because they are poorly aligned with local ecological functions, due to the lack of place-based knowledge.

All four farmers developed their own place-based (also known as vernacular[137]) knowledge through learning about the local ecosystems and making a conscious effort to adapt their farming activities, ceasing damaging practices and stopping degradation. All four articulate dismay and sadness about the legacy of past misunderstandings of, blindness to and ignorance of ecosystem functions. All identified the lack of knowledge about these ecological functions as one of the biggest obstacles to changing practices. While not all farmer knowledge may be appropriate or necessarily beneficial,[138] it has been recognised that: '[k]nowledge is a key driver of change in farming systems, both in relation to farming practices and in relation to biodiversity'.[139]

Vernacular knowledge is neither quickly nor easily attained; it takes a long time to understand even a little because places are complex and dynamic. All four farmers recognise that their knowledges are still only nascent. All write about Indigenous knowledge as a vital and authoritative source, and regret its ongoing

134 Massy (2017), 71, 133, 296–7 and 473.
135 Ibid, 227.
136 Bartel (2014; 2017).
137 Wagner (2007); Bartel (2014); Goldstein (2015).
138 Gill (2014); Bartel (2014).
139 Norton and Reid (2013), 17; Also Massy (2017), 52.

diminution and loss through dispossession and persistent colonisation.[140] In addition to his trial-and-error approach, Fenton 'sought to learn about and understand the land management techniques that the Aborigines employed'.[141] One of Purvis' strategies was fire, which he only belatedly recognised as a farming technique.[142] He appreciates that 'the blackfella burned' but his father's generation prohibited fire and this knowledge was 'lost'[143] to him.

Purvis expresses deep anger about the loss of the critical place-based knowledge of the dispossessed Indigenous owners and the delay this created in addressing the degradation of the land, saying that he had to 'learn what the blackfellas knew three generations back'.[144] He laments that he had to 'teach himself how to do this' and that it took him years of trying 'many avenues' instead of learning in a far shorter time from Indigenous Australians whose knowledge was place-based, authoritative and long-standing.

Purvis acknowledges that 'all the landscapes are slightly different, what applies here doesn't necessarily apply somewhere else'. His observation that there is no substitute for place-based learning and place-specific knowledge was at odds with government policy and industry practice that imposed a one-size-fits-all approach to farming. Other farmers in his region have eventually appreciated that the strategies Purvis used to turn his farming business around could work for them, and he says others have ultimately followed his lead.[145] While farmers may be conservative, they are also avid innovators and early adopters of new techniques.[146]

Vernacular knowledge and place attachment,[147] sense of place,[148] belonging,[149] biophilia, topophilia or endemophilia[150] are different ways of conceptualising relationships between people and place. Place attachment is defined as 'the emotional bonds between people and a particular place or environment'.[151] It serves as a useful frame to describe the love of place apparent in the four narratives. They love the landscapes in which they live and work. All of them point to the real and significant pressures experienced by Australian farmers, financially, legally and personally, in keeping their farms viable, let alone successful. What keeps them all there, they say, is their attachment to the land.

When Purvis inherited his father's property, its debts exceeded its commercial value.[152] He sees it as his job to 'arrest the decline' and 'stop the slide', realising

140 Massy (2017), 78, 123, 288, 480, 491, 505.
141 Fenton (2010), 9.
142 Also Massy (2017), 281ff.
143 Purvis (1999), 82.
144 Ibid.
145 ABC TV (2014, 2017).
146 Cross and Ampt (2016); Massy (2017), 116, 195, 201; Telesetsky (2017).
147 Bartel and Graham (2016).
148 Robertson et al (2000), 126.
149 Wright (2015).
150 See Bartel (2014); also Albrecht (2012).
151 Seamon (2014), 11.
152 Purvis (1986), 111.

that this is all he could achieve in a single lifetime.[153] He hopes his family will learn the knowledge of the land from him, and share his love for it: 'I love this land, I was born here. Why would I want to live anywhere else?'[154] Purvis connects place attachment to good farming practice.[155]

Fenton similarly urges farmers to relate to their properties beyond the paper title deeds, as a partnership rather than a subject–object relation: 'I personally prefer to see it as a privilege rather than a responsibility'.[156] His analysis is that 'most farmers' already do have an emotional attachment to the land, but that it may not be articulated in words: '[w]hat other reason could a farmer have for working 50–60 hours a week for not much money?'[157] That is, at least, his experience:

> I had a passion for the environmental work that I undertook. Without that passion, obsession, there is no way I could have persevered with the work over half a century, for it certainly took its toll, physically, financially and emotionally.[158]

His connection to the land also causes him to reflect on the colonisation and dispossession of 'local Aborigines', observing that: '[i]n barely 200 years, Europeans have succeeded in turning a healthy landscape into one that is clearly ailing'.[159] Fenton is 'convinced' that farmers must 'fundamentally change' their relation to land, learning from the place-based knowledge and place attachment that underpins the success of Indigenous land laws and economies:

> My own belief is that we should view the land as the Aborigines did. They regarded it with reverence, so they treated it with reverence. We must do the same. We need to feel an emotional attachment, maybe even a spiritual attachment, to the land. We need to nurture the land and its resources, not exploit them.[160]

Like Purvis, Fenton understands the role of time in the long project of amending for the environmental damage of past mistakes. His property is being worked to a 100-year plan that necessitates thinking about landscapes beyond our own lifetimes – 'if we are to achieve equilibrium' in the environment.[161]

The narratives all identify the disconnection between people and place, the dichotomous logic that separates humans and nature, and the loss of place-based

153 Purvis (1999), 84.
154 Ibid, 84.
155 Ibid, 82.
156 Fenton (2010), 6.
157 Ibid, 6.
158 Ibid, 150.
159 Ibid, 96.
160 Ibid, 6.
161 Ibid, 164.

knowledge, as major sources of persistent adverse environmental change. Their separate but similar experiences in different geographies and jurisdictions offer critical insights into the problems caused by a paradigm of 'placelessness' that underpins Anglo-Australian land laws.[162]

## 5.5 Breaking the impasse: reconciling humans to nature and property to the environment

The narratives of all four farmers reveal that their motivations and experiences have developed largely independently of, despite, and sometimes in contravention of, the law. Furthermore, their narratives indicate that severely degraded and almost non-functional landscapes would not have been improved had the law been followed. Landscape function, productivity and the lives of landholders have all benefitted from listening to the land, rather than to the law. Purvis accepts that past poor management 'could be excused on the basis of ignorance', but that this situation should be 'now long gone'.[163] However, the legal position remains lagging and incoherent, especially while environmental law and property are antagonistic.[164]

Current approaches may be undermining rather than promoting the achievement of environmental objectives. A focus on the restoration of isolated parcels may not be the ideal aim. Massy declares: 'we cannot return landscapes to a state that existed before the invasion of the Europeans',[165] and further, in describing others' experiences: 'the key was a functioning ecosystem – not some impossible dream of returning to a theoretical pre-1830s biota'.[166] Fenton's work was also not about conserving nature or restoring an environment somehow separate to humans, but about productive and viable farming practice. Neither Andrews nor Purvis has adopted restoration approaches, evident in their deployment of weeds as a curative. Purvis describes 'restorative' projects as folly: 'you can't ever have it like it was, but you realise all things are intertwined'.[167]

The farmers' narratives underscore the need for more diversity, nuance and flexibility in legal rules and categories, rather than their enforcement, including a relaxation of species categories.[168] Regulatory systems and attendant bureaucracies are often not well-equipped or designed to address a myriad of complex environmental issues at multiple scales. Approaches are generally a-spatial, and are state- or jurisdiction-centric, also known as top-down[169] and uni-level,[170] applying state-wide irrespective of geographical variability. Centralised and universalist approaches are also adopted for administration and

162 Graham (2011), 8.
163 Purvis (1986), 116.
164 Vanclay (2004); Beeton and Lynch (2012).
165 Massy (2017), 506.
166 Ibid, 249, 243, 301.
167 Purvis (1999), 82.
168 Bartel (2014; 2017).
169 Massy (2017), 406, 420.
170 Lockwood (2010). Also Winter (2005).

enforcement: the same system is applied to all, irrespective of biophysical heterogeneity and dynamism. Although consistency is appropriate in terms of the appearance of fairness and efficiency, because the environment is not homogenous, the application of a single rule and process that cannot be adapted to different and changing geographical conditions often renders it unviable and ineffective.[171]

The narratives demonstrate that although their farming practices were often neither legally nor culturally compliant – contravening both formal and informal rules – their practices were consistent with the overarching objectives of many well-intentioned environmental laws in terms of improving ecological outcomes. This compliance with the spirit (if not always the letter) of the law reveals their intrinsic (rather than extrinsic) motivation[172] in achieving positive environmental outcomes on private agricultural lands. Such departures were pragmatic[173] but they were also philosophical. They have recognised non-human agency and adapted reflexively to it, rather than attempting to bend nature to their own will, or that of the law. They have seen themselves as part of nature rather than apart from it.[174] The farmers' stories demonstrate how we may attempt a less anthropocentric engagement.[175] According to this de-centred and de-binarised view,[176] the agency of non-human nature is accorded greater recognition,[177] and human–nature relationships may be characterised as co-evolution.[178] The four narratives demonstrate the potential of a relational–material ontology for conservation on private property through reconciliation, that allows us to question the singularity of human agency and demonstrates instead how we are interconnected with a multitude of agential things.[179]

Materiality and relationality are each evidently critical for farmscaping – Massy refers to his 'continuous dialogue with my landscape and all its functions',[180] and that the landscape furthermore demonstrates its own intelligence and cognisance.[181] This presents obvious challenges to the subject–object dualism that is central to the institution of private property.[182] It is the latter conceit (and deceit) that also causes problems when extending the property institution to other natural resources through commodification and commensuration.[183] Similarly de-binarised nature-orientations have been found to be associated with more

171 Robards et al (2011); Barr and Devine-Wright (2012); Bartel (2014).
172 See Gunningham and Grabosky (1998).
173 See McDonald et al (2004).
174 Massy (2017), 313, 492.
175 Bartel (2017).
176 Bartel et al (2014).
177 Goodman (1999).
178 Massy (2017), 382.
179 Goodman (1999); Trentmann (2009); Bennett (2010); Connolly (2013); Bawaka Country et al (2016); Davies (2017).
180 Massy (2017), 310.
181 Ibid, 321.
182 See Davies (2016).
183 Bartel et al (2014).

environmentally friendly views, values and practices.[184] This could mean that ownership, long critiqued in ecological circles, is not necessarily obstructive. Fenton refers to his property as 'part of the natural environment', and states that the health of the environment simply was the health of his farm and *vice versa*.[185] Private land ownership may present unique opportunities to support regenerative agriculture and ecological reconciliation – by first acknowledging and prioritising its very possibility in the health of the land itself. Private land ownership can and could be reconceptualised and reimagined further to incorporate place-based perspectives.[186]

Anglo-Australian property law developed in part through the rejection of the place-based knowledge systems it attempted to replace.[187] If there is a way to restore place to property, to decolonise the placelessness of Anglo-Australian law, it involves the recognition, recovery and development of vernacular knowledge as an authoritative source of land-use rules and practices. Vernacular knowledge of diverse and dynamic places is the authoritative source informing these farmers' land-use practices, rather than homogenous laws abstracted from places and its biological particularities.

Rather than persisting with an idealised notion of 'restored' or pristine environments, and a one-size-fits-all approach to regulation, the farmers' narratives indicate that preferable outcomes can be achieved by integrating place-based landholder knowledge into private agricultural land-use practices.[188] The narratives demonstrate the potential of ecological reconciliation to transform the dichotomous structure and disingenuous erasure of humans from landscapes, underpinning the approaches of both reservation and restoration in attempting to reinstate and preserve 'historical' or 'classical' landscapes. Instead of imposing an essentialist paradigm, a more dynamic and place-based approach to environmental engagements is now required of law, that allows for a more 'open-ended'[189] and emergent diversity of future agricultural ecologies, including maximum biodiversity, rather than focusing primarily on pre-1788 environments.[190] Such practices are particularly pertinent for private land and regenerative agriculture, whose aims go beyond the 'restoration' of historical biodiversity in often partial and piecemeal parcels of land.[191] They are instead whole-of-property and landscape-scale approaches which meet multiple aims including, but not limited to, the re-establishment of biodiversity. Biodiversity conservation is integrated into a wider agenda of private land ownership that includes regenerating soil, water, energy, and social systems and functions,

184 Gosling and Williams (2010); Bartel and Graham (2016).
185 Fenton (2010), 2.
186 Bartel and Graham (2016); Massy (2017), 260.
187 See *Steel v Houghton* (1788) 1 BH. H. 51, 126 ER 32. See Graham (2011), Ch. 3.
188 Bartel (2014); England (2015); Godden and Cowell (2016).
189 Massy (2017), 65.
190 Miller and Hobbs (2007); Davis et al (2011); Head (2012).
191 See Campbell et al (2017).

encompassing community health and wellbeing, as well as sustaining an income and generating production.[192]

Were law reform endeavours to adopt a non-anthropocentric place-based paradigm, rather than focusing on compelling the 'restoration' of nature, then public policy objectives could become oriented towards the reconciliation of humans with nature,[193] and reconnecting agriculture and 'nature'.[194] This should not attempt to annihilate difference or alterity,[195] nor revert to neo-environmental determinism,[196] nor appropriate Indigenous laws and ontologies. Rather, Anglo-Australian legal approaches to ecological restoration on private agricultural land should focus on appreciating and fostering new and emergent co-becomings.[197]

Although it would certainly be challenging to develop a property regime that is sufficiently flexible and dynamic to support a plurality of emplaced and embedded objectives; a diversity of interactions across space and time; and a plethora of interventions – it is possible. This is evident from the long-term success of the original and enduring place-based laws of Australia – Indigenous laws – that are yet to be recognised by Anglo-Australian law and culture,[198] and which in most cases were forcefully supplanted.

## 5.6 Conclusions

Reconceiving the relationship between private property and the environment is a major challenge of our time. The scale of modern impacts serves (paradoxically) to remind us of our power, but also of our powerlessness – as we are also subject to the adverse consequences of anthropogenic environmental change. Recognising the role of humans as agents of our own demise as part of nature is one step towards a fundamental re-examination of the paradigm underpinning our approach to land use, as well as to land-use laws including both property and environmental law.

Ecological restoration is one construct that is being used to reimagine land management. While preferable to more exploitative human–nature relationships, restoration is generally aimed at reinstating pre-1788 habitats in Australia, which may be neither desirable nor feasible, has little relevance for broader-scale application, and perpetuates an anthropocentric approach that may ignore more fundamental questions.

This chapter has proffered an alternative – reconciliation. It is essentially a de-binarised variation of restoration that re-integrates humans with nature, and

192 Massy (2017), 302–3, 419.
193 Mace (2014).
194 Winter (2005).
195 Head and Gibson (2012); Plumwood (2006).
196 Wilcock et al (2013).
197 Bartel (2017).
198 See discussion of Graham (2008) in Davies (2016). See also Orts (1995). This is not to discount other arrangements such as Indigenous Protected Areas: Figgis et al (2005); Godden and Cowell (2016).

unsettles the subject–object dichotomy of conventional private property relations. Its potential has been inspired and demonstrated by regenerative agriculture and landholders whose land-use practices are place-based approaches that de-centre the human, where the environment is not an end point, but a relational–material co-becoming. Instead of restoration, they have achieved reconciliation – between human and nature, and between agricultural practices and Australian landscapes. The farmscape is the result. A similar way forward for the reform of legal and cultural discourses around the ownership and regulation of land use, and environmental law and management, is proposed – one that is dynamic, reflexive and place-based, that aims to reconnect humans with nature that is already there – rather than at humans restoring a 'nature' that is external and prior to us.

## References

### *Primary sources (law materials)*

Pastoral Land Act 1992 (NT)

*Steel v Houghton* (1788) 1 BH. H. 51, 126 ER 32

### *Secondary sources*

ABC TV (2005) 'Of Droughts and Flooding Rains – Parts 1 and 2' *Australian Story*, Australian Broadcasting Corporation, www.abc.net.au/austory/of-droughts-and-flooding-rains—part-1/9169558 and www.abc.net.au/austory/of-droughts-and-flooding-rains—part-2/9169550.

ABC TV (2014) 'Letter from Wooleen' *Australian Story*, Australian Broadcasting Corporation, www.abc.net.au/austory/letter-from-wooleen/5912566.

ABC TV (2015) 'The Battle for Tarwyn Park – Parts 1 and 2' *Australian Story*, Australian Broadcasting Corporation, www.abc.net.au/austory/the-battle-for-tarwyn-park/6441486 and www.abc.net.au/austory/the-battle-for-tarwyn-park—part-two/6461782.

ABC TV (2017) 'Splendour in the Grass' *Australian Story*, Australian Broadcasting Corporation, www.abc.net.au/austory/splendour-in-the-grass-teaser/8508044.

Vanessa Adams and Katie Moon (2013) 'Security and Equity of Conservation Covenants: Contradictions of Protected Area Policies in Australia' 30 *Land Use Policy* 114.

Afshin Akhtar-Khavari and Anastasia Telesetsky (2016) 'From Protection to Restoration: A Challenge for Environmental Governance' in D. Fisher (ed.) *Research Handbook on Fundamental Concepts of Environmental Law*, Edward Elgar.

Glenn Albrecht (2012) 'Psychoterratic Conditions in a Scientific and Technological World' in P. Kahn and P. Hasbach (eds) *Ecopsychology: Science, Totems, and the Technological Species*, MIT Press, 241.

Peter Andrews (2006) *Back from the Brink: How Australia's Landscape can be Saved*, ABC Books.

Peter Andrews (2008) *Beyond the Brink: Peter Andrews' Radical Vision for a Sustainable Australian Landscape*, ABC Books.

Jennifer Atchison (2015) 'Experiments in Co-existence: The Science and Practices of Biocontrol in Invasive Species Management' 47 *Environment and Planning A* 1697.

Australian Law Reform Commission (2015) *Traditional Rights and Freedoms – Encroachments by Commonwealth Laws*, Report 129, ALRC.
Elaine Barclay and Robyn Bartel (2015) 'Defining Environmental Crime: The Perspective of Farmers' 39 *Journal of Rural Studies* 188.
Robyn Bartel (2014) 'Vernacular Knowledge and Environmental Law: Cause and Cure for Regulatory Failure' 19 *Local Environment: The International Journal of Justice and Sustainability* 891.
Robyn Bartel (2017) 'Place-thinking: The Hidden Geography of Environmental Law' in V. Brooks and A. Philippopoulos-Mihalopoulos (eds) *Handbook of Research Methods in Environmental Law*, Edward Elgar, 159.
Robyn Bartel (2018) 'Place-speaking: Attending to the Relational, Material and Governance Messages of Silent Spring' 184 *Geographical Journal* 64.
Robyn Bartel and Elaine Barclay (2011) 'Motivational Postures and Compliance with Environmental Law in Australian Agriculture' 27 *Journal of Rural Studies* 153.
Robyn Bartel and Nicole Graham (2016) 'Property and Place Attachment: A Legal Geographical Analysis of Biodiversity Law Reform in New South Wales' 54 *Geographical Research* 267.
Robyn Bartel, Paul McFarland and Colin Hearfield (2014) 'Taking a De-Binarized Envirosocial Approach to Reconciling the Environment vs Economy Debate: Lessons from Climate Change Litigation for Planning in NSW, Australia' 85 *Town Planning Review* 67.
Stewart Barr and Patrick Devine-Wright (2012) 'Resilient Communities: Sustainabilities in Transition' 17 *Local Environment: The International Journal of Justice and Sustainability* 525.
Bawaka Country et al (2013) 'Caring as Country: Towards an Ontology of Co-becoming in Natural Resource Management' 54 *Asia Pacific Viewpoint* 185.
Bawaka Country et al (2016) 'Co-becoming Bawaka: Towards a Relational Understanding of Place/Space' 40 *Progress in Human Geography* 455.
Robert Beeton and Jasmine Lynch (2012) 'Most of Nature: A Framework to Resolve the Twin Dilemmas of the Decline of Nature and Rural Communities' 23 *Environmental Science and Policy* 45.
Jane Bennett (2010) *Vibrant Matter: A Political Ecology of Things*, Duke University Press.
Marc Bekoff (2014) *Rewilding Our Hearts: Building Pathways of Compassion and Coexistence*, New World Library.
Fikret Berkes (2003) 'Rethinking Community-based Conservation' 18 *Conservation Biology* 621.
Karl J. Bonnedahl (2017) 'From Exploitation and Expansion to Evolutionary Coexistence: A New Realism for Life Beyond the Anthropocene' in P. Heikkurinen (ed.) *Sustainability and Peaceful Coexistence for the Anthropocene*, Routledge, 162.
Irus Braverman (2012) *Zooland*, Stanford University Press.
Andrew Campbell, Jason Alexandra and David Curtis (2017) 'Reflections on Four Decades of Land Restoration in Australia' 39 *Rangeland Journal* 405.
Neil H. Carter and John D.C. Linnell (2016) 'Co-adaptation is Key to Existing with Large Carnivores' 31 *Trends in Ecology and Evaluation* 575.
Gerado Ceballos et al (2015) 'Accelerated Modern Human-induced Species Losses: Entering the Sixth Mass Extinction' 1 *Science Advances* e1400253.
William Connolly (2013) 'The "New Materialism" and the Fragility of Things' 41 *Millennium: Journal of International Studies* 399.
Richard T. Corlett (2016) 'Restoration, Reintroduction, and Rewilding in a Changing World' 31(6) *Trends in Ecology and Evolution* 453.

Rebecca Cross and Peter Ampt (2016) 'Exploring Agroecological Sustainability: Unearthing Innovators and Documenting a Community of Practice in Southeast Australia' 30 *Society and Natural Resources* 585.

Margaret Davies (2016) 'Material Subjects and Vital Objects – Prefiguring Property and Rights for an Entangled World' 22(2) *Australian Journal of Human Rights* 37.

Margaret Davies (2017) *Law Unlimited: Materialism, Pluralism and Legal Theory*, Routledge.

Mark Davis et al (2011) 'Don't Judge Species on Their Origins' 474(7350) *Nature* 153.

Uzuazo Etemire and Menes Abinami Muzan (2017) 'Governance and Regulatory Strategies Beyond the State: Stakeholder Participation and the Ecological Restoration of Ogoniland' 26 *Griffith Law Review* 275.

Brendan Edgeworth (2017) *Butt's Land Law* 7th edn, Lawbook Company.

Robert Elliot (1985) *Faking Nature: The Ethics of Environmental Restoration*, Routledge.

Phillipa England (2015) 'Conservation Covenants: Are they Working and What Have we Learned?' 34 *University of Tasmania Law Review* 92.

John Fenton (2010) *The Untrained Environmentalist: How an Australian Grazier Brought his Barren Property Back to Life*, Allen and Unwin.

Penelope Figgis, Doug Humann and Michael Looker (2005) 'Conservation on Private Land in Australia' 15 *Parks* 19.

James A. Fitzsimons and C.B. Carr (2014) 'Conservation Covenants on Private Land: Issues with Measuring and Achieving Biodiversity Outcomes in Australia' 54 *Environmental Management* 606.

Jodi Frawley and Iain McCalman (eds) (2014) *Rethinking Invasion Ecologies from the Environmental Humanities*, Routledge.

Bill Gammage (2011) *The Biggest Estate on Earth*, Allen and Unwin.

Rupert Gerritsen (2008) *Australia and the Origins of Agriculture*, Archaeopress.

Nicholas Gill (2014) 'Making Country Good: Environmental Change and Stewardship in Central Australian Pastoral Culture' 39 *Transactions of the Institute of British Geographers* 277.

Lee Godden and Stuart Cowell (2016) 'Conservation Planning and Indigenous Governance in Australia's Indigenous Protected Areas' 24(5) *Restoration Ecology* 692.

Gondwana Link (2014) *Gondwana Link Whole of Link Ecological Guide*, Gondwana Link.

Diane Goldstein (2015) 'Vernacular Turns: Narrative, Local Knowledge, and the Changed Context of Folklore' 128 *Journal of American Folklore* 125.

David Goodman (1999) 'Agro-food Studies in the "Age of Ecology": Nature, Corporeality, Bio-politics' 39 *Sociologia Ruralis* 17.

Elizabeth Gosling and Kathryn Williams (2010) 'Connectedness to Nature, Place Attachment and Conservation Behaviour: Testing Connectedness Theory Among Farmers' 30 *Journal of Environmental Psychology* 298.

Nicole Graham (2011) *Lawscape: Property, Environment, Law*, Routledge.

Nicole Graham (2014) 'This is Not a Thing: Land, Sustainability and Legal Education' 26 *Journal of Environmental Law* 395.

Nicole Graham (2016) 'The ALRC Freedoms Inquiry, Real Property and Environmental Law' 31 *Australian Environment Review* 364.

Neil Gunningham (2002) 'Beyond Compliance: Next Generation Environmental Regulation' 36 *Journal of Law and Society* 145.

Neil Gunningham (2009) 'The New Collaborative Environmental Governance: The Localization of Regulation', Current Issues in Regulation Conference, Australian Institute of Criminology, Melbourne.

Neil Gunningham and Peter Grabosky (1998) *Smart Regulation: Designing Environmental Policy*, Clarendon Press.

Marcus Hall (2010) *Restoration and History: The Search for a Usable Environmental Past*, Routledge.

Mathew Hardy et al (2017) 'Exploring the Permanence of Conservation Covenants' 10 *Conservation Letters* 221.

Lesley Head (2012) 'Decentring 1788: Beyond Biotic Nativeness' 50 *Geographical Research* 166.

Lesley Head and Chris Gibson (2012) 'Becoming Differently Modern: Geographic Contributions to Generative Climate Politics' 36 *Progress in Human Geography* 699.

Lesley Head, Jennifer Atchison and Catherine Phillips (2015) 'The Distinctive Capacities of Plants: Re-thinking Difference via Invasive Species' 40 *Transactions of the Institute of British Geographers* 399.

Pasi Heikkurinen (2017) 'Introduction' in P. Heikkurinen (ed.) *Sustainability and Peaceful Coexistence for the Anthropocene*, Routledge.

Eric Higgs (1997) 'What is Good Ecological Restoration?' 11 *Conservation Biology* 338.

Rickard Hobbs (2013) 'Foreword' in D. Norton and N. Reid (eds) *Nature and Farming: Sustaining Native Biodiversity in Agricultural Landscapes*, CSIRO Publishing.

Richard Hobbs et al (2006) 'Novel Ecosystems: Theoretical and Management Aspects of the New Ecological World Order' 15 *Global Ecology and Biogeography* 1.

Richard Hobbs and Viki Cramer (2008) 'Restoration Ecology: Interventionist Approaches for Restoring and Maintaining Ecosystem Function in the Face of Rapid Environmental Change' 33 *Annual Review of Environment and Resources* 39.

Richard Howitt and Sandie Suchet-Pearson (2006) 'Rethinking the Building Blocks: Ontological Pluralism and the Idea of "Management"' 88 *Geografiska Annaler B* 323.

Peter Howden et al (1998) 'Working with the Grain: Farming Styles Amongst Australian Broadacre Croppers' 8 *Rural Society* 109.

Daniel Imhoff (2003) *Farming with the Wild: Enhancing Biodiversity on Farms and Ranches*, Sierra Club Books.

William R. JordanIII (2003) *The Sunflower Forest: Ecological Restoration and the New Communion with Nature*, University of California Press.

Eric Katz (1992) 'The Big Lie: Human Restoration of Nature' 12 *Research in Philosophy and Technology* 231.

Jozef Keulartz and Cor van der Weele (2009) 'Between Nativism and Cosmopolitanism: Framing and Reframing in Invasion Biology' in M. Drenthen, J. Keulartz and J. Proctor (eds) *New Visions of Nature: Complexity and Authenticity*, Springer, 237.

Bruce Lindsay (2016) 'Legal Instruments in Private Land Conservation: The Nature and Role of Conservation Contracts and Conservation Covenants' 24 *Restoration Ecology* 698.

Michael Lockwood (2010) 'Good Governance for Terrestrial Protected Areas: A Framework, Principles and Performance Outcomes' 81 *Journal of Environmental Management* 754.

Thomas E. Lovejoy (2016) 'Conservation Biology: The Importance of Wilderness' 26 *Current Biology* R1226.

Tim Low (2017) *The New Nature*, Penguin Books.

Richard Mabey (2010) *Weeds: How Vagabond Plants Gatecrashed Civilisation and Changed the Way We Think About Nature*, Profile Books.

J.B. MacKinnon (2013) *The Once and Future World: Nature as It Was, as It Is, and as It Could Be*, Random House.

Emma Marris (2011) *Rambunctious Garden: Saving Nature in a Post-wild World*, Bloomsbury.

Charles Massy (2017) *Call of the Reed Warbler*, University of Queensland Press.

Georgina Mace (2014) 'Whose Conservation?' 345 *Science* 1558.

Jan McDonald (2010) 'Mapping the Legal Landscape of Climate Change Adaptation' in T. Bonyhady, A. Macintosh and J. McDonald (eds) *Adaptation to Climate Change: Law and Policy*, Federation Press, 1.

Adrian McDonald et al (2004) 'Rivers of Dreams: On the Gulf between Theoretical and Practical Aspects of an Upland River Restoration' 29 *Transactions of the Institute for British Geographers* 257.

Tein McDonald, Justin Jonson and Kingsley Dixon (2016a) 'National Standards for the Practice of Ecological Restoration in Australia' 24 (S1) *Restoration Ecology* S4.

Tein McDonald et al (2016b) *International Standards for the Practice of Ecological Restoration – Including Principles and Key Concepts*, Society for Ecological Restoration.

Gary Meffe, Ronald Carroll and Martha Groom (2006) 'What is Conservation Biology?' in M. Groom, G. Meffe and R. Carroll (eds) *Principles of Conservation Biology*, Sinauer Associates, 3.

James Miller (2006) 'Restoration, Reconciliation and Reconnecting with Nature Nearby' 127 *Biological Conservation* 356.

James Miller and Richard Hobbs (2007) 'Habitat Restoration – Do We Know What We're Doing?' 15 *Restoration Ecology* 382.

Thaddeus Miller, Ben Minteer and Leon-C. Malan (2011) 'The New Conservation Debate: The View from Practical Ethics' 144 *Biological Conservation* 948.

George Monbiot (2013) *Feral: Searching for Enchantment on the Frontiers of Rewilding*, Allen Lane.

Cameron Muir (2014) *The Broken Promise of Agricultural Progress: An Environmental History*, Routledge.

David Norton and Nicholas Reid (2013) *Nature and Farming: Sustaining Native Biodiversity in Agricultural Landscapes*, CSIRO Publishing.

Tao Orion (2015) *Beyond the War on Invasive Species*, Chelsea Green.

Eric Orts (1995) 'Reflexive Environmental Law' 89 *Northwestern Law Review* 1227.

Fred Pearce (2015) *The New Wild: Why Invasive Species will be Nature's Salvation*, Icon Books.

Bruce Pascoe (2014) *Dark Emu*, Magabala Books.

Val Plumwood (2006) 'The Concept of a Cultural Landscape: Nature, Culture and the Agency of Land' 11 *Ethics and the Environment* 115.

Michel Prieur (2012) 'Non-regression in Environmental Law' 5 *SAPIENS Surveys and Perspectives Integrating Environment and Society* 53.

Productivity Commission (2002) *Pastoral Leases and Non-Pastoral Land Use*, Australian Government.

Productivity Commission (2016) *Regulation of Australian Agriculture*, Australian Government.

Jebediah Purdy (2015) *After Nature*, Harvard University Press.

Bob Purvis (1999) 'I Love this Land, I Was Born Here' in J. Sinatra and P. Murphy (eds) *Listen to the People, Listen to the Land*, Melbourne University Press.

J.R. Purvis (1986) 'Nurture the Land: My Philosophies of Pastoral Management in Central Australia' 8 *Australian Rangelands Journal* 110.

Joanna Quinn (2009) *Reconciliation(s): Transitional Justice in Postconflict Societies*, McGill – Queen's University Press.

Benjamin J. Richardson (2015) 'Reclaiming Nature: Eco-Restoration of Liminal Spaces' 2 *Australian Journal of Environmental Law* 1.

Benjamin J. Richardson (2017) 'Restoring Layered Geographies: Ecology, Society and Time' 26 *Griffith Law Review* 154.

Martin Robards et al (2011) 'The Importance of Social Drivers in the Resilient Provision of Ecosystem Services' 21 *Global Environmental Change* 522.

Margaret Robertson et al (2000) 'Environmental Narratives and the Need for Multiple Perspectives to Restore Degraded Landscapes in Australia' 6 *Ecosystem Health* 119.
Michael Rosenzweig (2003) *Win Win Ecology – How the Earth's Species Can Survive in the Midst of Human Enterprise*, Oxford University Press.
Sadie Ryan and Joel Hartter (2012) 'Beyond Ecological Success of Corridors: Integrating Land Use History and Demographic Change to Provide a Whole Landscape Perspective' 30 *Ecological Restoration* 320.
Joseph Sax (2008) 'Environmental Law Forty Years Later: Looking Back and Looking Ahead' in M. Jeffery, J. Firestone and K. Bubna-Litic (eds) *Biodiversity Conservation, Law and Livelihoods*, Cambridge University Press.
David Seamon (2014) 'Place Attachment and Phenomenology' in L. Manzo and P. Devine-Wright (eds) *Place Attachment: Advances in Theory, Methods and Applications*, Routledge, 11.
Hanna Siurua (2006) 'Nature Above People: Rolston and "Fortress" Conservation in the South' 11 *Ethics and the Environment* 71.
Society for Ecological Restoration International (2004) *The SERI Primer on Ecological Restoration*, Science and Policy Working Group.
David Smith (2000) *Natural Gain in the Grazing Lands of South Australia*, UNSW Press.
Sean M. Smukler et al (2010) 'Biodiversity and Multiple Ecosystem Functions in an Organic Farmscape' 139 *Agriculture, Ecosystems and Environment* 80.
Anastasia Telesetsky (2017) 'Eco-restoration, Private Landowners and Overcoming the Status Quo Bias' 26 *Griffith Law Review* 248.
Anastasia Telesetsky, An Cliquet and Afshin Akhtar-Khavari (2017) *Ecological Restoration in International Environmental Law*, Routledge.
Frank Trentmann (2009) 'Materiality in the Future of History: Things, Practices, and Politics' 48 *Journal of British Studies* 283.
Frank Vanclay (2004) 'Social Principles for Extension to Assist in the Promotion of Natural Resource Management' 44 *Australian Journal of Experimental Agriculture* 213.
Wolfgang Wagner (2007) 'Vernacular Science Knowledge: Its Role in Everyday Life Communication' 16 *Public Understanding of Science* 7.
Dionne Walsh (2009) *A Maximum Sustainable Stocking Rate System in Central Australia: Woodgreen Station, NT.* DKCRC Working Paper 54, The Central Australian Grazing Strategies Project Working Paper Series, Desert Knowledge CRC, Alice Springs.
Charles Warren (2007) 'Perspectives on the Alien Versus Native Species Debate: A Critique of Concepts, Language, and Practice' 31 *Progress in Human Geography* 427.
Dierdre Wilcock, Gary Brierley and Richard Howitt (2013) 'Ethnogeomorphology' 37 *Progress in Physical Geography* 573.
Michael Williams (2006) *Deforesting the Earth: From Prehistory to Global Crisis*, Chicago University Press.
Michael Winter (1991) 'Agriculture and Environment: The Integration of Policy?' 18 *Journal of Law and Policy* 48.
Michael Winter (2005) 'Geographies of Food: Agro-food Geographies – Food, Nature, Farmers and Agency' 5 *Progress in Human Geography* 609.
Mark Woods (2017) *Rethinking Wilderness*, Broadview Press.
Sarah Wright (2015) 'More-than-human, Emergent Belongings: A Weak Theory Approach' 39 *Progress in Human Geography* 391.
Edward O. Wilson (2016) *Half-Earth: Our Planet's Fight for Life*, Liveright Publishing.
George Wuerthner, Eileen Crist and Tom Butler (2014) *Keeping the Wild: Against the Domestication of Earth*, Island Press.

# 6 Linking restoration science and law

*An Cliquet and Kris Decleer*

## 6.1 Introduction

The relationship between restoration science and law is one of the key themes of this book, and the focus of this chapter. The relationship involves mutual influences. On the one hand, restoration science is one of the fundamental foundations for restoration law.[1] On the other hand, restoration legislation plays an important role in inducing restoration and to put scientific insights into practice through restoration obligations.

This two-edged relationship leads to the following questions: to what extent has scientific knowledge been incorporated in restoration law? How should science underpin legal obligations? To what extent is legal background knowledge being incorporated in restoration ecology literature? Linking restoration science and law poses several challenges. This chapter addresses three of these challenges by giving some concrete examples of these issues:

1 Conflicting views amongst scientists can render legal decision-making and implementation more difficult.
2 The lack of understanding by lawyers of the ecological complexities that underpin ecological conservation and restoration can undermine the implementation of restoration obligations and ultimately of restoring degraded ecosystems.
3 The absence of legal expertise in ecological research can lead to a lack of understanding of legal instruments and disputable recommendations by restoration ecologists.

We do not attempt to give a definitive answer to these questions, but rather work with anecdotic examples. Although we recognize the methodological limitations in our research, we can still draw conclusions and recommendations from these examples. Also, in advancing ecological restoration on the ground, not only linking restoration science with law is necessary. Other disciplines, including

1 More broadly speaking, science is an essential input for environmental decision-making: see Biber (2012).

political, social and economic disciplines, are equally important. We will, however, focus on legal and policy issues and science, and not address the other issues further. And although we advocate for a science-based law and policy, we acknowledge the importance of lay knowledge[2], of public participation[3] and of environmental economics in ecological restoration governance.[4]

The chapter provides a review of the current literature, using insights from restoration science and legal analysis, rather than any empirical fieldwork. On the first issue evaluated in this chapter, we have chosen two examples of conflicting opinions expressed by restoration ecologists in a series of publications concerning 'novel ecosystems' and 'standards' for ecological restoration. Both discussions have important legal implications. For the second issue, namely the lack of understanding of ecological complexity by lawyers, we undertook more extensive research that builds on our published article in the *Griffith Law Review*.[5] The findings of this research are for a large part reproduced in this chapter.[6] The research examines the scientific literature in order to explain several legally relevant ecological concepts. We focused on the Convention on Biodiversity (CBD)[7] and subsequent policy documents and instruments associated with this treaty. At the regional level, we assessed the European Union (EU) Habitats Directive[8] guidelines, and other documents from the European Commission and some cases by the European Court of Justice. For the third issue, namely the lack of legal expertise in ecological studies, we used a limited number of interdisciplinary articles on (conservation) science and law to support our general comments and observations. We use one example of a scientific article in which a statement is made about law, without using legal background knowledge.

The structure of the chapter is as follows. First, we discuss the three issues identified above, namely: scientific uncertainties and law; the need for lawyers to understand certain ecological concepts and ecological complexities; and the need for legal methodology in ecological research. A final part includes some recommendations on how to better align restoration science and law, advocating for more interdisciplinary research to allow better implementation of restoration obligations.

2 See Richardson and Lefroy (2016).
3 See the special issue of *Restoration Ecology* (volume 26, April 2018) and Swart et al (2018) for a synthesis article of this special issue.
4 See Iftekhar et al (2017).
5 Cliquet and Decleer (2017).
6 In Cliquet and Decleer (2017) we discuss three concepts (extinction debt, ecological trap and dark diversity). We limit this chapter to two concepts: extinction debt and ecological trap.
7 Convention on Biological Diversity (CBD) (1992), 31 ILM 818.
8 Directive 92/43/EEC of 21 May 1992 on the Conservation of Natural Habitats and of Wild Fauna and Flora, *OJ L* 206, 22 July 1992 (Habitats Directive).

## 6.2 Scientific uncertainties and law

Restoration ecology, as the principal field of science underpinning ecological restoration activities, is a relatively young discipline. In restoration projects many questions may arise, including: what is the restoration target and what (ecological) constraints are present?; which restoration measures are appropriate and can be applied?; how should the restoration success be monitored?[9] More generally speaking, while science is essential to environmental law, science cannot always provide clear answers to the questions that environmental decision makers ask. This gap is in large part because of the significant uncertainties that surround much environmental science.[10] Rather than elaborating on this issue from a general perspective, we choose to give two examples of ongoing scientific discussions. The first discussion relates to the so-called 'novel ecosystems', and the second discussion concerns the appropriate scientific standards for ecological restoration activities.

### *6.2.1 Novel ecosystems*

The analytical concept of novel ecosystems was developed by scientists as a way to describe observed fundamental ecosystem changes as a consequence of human influences, such as climate change, invasive alien species or severe environmental degradation. The concept suggests that in some of these circumstances, restoration efforts should no longer be targeted to a historical reference situation to the extent possible, but rather that decision-makers should embrace new opportunities and accept novel species combinations and altered ecosystems that are more resilient to environmental change.[11] Hobbs and others define novel ecosystems as 'a system of abiotic, biotic and social components (and their interactions) that, by virtue of human influence, differ from those that prevailed historically'.[12] The concept has also made quite an impression among environmental lawyers, including some contributors to this book, such as Phillipa McCormack, and Robyn Bartel and Nichole Graham, who are sceptical of the value of historical baselines as a framework for environmental restoration in a rapidly changing planet.

The concept of novel ecosystems, however, has also raised many concerns and criticisms among restoration scientists and practitioners,[13] and triggered vivid debates on the implications for management and restoration.[14] The main concern with the concept is its basic assumption that restoration towards historical conditions (i.e. with less or no ecosystem degradation) is difficult, if not impossible. This could create the impression that novel ecosystems are 'the new normal' and

9 Valkó (2018).
10 Biber (2012).
11 See on novel ecosystems, for instance Hobbs, Higgs and Harris (2009); Hobbs, Higgs and Hall (2013).
12 Hobbs, Higgs and Hall (2013).
13 See for example Truitt et al (2015); Murcia et al (2014).
14 For a vivid overview of this conflicting debate, see Woodworth (2013).

become the 'easy way forward' for decision-makers. It is argued that in spite of degradation, many ecosystems can still be restored to some extent.[15] Even with further environmental change intensifying in the Anthropocene, critics of the novel ecosystems hypothesis believe that most places around the world are likely to be only incrementally different from today, and so a complete break with their environmental history is unnecessary.[16] It should not automatically be assumed that species or assemblages cannot adapt to rapid changes.[17] The effects of biological and ecological change are not distributed evenly and may vary tremendously at local and regional scales. Thus, some areas may be relatively resistant to change, and their restoration with a focus on historically determined goals should still be viable.[18] Even when novel ecosystems involving new species assemblages appear to be inevitable, the use of historical information may still be significant as a source of context and constraint in shaping the goals of restoration projects. Historical knowledge will play a key role in restoration, for example in improving understanding of range shifts, species interactions and adaptive capacity, regardless of the extent to which it is used as the basis of goal-setting.[19]

What could be the legal and policy implications of the concept of novel ecosystems? This has been studied only rarely.[20] We will raise just a few issues here. From a legal point of view, there is lack of clarity regarding the meaning of the concept. If the concept is understood as ecosystems being changed by human activities, then semi-natural habitats, as we find them in most of Europe, for instance, can also be considered as 'novel' ecosystems;[21] already, such 'novel' ecosystems are legally recognized and protected as 'semi-natural habitats' in the EU Habitats Directive.[22] The whole concept of novel ecosystems might thus not be so 'novel' at all. Even authors advocating this concept recognize that all ecosystems can be considered 'novel' when placed in the appropriate temporal context.[23] However, these researchers make a distinction between semi-natural habitats and the current ensemble of emerging novel ecosystems, as characterized by new species combinations because of exotic species invasions or climate change. But it is unclear how much change should have occurred before a system can be qualified as 'novel' and therefore would no longer qualify under existing legal definitions of 'ecosystems' (such as defined in the CBD[24]), or the definition of 'natural habitats' (such as defined in the Habitats Directive[25]).

15 Murcia et al (2014).
16 Corlett (2016).
17 Murcia et al (2014).
18 Keenleyside et al (2012).
19 Ibid.
20 See on policy: Bridgewater and Yung (2013); see on legal aspects: Telesetsky, Cliquet and Akhtar-Khavari (2017), Chs 2 and 11; Cliquet (forthcoming, 2019).
21 See also Backstrom et al (2018).
22 See definition of 'natural habitats' in article 1(b), Habitats Directive: natural habitats means terrestrial or aquatic areas distinguished by geographic, abiotic and biotic features, whether entirely natural or semi-natural.
23 Hobbs, Higgs and Harris (2009).
24 Article 2, CBD.
25 Article 1(b), Habitats Directive.

Another unclear implication of the concept of novel ecosystems relates to existing conservation and restoration obligations in law. Are existing laws unable to cope with these changes, and to what extent can this be generalized? Advocates of the novel ecosystems concept call for more flexibility in law.[26] However, we are cautious of claims that nature conservation and restoration legislation is too static to deal with environmental changes. For instance, the EU Habitats Directive has sometimes been depicted as rigid and static; yet a recent evaluation of the Directive suggests that it is able to cope with environmental changes such as climate change.[27]

Furthermore, there are concerns that relate to policy aspects of restoration and the implementation of restoration obligations. There is a risk that the concept can be used or interpreted in a way to undermine or question the usefulness of restoration. This scepticism could easily be abused by vested interests as a means of weakening the legal obligations to conserve and restore ecosystems.[28] If novel ecosystems are regarded as the 'new normal', this could probably reduce the willingness of governments or other stakeholders to undertake restoration efforts towards a historical trajectory of ecosystems, or it could lead them to lower the quality standards for restoration. So far, there is no proof of a point of no return, of ecological thresholds that prevent restoration. It is rather political and financial thresholds that will hamper restoration projects.[29] If the loss of original keystone species and habitats is seen as inevitable, this could easily lead to a weakening of the protection system of these habitats and species. It might lead to declassification of protected areas, as the keystone species and habitats for which the area has been designated were lost and cannot be restored. However, species could as well disappear because of the lack of sufficient management measures, or because of fragmentation and other negative impacts on biodiversity. Increasing regulatory flexibility often opens the door to potential unscrupulous agendas.[30]

### *6.2.2 Standards*

Due to the lack of definitions and clarity of the legal restoration obligations and targets in international law[31], the development of principles and standards can be a very useful instrument to complement and clarify the legal requirements. Principles and standards can also help to define different types of restoration projects (e.g. post-mining remediation versus landscape restoration), and ensure that these projects are implemented in an appropriate way and according to the best available techniques. This is one of the main conclusions of the landmark book by Telesetsky and others,[32] who conducted extensive legal research on ecological restoration in environmental law.

26 See for instance Bridgewater and Yung (2013).
27 See for example Trouwborst (2015).
28 See also Simberloff, Murcia and Aronson (2015).
29 Murcia et al (2014).
30 See also similar conclusions in Craig (2010).
31 See Telesetsky, Cliquet and Akhtar-Khavari (2017).
32 Ibid, Ch. 12 on future directions for ecological restoration law.

Several key principles for ecological restoration for protected areas have been put forward by Keenleyside and others[33]: *effective* ecological restoration for protected areas re-establishes and maintains the values of a protected area; *efficient* ecological restoration for protected areas maximizes beneficial outcomes while minimizing costs in time, resources and effort; *engaging* ecological restoration for protected areas collaborates with partners and stakeholders, promotes participation and enhances visitor experience. These principles are complemented by guidelines and best practices. Suding et al proposed four principles for ecological restoration (restoration increases ecological integrity; restoration is sustainable in the long term; restoration is informed by the past and future; restoration benefits and engages society).[34]

Restoration standards have been developed at both national and global levels. In 2016 national standards for ecological restoration have been developed for Australia by the Society for Ecological Restoration Australasia (SERA), in collaboration with various partners.[35] These served as an important source of inspiration for international standards for restoration that were adapted by SER International in late 2016,[36] and released during the 13th Conference of Parties of the Biodiversity Convention in December 2016. The International SER Standards depart from six key concepts, including: ecological restoration practice is based on an appropriate local native reference ecosystem, taking environmental change into account; and restoration seeks 'highest and best effort' progression towards full recovery. The Standards develop a five-star performance rating system, in which the highest rating, five stars, means a status where the ecosystem is on a self-organizing trajectory to full recovery (based on an appropriate local native reference ecosystem). This is considered the 'gold standard' to which all ecological restoration projects aim, insofar as is possible.

The foregoing principles and standards are welcome developments, but so far there are no legal obligations on states to actually use them. So, there might be a need to develop additional legal instruments that refer to scientifically accepted principles and standards, either as recommended guidance or as firm legal standards. As the legal obligation for restoration in international law is evolving from an obligation of means to an obligation of result, performance standards, such as the SER Standards, could prove to be very useful as guidance to decide what qualifies as ecological restoration for legal purposes.[37] However, standards should be able to cope with diversity: restoration in semi-natural habitats in Europe can pose different challenges for restoration than in North America or Australia.

As restoration science is still a young and evolving science, there might not be a scientific agreement on the appropriate standards. A recent article in *Restoration Ecology* by Higgs and others has given some critical comments on the SER

33 Keenleyside et al (2012).
34 Suding et al (2015).
35 McDonald, Jonson and Dixon (2016).
36 McDonald et al (2016).
37 Telesetsky, Cliquet and Akhtar-Khavari (2017).

Standards, which led to a response and in turn a counter response.[38] Without going into the detail, the basic argument from Higgs and his co-researchers is that there should be a principles-first approach (principles that can generate guidelines and be further refined in best practices and standards). It seems that Higgs and others have most problems with the five-star recovery system, which they view as too rigid, especially for heavily degraded environments. In response, Gann and others argue that the SER Standards operationalize the principles of ecological restoration (effective, efficient and engaging) and that the Standards allows for flexibility and openness in the implementation of principles.

To us, this discussion on principles and standards seems to a certain extent a semantics discussion. From an international environmental law perspective, we are familiar with general principles of environmental law, such as the precautionary principle, and specific rules. Both can include standards, be it that the standards that are embodied in a general principle will be of a more general nature.[39] The term 'standards' is sometimes used in the sense of performance standards (e.g. emission standards or environmental quality standards), but is also sometimes used in a more general way, relating to environmental rules, that can be either binding regulations or voluntary standards (such as guidelines and recommendations).

The SER Standards of 2016 are not intended by the SER to be a final or definitive statement, but rather a framework for ongoing refinement and modification according to new scientific insights. At this stage, therefore, it might be too soon to include these Standards in legislation. Such a step requires further discussion on whether and how principles and standards should be included in or referred to in law, which principles and standards should be acknowledged, and at what level of detail, among a variety of issues to resolve. Legal scholars must participate in such discussions.

Both the examples discussed above, on novel ecosystems and on standards for restoration, demonstrate that the science underpinning restoration activities is young and evolving. The implications for ecological restoration law are still unclear: should the law embrace the concept of novel ecosystems while it remains in such fervent debate among restoration ecologists? Should performance standards be incorporated in law? Undoubtedly there will always be some discussions amongst scientists, so the fact that there is disagreement, as also seen in other areas such as climate change science, should not be reason to postpone action. The urgency of the current environmental crisis, leading to a sixth mass extinction, cannot wait for a final answer from scientists. Environmental law principles, such as the precautionary principle, behove policy makers to act now rather than wait for irreparable and costly environmental harm to possibly materialize.

38 Higgs et al (2018a); response by Gann et al (2018); and counter-response by Higgs et al (2018b).

39 But even on the distinction between principles and rules there are discussions amongst legal scholars, see for instance Beyerlin (2007); Sands et al (2012).

## 6.3 Ecological concepts in restoration law

International obligations and targets on restoration are far from being met, in spite of restoration measures being increasingly taken worldwide.[40] This is partly caused by a lack of ambition in the restoration measures taken, which in turn is owing to the lack of political will, lack of finance, lack of staff and conflicts with stakeholders.[41] But there is also a lack of understanding of the ecological complexities that form the basis of taking restoration measures. In particular, as a consequence of the delayed effects of detrimental human activities and the ecological changes they have on biodiversity, it might (wrongly) seem that conservation and restoration obligations and targets are fulfilled, creating a false sense of security. In reality, the conservation status might be far worse off, and more restoration measures are in order. For instance, focal species that are still present in a certain area can lead to the false impression that the conservation measures that were taken are sufficient. However, it could be that the presence of these species is not sustainable and additional restoration measures could be required. In order to better understand and implement the restoration obligations under nature conservation laws, it seems indispensable that lawyers, judges and policy makers should be aware of certain ecological concepts that are important for ecological restoration.

We examine two ecological concepts, namely extinction debt and ecological trap. Although ecologically highly relevant, they are neglected by, or even absent from, legal primary sources and the scholarly law literature. These concepts all relate to the extinction of species, and they contribute to our knowledge about underlying processes that can lead to loss of biodiversity.

### *6.3.1 Ecological concepts*

The concept of 'extinction debt' refers to the often-substantial time delay between the impact of environmental changes and pressures on a habitat specialist species and the moment the species goes extinct.[42] Species suffering from extinction debt are sometimes called 'the walking dead'. Extinction debts can arise because 1) individuals of a species with a long generation time may survive long after the habitat quality change; 2) populations may have become so small that stochastic catastrophic events sooner or later make them surpass their extinction threshold; and 3) metapopulations may survive long after that connectivity has decreased if colonization–extinction dynamics are slow.[43] Especially in highly fragmented landscapes, there are many relict species at risk of an extinction debt.[44]

In order to avoid ultimate extinctions and the gradual decrease in species richness of a focal habitat or area, it is crucial that nature conservation laws and

40 Secretariat of the Convention on Biological Diversity (2014), 101; Tittensor et al (2014).
41 See also Cortina, Decleer and Kollmann (2016).
42 Kuussaari et al (2009).
43 Hylander and Ehrlén (2013).
44 On this aspect, see Krauss et al (2010).

policies fully recognize the 'extinction debt' concept and its causes. It is not sufficient to simply record that a focal species is still present in an area; it is also necessary to monitor its population size over time and assess the risk for extinction over the long term, and its causes.

It is also crucial that the effect of any time delay in measures taken to tackle biodiversity loss is taken into account by policy makers. The urgency to take additional measures for effects that are currently not visible is challenging, as legal processes for taking nature-conservation and restoration measures are usually time consuming. By the time environmental monitoring reports, which are commonly a periodic legal requirement, show the extinction risk, and by the time any additional conservation or restoration measures are implemented, it might be too late to save the species from extinction.

The 'ecological trap' concept refers to the presence of poor-quality habitats that make it difficult or impossible for a species to successfully reproduce and settle a sustainable population.[45] This may affect the size and the functioning of the regional metapopulation as a whole. The concept is linked to the extinction debt concept in the sense that the presence of a trap in a landscape will gradually drive a local population to extinction. A well-known example is meadow birds breeding in agricultural fields, where their nests are destroyed by the mowing activities of the farmer. When reproduction has successively failed over several years and the parent birds have died, the local population goes extinct. Offspring of animals from neighbouring areas may colonize these unsuitable agricultural fields, in the end risking the extinction of metapopulations if the causes of reproduction failure are widespread. Common approaches to population modelling may introduce faulty assumptions that mask the effects of ecological traps and lead to overly optimistic predictions about population persistence.[46]

The lesson for conservation laws is that it is not sufficient to simply record the presence of focal species in an area. In order to be able to anticipate the presence of ecological traps, it is essential to have some knowledge of the reproductive success of the focal species or a set of specialist species from a focal habitat. It is important to note that ecological traps may also occur in areas where ecological restoration measures have been taken, as long as not all conditions for successful reproduction have been achieved.[47]

### *6.3.2 Absence of these concepts in international and EU law*

The above-mentioned concepts are largely unknown in legal literature and largely absent from legal sources. Neither important international conventions (such as the Biodiversity Convention) nor regional instruments (such as the EU Habitats Directive) explicitly refer to these concepts in their primary legal texts. However, in subsidiary policy documents and guidance instruments developed under the

45 Battin (2004).
46 Ibid.
47 Cristescu et al (2013); Severns (2011).

auspices of these instruments, reference to some of these concepts can occasionally be found.[48] Implicitly, the precautionary principle can also play a key role in the application of the foregoing ecological concepts. The precautionary principle can be invoked to avoid extinction, even when there is a lack of full scientific certainty that there is an extinction debt and that species will go extinct. In the framework of the CBD, the precautionary approach is linked to adaptive management:[49] this allows for decision makers to respond to new knowledge and changed circumstances. This response might include monitoring the risk of extinction debt and ecological traps, and adapting conservation and restoration measures to this new knowledge. Although the concepts as such are mostly absent in law, in the next part we will identify the possible measures to implement them.

### *6.3.3 Measures to prevent extinction debt and ecological traps*

In order to identify whether international and EU nature conservation law is capable of dealing with extinction debt and ecological traps, we see the following measures as crucial: 1) adequate monitoring; 2) ambitious restoration site-specific objectives and measures; and 3) measures for habitat connectivity. Adequate monitoring is crucial in order to assess if there is a risk for an extinction debt or ecological trap condition. If this is the case, then two types of measures are required in order to prevent extinction: 1) sufficiently ambitious restoration in terms of the size and quality of the habitat to be restored; and 2) measures to improve habitat connectivity to the extent that proper meta-population functioning is achieved in a sustainable way. In the following sections we assess the adequacy of the law to accommodate these types of measures.

#### *Monitoring*

Monitoring and data collecting are of crucial importance. However, we should be aware that data collecting as such is not enough, and on its own will not stop further biodiversity loss.[50] It is equally important that monitoring is included in an adaptive management process. If monitoring shows that restoration has been inadequate, additional conservation and restoration measures should be taken. Monitoring should focus on the early detection of the occurrence of an extinction debt or ecological trap condition for habitat-specific species. Monitoring should thus go beyond a simple 'ticking the box' ('species X is present in area Y, so the legal obligations are fulfilled'), and look specifically at issues such as population dynamics and reproduction success, in both conservation and restoration areas. Special attention should go to 'umbrella species' as indicators for the overall integrity of the ecosystem and the local pool of habitat-specific species.[51]

48 For examples, see Cliquet and Decleer (2017).
49 CBD (2004) COP Decision VII/11, Ecosystem approach, Annex 1, Principle 6, Implementation Guideline 6.2; see Morgera (2017), 75–77.
50 Ellison (2016).
51 On the usefulness of umbrella species, see Roberge and Angelstam (2004).

Most international and regional conservation laws include the obligation for monitoring (see e.g. Article 7 of the CBD; Article 11 of the Habitats Directive). In the CBD's short-term action plan on ecosystem restoration,[52] monitoring is identified as one of the key actions for state parties.[53] At the EU level, there are monitoring and reporting obligations under the Habitats Directive on the conservation status of habitats and species in a Member State across the present biogeographical regions. This obligation provides possibilities to take the assessment of extinction debt and ecological traps into account, even though it is not explicitly provided in the reporting format[54] and explanatory guidelines.[55] In practice, monitoring under the Habitats Directive might still fail to detect extinction debt or the presence of ecological traps: the monitoring is done at the Member State level, and is limited to the habitats and species listed in the Habitats Directive, which is only a fraction of habitats and species present in Europe. Furthermore, it seems that Member States in their assessment of Annex I habitats[56] focus mainly on the presence of a selection of typical plant species, and neglect the vast majority of biodiversity, namely the typical fauna. Most of the typical fauna have high demands for habitat size, habitat quality and landscape configuration, and are therefore often more vulnerable to habitat loss and degradation, and loss of connectivity. However, to detect species with risks for an extinction debt or ecological trap condition, and in order to enable timely implementation of effective counter measures, monitoring at site level is desirable. Monitoring in individual sites is required to enable a reliable assessment of the conservation degree of the sites in the so-called 'Standard Data Form' for each site. However, in the European Commission's Decision on the Standard Data Form[57] there is no explicit mention of extinction debt or ecological traps, although Member States do have to indicate the degree of isolation of the population present on the site in relation to the natural range of the species.[58]

Although general monitoring obligations in international and EU nature conservation law do not explicitly call for monitoring of extinction debt and ecological traps, they do not exclude this either. More concrete guidelines on monitoring, taking into account extinction debt and ecological traps, are necessary.

### *Ambition level of restoration goals and measures*

In spite of international legal obligations for restoration, states are not taking enough steps to implement their responsibilities. Although the number and extent of restoration projects are growing worldwide, including many being initiated by

52 CBD (2016) Decision XIII/5. Ecosystem restoration: short term-action plan.
53 Ibid, Annex, para 11.
54 Reporting under Article 17 of the Habitats Directive. Report format for the period 2013–2018. Final version – November 2016. http://cdr.eionet.europa.eu/help/habitats_art17.
55 DG Environment (2017).
56 European Commission (2013a).
57 European Commission (2011).
58 This assessment is limited to certain species (species mentioned in article 4 of the Birds Directive and Annex 2 of the Habitats Directive).

non-state actors, this trend has thus far been largely insufficient to actually halt or reverse the detrimental state biodiversity is in. Yet, the overall negative trend is documented in a large amount of literature pointing to biodiversity loss and the lack of conservation or restoration measures.[59] The scientific insights relating to concepts such as extinction debt and ecological traps behove decision makers to allow for even more caution. What might appear as successful conservation measures, because of the presence or return of certain species, might not take into account the delayed effects of biodiversity loss.

In order to reverse negative trends, restoring or maximizing habitat size and restoring habitat quality are both important. Maximizing habitat size (habitat quantity) can be achieved by designating additional or expanding existing protected areas. For instance, several Conference of the Parties' (COP) decisions under international biodiversity conventions, including the CBD, urge states to strengthen the ecological networks as a way to face the challenges of climate change.[60]

Restoring habitat quality requires adequate conservation and restoration measures. Although some international environmental laws include restoration obligations, they are not particularly ambitious or articulated precisely enough to make states measurably accountable. One way to overcome these limitations would be to recast conservation and restoration obligations as an obligation of result. Consider the EU Habitats Directive. The general aim of the Habitats Directive[61] is to contribute towards ensuring biodiversity through the conservation of natural habitats and their wild fauna and flora in the European territory of the Member States. This is an outcomes-focused obligation.[62] Measures taken pursuant to this Directive shall be designed to maintain or restore, at favourable conservation status, natural habitats and species of wild fauna and flora of EU interest.[63] The measures to obtain this obligation include both the designation and protection of an ecological network in the EU, as the strict protection of species. It appears from these obligations that more measures will be necessary where a species has an unfavourable conservation status. The concept of 'favourable conservation status' represents a high level of ambition set by the Habitats Directive.[64] In the European Commission's guidelines on species protection, the legal concept is clarified: the fact that a habitat or species is not threatened (i.e. not faced by any direct extinction risk) does not necessarily mean that it has favourable conservation status. The target of the Directive is defined in a positive way, as a

59 See for example Secretariat of the Convention on Biological Diversity (2014); Tittensor et al (2014); European Environment Agency (2015); European Commission (2015).

60 For references to the COP decisions, see Telesetsky, Cliquet and Akhtar-Khavari (2017) 275–278.

61 Article 2 (1), Habitats Directive.

62 European Commission (2000) 18.

63 Article 2 (2), Habitats Directive.

64 See article 1 of the Habitats Directive for a definition of a favourable conservation status.

'favourable' situation to be reached and maintained, which needs to be defined based on the best available knowledge. Therefore, the obligation of a Member State is more than just avoiding extinction; all measures taken under the Directive must aim to reach or maintain a favourable conservation status.[65] The guidelines also mention that '[i]t is important to note that the assessment of conservation status not only includes an element of "diagnosis" based on current conditions, but also an important element of "prognosis" (foreseeable future) based on influences'.[66] In spite of the definition, and the guidance by the Commission, the concept of favourable conservation status still leaves room for interpretation and needs further clarification.[67]

In order for governments to reach a high ambition level, such as 'favourable conservation status', the conservation and restoration measures should take into account the potential delayed effects of extinction debt and ecological traps. This implies that detrimental human activities should be restricted or modified. These restrictions can be imposed within protected areas, or can apply to the whole territory (e.g. species protection measures). Defining which measures are to be taken should be based on monitoring that takes into account issues such as reproduction success. In the example of an ecological trap given above – meadow birds breeding in agricultural fields where the nests are destroyed by the mowing activities of the farmer – this should entail that mowing activities are restricted.[68]

In spite of the restoration obligations in the EU, as an outcome of a results-based obligation, the implementation of EU legislation has so far not been able to stop the degradation of nature. More guidelines from the European Commission to raise the ambitiousness of restoration measures remain necessary, as well as more enforcement at the national and EU levels. Also, at the international level, more concrete guidelines for ecological restoration to help states in implementing their restoration obligations would be useful.[69]

### *Restoration of connectivity*

Connectivity is recognized as an important measure in nature conservation and restoration.[70] Restoring connectivity can allow movement of species for genetic interchange and help species adapt to climate change.[71] Restoring ecological stepping stones between protected areas can help to ensure that migratory species have resting and feeding locations in order to ensure their safe passage. A

65 European Commission (2007) 9–10.
66 Ibid, 10.
67 For an analysis of 'Favourable conservation status', see Epstein et al (2016); Epstein (2016); Trouwborst, Boitani and Linnell (2017).
68 See for instance a case for the European Court of Justice on the European hamster: Case C-383/09 *Commission v France* (2011).
69 Cortina et al (2016).
70 On the importance of connectivity see inter alia: Bennett (1998, 2003); Bennett and Mulongoy (2003); Lausche et al (2013).
71 See Heller and Zavaleta (2009) 14–32; Cross et al (2013) 307–329.

landscape mosaic can link various habitats into a viable and more functional ecosystem.[72] A decrease of connectivity can lead to lower recolonization and immigration rates, affecting the equilibrium in a meta-population. These processes that affect meta-population functioning will often unfold over a period of time before they become manifest and obvious, which thereby creates the risk of an extinction debt.[73] Restoring connectivity between fragmented areas can enable proper meta-population functioning and thus counter extinction debt and ecological traps.

Measures for habitat connectivity are already included in international law.[74] Virtually all global and regional legal instruments dealing with biodiversity, climate change and natural resources require states to provide for adequate wildlife connectivity conservation.[75] Although several international legal instruments mention networks of protected areas, and some also refer to habitat corridors, few states have actually tried to implement them.[76] The focus of nature conservation law is mostly on the designation and protection of core protected areas, but often without specific measures for connecting them. In the COP decisions of several international and regional nature conservation conventions, attention has been given to the importance of connectivity,[77] including the Ramsar Convention,[78] the Bonn Convention on migratory species,[79] the European Bern Convention[80] and the CBD.[81] Connectivity is also part of the Aichi Targets. Although initiatives exist to develop corridors, there is still not sufficient connection between protected

72 Keenleyside et al (2012) 14.
73 See Hylander and Ehrlén (2013).
74 See more extensively on legal aspects of connectivity: Lausche et al (2013); see specifically on restoration of connectivity: Telesetsky, Cliquet and Akhtar-Khavari (2017), 247–255.
75 Lausche et al (2013), 175.
76 Gillespie (2007), 151.
77 See more in detail: Telesetsky, Cliquet and Akhtar-Khavari (2017), 247–255.
78 Convention on Wetlands of International Importance especially as Waterfowl Habitat, Iran, 2 February 1971 (1972), 11 *ILM* 963; Resolution XI.14 (2012) Climate change and wetlands: implications for the Ramsar Convention on Wetlands.
79 Convention on the Conservation of Migratory Species of Wild Animals, 23 June 1979 (1979), 19 *ILM* 15; Resolution 10.3 (2011) The role of ecological networks in the conservation of migratory species; Resolution 10.19 (2011) Migratory species conservation in the light of climate change; Resolution 11.25 (2014) Advancing ecological networks to address the needs of migratory species; Resolution 11.26 (2014) Programme of work on climate change and migratory species.
80 Convention on the Conservation of European Wildlife and Natural Habitat, 19 September 1979, ETS No. 104; Recommendation No. 25 on the conservation of natural areas outside protected areas proper (1991); Recommendation No. 135 on addressing the impacts of climate change on biodiversity (2008); Recommendation No. 143 on further guidance for Parties on biodiversity and climate change (2009); Recommendation No. 180 on improving the conservation of nature outside protected areas proper (2015).
81 CBD (2004) Programme of Work on Protected Areas: Goal 1.2; CBD (2010) Decision X/31. Protected areas; CBD (2010) Decision X/33. Biodiversity and climate change.

areas.[82] Few national biodiversity strategies and action plans explicitly address the connection or integration of protected areas into wider landscapes and seascapes.[83]

Articles 3 and 10 of the EU Habitats Directive ask Member States to encourage the management of features of the landscape that are of major importance for wild fauna and flora. These articles must be read together with the obligation requiring Member States to achieve favourable conservation status. Because a large number of habitats and species in the EU have an unfavourable status of conservation due to fragmentation and other human impacts, connectivity measures will be essential to reach favourable conservation status. According to the Commission, the concept of favourable conservation status is not limited to the Natura 2000 network or to the species protected by this network.[84] Natura 2000 is the European ecological network of special areas of conservation, designated under the Habitats Directive and the special protection areas classified under the EU Birds Directive.[85] Member States should promote the implementation of connectivity measures where these are required to maintain or restore an area to favourable conservation status, irrespective of their contribution to the coherence of the Natura 2000 network.[86]

In conclusion, connectivity has been recognized in the framework of international and regional nature conservation instruments as a way to cope with the effects of climate change and landscape fragmentation. However, it often lacks implementation on the ground, thus increasing the risks of extinction debt and ecological traps.

## 6.4 Legal expertise in restoration science

As demonstrated in the previous section, it is important that lawyers are familiar with ecological concepts, in order to grasp the underlying ecological complexities in the implementation or enforcement of restoration obligations. But it is equally important that scientists are aware of policies and legal requirements on restoration,[87] and also are conscious of the potential impact of their statements on policy and law. This is especially true if they make statements or give recommendations about law.

Scientific literature on ecological restoration rarely refers to the legal context. We did not conduct a systematic screening, but a cursory screening of key journals such as *Restoration Ecology* for the past year showed only a handful of papers relating to law (including those on restoration standards cited above). Most of these articles make only passing mention of the law without any assessment thereof.

82 Leadley et al (2014).
83 Ibid.
84 European Commission (2007).
85 Directive 2009/147/EC of the European Parliament and of the Council of 30 November 2009 on the conservation of wild birds, OJ L 20, 26 January 2010 (Birds Directive).
86 Kettunen et al (2007).
87 Zedler (2018).

We will use one article as an example where recommendations on legal instruments were made. Pettorelli et al[88] make policy recommendations for rewilding, including recommendations on revising the existing legislation. Referring to Hobbs and others,[89] they state that revision of environmental policies and legislation that currently focus on existing or historical assemblages may be required for rewilding to fully reach its conservation potential. More concretely, they assess the EU Habitats and Birds Directives. According to the authors 'these directives are based on a "compositionalist" paradigm, predicated on the preservation of particular species assemblages and habitats types' and therefore 'rewilding projects focused on ecosystem processes and embracing uncertain outcomes could be difficult to accommodate within this policy framework'.[90] In their conclusions they further state that:

> rethinking of the key pieces of legislation shaping biodiversity conservation and land use in countries, such as the Birds and Habitats Directives in the EU, could facilitate the development and testing of novel environmental management funding mechanisms focused on payments for the delivery of desired ecosystem services, based on measurable outcomes rather than prescriptive management measures.[91]

It is not our intention to critically assess the article as a whole. Yet, on the Birds and Habitats Directives, we notice the authors give a rather limited and one-sided interpretation of these directives. These directives do not exclude the protection of ecological functioning and processes,[92] which are key indicators for the conservation status of protected sites (see Standard Data Form).[93] The existing Natura 2000 network includes rewilding areas, such as the Southern Carpathians rewilding area, and the European Commission has even produced a guidance document on wilderness and Natura 2000.[94]

It should thus be clear that scientists' commentaries on environmental legislation can lack the necessary nuanced detail to understand their strengths and limitations. The above example also fails to cite relevant secondary legal literature, in which one can find extensive interpretations of these directives.[95] There is also no reference to the 'fitness check'[96] on these directives, which concluded after a thorough study that there is no need for amending the directives.[97] Also, there is

88 Pettorelli et al (2018).
89 Hobbs, Higgs and Harris (2009).
90 Pettorelli et al (2018), 1121.
91 Ibid, 1122.
92 See Dodd et al (2010).
93 European Commission (2011).
94 European Commission (2013b).
95 See Born et al (2015).
96 Fitness checks are comprehensive policy evaluations assessing whether the regulatory framework for a policy sector is 'fit for purpose'; see http://ec.europa.eu/environment/nature/legislation/fitness_check/docs/Mandate%20for%20Nature%20Legislation.pdf.
97 Milieu, IEEP and ICF (2016); European Commission (2016).

no reference to the essential guidance document of the European Commission on wilderness and Natura 2000.

This example shows that it is important for scientists to consult legal literature, especially when making statements about the quality of environmental law. Legal articles are often published in legal books or legal journals that are not included in the web-of-science or open access. Co-authored publications are still unusual in the legal academic world. However, the emergence of some multidisciplinary environmental journals in recent years, such as *Global Environmental Change* and the *Anthropocene Review*, should help to counteract the segmentation of knowledge into disciplines that has hampered environmental restoration and others facets of environmental governance. But more collaboration among scientists and lawyers is urgently needed, as we will discuss further in the next part.

## 6.5 Better aligning of science and law for restoration

Cooperation between lawyers and restoration ecologists is seen as a precondition for future development and implementation of international restoration law[98] and as a way to advance restoration on the ground. However, interdisciplinary research or literature between restoration scientists and lawyers is scarce. Chapron and others give the example of the 27th International Conference on Conservation Biology in 2015, in which only a handful of the 900 talks had a legal focus.[99] A similar observation can be made of the annual meeting of the British Ecological Society in 2017.[100] Likewise, there were no presentations by ecologists or experts in cognate environmental disciplines at the IUCN Academy for Environmental Law colloquium in 2016.[101] At the (world) conferences for ecological restoration, there have been some legal presentations, and a specific legal symposium 'Legal mechanisms for restoration across the world' at the SER World conference on ecological restoration in 2015.[102]

Trouwborst and others have already acknowledged that there is confusion, ignorance and misinterpretation amongst conservationists regarding international wildlife law. They call for increased cooperation between international wildlife lawyers and non-law conservation professionals.[103] Cooperation between conservationist and wildlife lawyers can especially be interesting in order to clarify the implications of international instruments for given issues, to identify gaps and

98 Telesetsky, Cliquet and Akhtar-Khavari (2017).
99 Chapron et al (2017).
100 See conference programme: BES, GFÖ, NecoV and EEF, Ecology across borders, 11–14 December 2017, Ghent; www.britishecologicalsociety.org/events/annual-meeting-2017/ghent-2017.
101 Chapron et al (2017).
102 SER 2015 World Conference on Ecological Restoration. Towards Resilient Ecosystems: Restoring the Urban, the Rural and the Wild, Manchester, UK, 23–27 August 2015.
103 Trouwborst et al (2017).

inconsistencies, as well as to identify avenues toward improving the application of the law, or the law itself.[104]

There are several benefits for better aligning ecology and law. First, lawmakers can use ecology to refine existing laws so that they are more effective in achieving current goals, or to adopt more appropriate environmental goals.[105] Second, law is unique because of its binding and enforceable character, putting constraints on human impacts on the environment, compared with non-binding tools such as the advice of scientific publications.[106] Yet scientific knowledge is often essential to demonstrate compliance with the law, such as to show the conservation status of a species or to understand threats to its survival. Scientific evidence is increasingly being used in courtrooms, and judges refer to scientific information. Notable examples from the European Court of Justice are the *Hamster* case,[107] the *Finnish wolf hunt* case,[108] and the recent *Białowieża* case.[109] Scientists should be aware of the role they can play in environmental decision-making and environmental enforcement.

The three issues we discussed in this chapter can only strengthen this call for greater cooperation between scientists and lawyers. Specifically regarding science and law on ecological restoration, we suggest the following recommendations for ecologists and lawyers.

Both the publications on novel ecosystems and the article we discussed on rewilding make simple proposals that legislation should be revised, but without evidence that the researchers consulted legal scholars or evaluated specific legal instruments, jurisprudence from courts or secondary legal literature. For restoration ecologists, it is important that they have basic knowledge of laws, guidance documents, and commentary by legal scholars and judicial decisions. They should be more involved in policy processes, or at least be aware of them, and of the potential political consequences of their statements. The novel ecosystems concept is a good example. The proponents of novel ecosystems seem to create the feeling that ambitious restoration targets are irrelevant when restoring a heavily degraded ecosystem to a historical reference situation becomes impossible, and legislation should be adapted to this. Woodworth calls this 'an engineering tradition that suggests we can turn nature into a designer shopping mall of human-oriented functions and services'.[110] Although the scientific publications on novel ecosystems are mostly more nuanced, this is the 'take-home' message that lingers on. Scientists might think that their publications and statements do not have an immediate impact on decisions taken by politicians; but they can have an impact, and taken away from their 'safe' context of scientific journals, they can lead to unwanted consequences. We do not advocate for restrictions being placed on the

104 Ibid.
105 Freyfogle (2006).
106 Chapron et al (2017).
107 Case C-383/09 *Commission v France* (2011).
108 Case C-342/05, *Commission v Finland* (2007).
109 Case C-441/17, *Commission v Poland* (Białowieża Forest) (2018).
110 Woodworth (2013).

freedom of scientific research, but we plea for more consideration of the possible abuse of statements and concepts, especially those that could possibly undermine strong legislation.

With regard to the academic discussions on standards for ecological restoration, we advocate more cooperation between ecologists and lawyers, especially for consideration of the question whether or not such standards should be articulated into legal formula. The SER standards exist at the international level, with unanswered questions about how they should be expressed and elaborated in global, regional, national and local restoration governance. The development of legal standards for ecological restoration at these levels will benefit from some comparison of progress made in other fields of environmental law, where well-developed standards exist in many domains, such as for water and air quality.

Likewise, in order for the conservation and restoration laws to be adequately implemented, it is crucial that policy makers, lawyers and judges are aware of ecological concepts and ecological complexities. We demonstrated this with reference to the concepts of extinction debt and ecological traps, and advocated that it is essential that these concepts are taken into account in the implementation of nature conservation and restoration laws and policies. We recommended several measures to avoid further species loss: monitoring beyond the simple presence/absence of habitat-specific species in a site, and the need for a good understanding of meta-population functioning in relation to habitat size, habitat quality and connectivity; an increase of habitat size and quality and better connectivity as preventive measures to avoid future species loss; and no loss in size, quality and connectivity. For species that are typical for a certain ecosystem but that have already disappeared in a given site, measures that enable resettlement are recommended in order to restore the typical species assemblage and proper ecosystem functioning. After species return, monitoring is needed in order to identify emerging threats that might lead to new extinction debt or ecological trap conditions.

Rather than adopting further new legislation at international or EU level, we recommend taking into account these concepts in the implementation of existing legal instruments. Although extinction debt or ecological traps are not explicitly mentioned in the CBD or Habitats Directive, the existing provisions are broad enough to encompass the recommended actions that we propose in this chapter. However, at the level of implementation, these concepts are not, or insufficiently, taken into account. Here again, cooperation between scientists and lawyers can be very valuable, for instance via the development of more specific guidelines on ecological restoration, such as through a COP decision under the CBD or European Commission guidelines.

## References

### *Primary sources (law materials)*

Case C-383/09*Commission v France* (2011) ECLI:EU:C:2011:369
Case C-342/05, *Commission v Finland* (2007) ECLI:EU:C:2007:341.
Case C-441/17, *Commission v Poland* (Białowieża Forest) (2018). ECLI:EU:C:2018:255.

Convention on Biological Diversity, 5 June 1992 (1992) 31 ILM 818.
Convention on Wetlands of International Importance especially as Waterfowl Habitat, 2 February 1971 (1972) 11 ILM 963.
Convention on the Conservation of Migratory Species of Wild Animals, 23 June 1979 (1979) 19 ILM 15.
Convention on the Conservation of European Wildlife and Natural Habitat, 19 September 1979, ETS No. 104.
Directive 92/43/EEC of 21 May 1992 on the Conservation of Natural Habitats and of Wild Fauna and Flora, OJ L 206, 22 July 1992
Directive 2009/147/EC of the European Parliament and of the Council of 30 November 2009 on the conservation of wild birds, OJ L 20, 26 January 2010.

### Secondary sources

Anna C. Backstrom et al (2018) 'Grappling with the Social Dimensions of Novel Ecosystems' 16 *Frontiers in Ecology and the Environment* 109.
James Battin (2004) 'When Good Animals Love Bad Habitats: Ecological Traps and the Conservation of Animal Populations' 18 *Conservation Biology* 1482.
Andrew F. Bennett (1998, 2003) *Linkages in the Landscape: The Role of Corridors and Connectivity in Wildlife Conservation*, IUCN.
Graham Bennett and Kalemani Mulongoy (2003) *Review of Experience with Ecological Networks, Corridors and Buffer Zones*, CBD Technical Series No. 23, Secretariat of the Convention on Biological Diversity.
Ulrich Beyerlin (2007) 'Different Types of Norms in International Environmental Law. Policies, Principles and Rules' in D. Bodansky, J. Brunnée and E. Hey (eds) *The Oxford Handbook of International Environmental Law*, Oxford University Press, 425.
Eric Biber (2012) 'Which Science? Whose Science? How Scientific Disciplines Can Shape Environmental Law' 79 *The University of Chicago Law Review* 471.
Charles-Hubert Born et al (eds) (2015) *The Habitats Directive in its EU Environmental Law Context: European Nature's Best Hope?* Routledge.
Peter Bridgewater and Laurie Yung (2013) 'The Policy Context: Building Laws and Rules that Embrace Novelty' in R. Hobbs, E. Higgs and C. Hall (eds) *Novel Ecosystems: Intervening in the New Ecological World Order*, Wiley-Blackwell, 272.
Guillaume Chapron et al (2017) 'Bolster Legal Boundaries to Stay Within Planetary Boundaries' *Nature Ecology and Evolution* 1, article 0086, doi:10.1038/s41559-017-0086.
An Cliquet (forthcoming, 2019) 'Boosting Ecological Restoration as a Legal Duty in the Anthropocene' in M. Lim (ed.) *Charting Environmental Law Futures in the Anthropocene*, Springer.
An Cliquet and Kris Decleer (2017) 'Halting and Restoring Species Loss: Incorporating the Concepts of Extinction Debt, Ecological Trap and Dark Diversity into Conservation and Restoration Law' 26 *Griffith Law Review* 178.
Richard T. Corlett (2016) 'Restoration, Reintroduction, and Rewilding in a Changing World' 31 *Trends in Ecology and Evolution* 453.
Jordi Cortina-Segarra, Kris Decleer and Johannes Kollmann (2016) 'Biodiversity: Speed Restoration of EU Ecosystems' 535 *Nature* 231.
Robin K. Craig (2010) '"Stationary is Dead"—Long Live Transformation: Five Principles for Climate Change Adaptation Law' 34 *Harvard Environmental Law Review* 9.

Romane H. Cristescu et al (2013) 'Potential "Ecological Traps" of Restored Landscapes: Koalas Phascolarctos Cinereus Re-occupy a Rehabilitated Mine Site' 8 *PLOS One* 1.

Molly S. Cross et al (2013) 'From Connect-the-dots to Dynamic Networks: Maintaining and Enhancing Connectivity as a Strategy to Address Climate Change Impacts on Wildlife' in J.F. Brodie, E.S. Post and D.F. Doak (eds) *Wildlife Conservation in a Changing Climate*, The University of Chicago Press, 307.

DG Environment (2017) *Reporting under Article 17 of the Habitats Directive: Explanatory Notes and Guidelines for the Period 2013–2018*, European Commission.

Andrew Dodd et al (2010) 'Protected Areas and Climate Change. Reflections from a Practitioner's Perspective' 6 *Utrecht Law Review* 141.

Aaron M. Ellison (2016) 'It's Time to Get Real About Conservation' 538(7624) *Nature* 141.

Yaffa Epstein (2016) 'Favourable Conservation Status for Species: Examining the Habitats Directive's Key Concept through a Case Study of the Swedish Wolf' 28 *Journal of Environmental Law* 221.

Yaffa Epstein, José Vicente López-Bao and Guillaume Chapron (2016) 'A Legal-ecological Understanding of Favorable Conservation Status for Species in Europe' 9 *Conservation Letters* 81.

European Commission (2000) *Managing Natura 2000 Sites: The Provisions of Article 6 of the 'Habitats' Directive 92/43/EEC*, Office for Official Publications of the European Communities.

European Commission (2007) *Guidance Document on the Strict Protection of Animal Species of Community Interest under the Habitats Directive 92/43/EEC.*

European Commission (2011) *Commission Implementing Decision of 11 July 2011 Concerning a Site Information Format for Natura 2000 Sites (Notified Under Document C(2011) 4892)*, OJ L 198, 30.7.2011.

European Commission (2013a) *Interpretation Manual of European Union Habitats – EUR28.*

European Commission (2013b) *Guidelines on Wilderness in Natura 2000. Management of Terrestrial Wilderness and Wild Areas within the Natura 2000 Network. Technical Report – 2013–2069.*

European Commission (2015) *The State of Nature in the EU. Reporting under the EU Habitats and Birds Directives 2007–2012.*

European Commission (2016) *Commission Staff Working Document. Fitness Check of the EU Nature Legislation (Birds and Habitats Directives)*. SWD(2016) 472 final.

European Environment Agency (2015) *State of Nature in the EU. Results from Reporting under the Nature Directives 2007–2012*, EEA Technical Report No. 2/2015.

Eric Freyfogle (2006) 'Conservation Biology and Law: Only a Start' 20 *Conservation Biology* 679.

George Gann et al (2018) 'The SER Standards: A Globally Relevant and Inclusive Tool for Improving Restoration Practice – A Reply to Higgs et al' 26 *Restoration Ecology* 426.

Alexander Gillespie (2007) *Protected Areas and International Environmental Law*, Martinus Nijhoff.

Nicola Heller and Erica Zavaleta (2009) 'Biodiversity Management in the Face of Climate Change: A Review of 22 Years of Recommendations' 142 *Biological Conservation* 14.

Eric Higgs et al (2018a) 'On Principles and Standards in Ecological Restoration' 26 *Restoration Ecology* 399.

Eric Higgs et al (2018b) 'The Evolution of Society for Ecological Restoration's Principles and Standards—Counter-response to Gann et al' 26 *Restoration Ecology* 431.

Richard Hobbs and Jim Harris (2001) 'Restoration Ecology: Repairing the Earth's Ecosystem in the New Millennium' 9 *Restoration Ecology* 239.

Richard Hobbs, Eric Higgs and Jim Harris (2009) 'Novel Ecosystems: Implications for Conservation and Restoration' 24 *Trends in Ecology and Evolution* 99.

Richard Hobbs, Eric Higgs and Carol Hall (2013) 'Defining Novel Ecosystems' in R. Hobbs, E. Higgs and C. Hall (eds) *Novel Ecosystems: Intervening in the New Ecological World Order*, Wiley-Blackwell.

Kristoffer Hylander and Johan Ehrlén (2013) 'The Mechanisms Causing Extinction Debts' 28 *Trends in Ecology and Evolution* 341.

Md S. Iftekhar et al (2017) 'How Economics Can Further the Success of Ecological Restoration' 31 *Conservation Biology* 261.

Karen Keenleyside et al (2012) *Ecological Restoration for Protected Areas: Principles, Guidelines and Best Practices*, IUCN.

Marianne Kettunen et al (2007) *Guidance on the Maintenance of Landscape Features of Major Importance for Wild Flora and Fauna – Guidance on the Implementation of Article 3 of the Birds Directive (79/409/EEC) and Article 10 of the Habitats Directive (92/43/EEC)*, Institute for European Environmental Policy.

Jochen Krauss et al (2010) 'Habitat Fragmentation Causes Immediate and Time-delayed Biodiversity Loss at Different Trophic Levels' 13 *Ecology Letters* 597.

Mikko Kuussaari et al (2009) 'Extinction Debt: A Challenge for Biodiversity Conservation' 10 *Trends in Ecology and Evolution* 564.

Barbara Lausche et al (2013) *The Legal Aspects of Connectivity Conservation. A Concept Paper*, IUCN.

Paul Leadley et al (2014) *Progress Towards the Aichi Biodiversity Targets: An Assessment of Biodiversity Trends, Policy Scenarios and Key Actions*, Secretariat of the Convention on Biological Diversity.

Tein McDonald, Justin Jonson and Kingsley Dixon (2016) 'National Standards for the Practice of Ecological Restoration in Australia' 24(S1) *Restoration Ecology* S4.

Tein McDonald et al (2016) *International Standards for the Practice of Ecological Restoration – Including Principles and Key Concepts*, Society for Ecological Restoration.

Milieu, IEEP and ICF (2016) *Evaluation Study to Support the Fitness Check of the Birds and Habitats Directives.*

Elisa Morgera (2017) 'The Ecosystem Approach and the Precautionary Principle' in E. Morgera and J. Razzaque (eds) *Biodiversity and Nature Protection Law*, Edward Elgar.

Carolina Murcia et al (2014) 'A Critique of the "Novel Ecosystem" Concept' 10 *Trends in Ecology and Evolution* 548.

Nathalie Pettorelli et al (2018) 'Making Rewilding Fit for Policy' 55 *Journal of Applied Ecology* 1114.

Benjamin J. Richardson and Ted Lefroy (2016) 'Restoration Dialogues: Improving the Governance of Ecological Restoration' 24 *Restoration Ecology* 668.

Jean-Michel Roberge and Per Angelstam (2004) 'Usefulness of the Umbrella Species Concept as a Conservation Tool' 18 *Conservation Biology* 76.

Philippe Sands et al (2012) *Principles of International Environmental Law*, Cambridge University Press.

Secretariat of the Convention on Biological Diversity (2014) *Global Biodiversity Outlook 4.*

Paul Severns (2011) 'Habitat Restoration Facilitates an Ecological Trap for a Locally Rare, Wetland-restricted Butterfly' 4 *Insect Conservation and Diversity* 184.

Daniel Simberloff, Caroline Murcia and James Aronson (2015) '"Novel Ecosystems" are a Trojan Horse for Conservation', *Ensia*, 21 January, http://ensia.com/voices/novel-ecosystems-are-a-trojan-horse-for-conservation.

Katharine Suding et al (2015) 'Conservation: Committing to Ecological Restoration' 348 (6235) *Science* 638

Jac A.A. Swart et al (2018) 'Involving Society in Restoration and Conservation' 26(S1) *Restoration Ecology* 53.

Anastasia Telesetsky, An Cliquet and Afshin Akhtar-Khavari (2017) *Ecological Restoration in International Environmental Law*, Routledge.

Derek Tittensor et al (2014) 'A Mid-term Analysis of Progress Toward International Biodiversity Targets' 346 *Science* 241.

Arie Trouwborst (2015) 'The Habitats Directive and Climate Change. Is the Law Climate-proof?' in C.-H. Born et al (eds) *The Habitats Directive in its EU Environmental Law Context: European Nature's Best Hope?*, Routledge.

Arie Trouwborst et al (2017) 'International Wildlife Law: Understanding and Enhancing Its Role in Conservation' 67 *BioScience* 784.

Arie Trouwborst, Luigi Boitani and John Linnell (2017) 'Interpreting "Favourable Conservation Status" for Large Carnivores in Europe: How Many are Needed and How Many are Wanted?' 26(1) *Biodiversity and Conservation* 37.

Amy M. Truitt et al (2015) 'What is Novel About Novel Ecosystems: Managing Change in an Ever-Changing World' 55 *Environmental Management* 1217.

Orsolya Valkó (2018) 'Concepts, Challenges, and Emerging Themes of Restoration Ecology' 32(4) *Conservation Biology* 966.

Paddy Woodworth (2013) *Our Once and Future Planet*, The University of Chicago Press.

Joy B. Zedler (2018) 'Communicating Useful Results from Restoration Ecology Research' 26 *Restoration Ecology* 395.

# 7 The green financing of ecosystem restoration

*Froukje Maria Platjouw*

## 7.1 Financialisation in an era of ecosystem degradation

### *7.1.1 Introduction*

Many ecosystems worldwide are now gravely threatened.[1] Recognising the threats of ecosystem degradation, in 2000, in his Millennium Report to the United Nations General Assembly, Kofi Annan emphasised the growing burden that degraded ecosystems place on human wellbeing and economic development, and the opportunity that well-managed ecosystems provide for meeting the goals of poverty eradication and sustainable development.[2] Ever-growing economic demands being placed on increasingly degraded ecosystems seriously diminish the prospects for sustainable development. Productive ecosystems, with an array of services and benefits, provide people with life-sustaining resources and options they can use as insurance in the face of natural catastrophes or social upheaval. While well-managed ecosystems reduce risks and vulnerability, poorly managed systems can exacerbate them by increasing risks of flood, droughts, crop failure, disease, and other adversities. Recently, the Intergovernmental Science-Policy Platform on Biodiversity and Ecosystem Services (IPBES) concluded that degradation of the Earth's lands and waters through human activities is negatively impacting the well-being of at least 3.2 billion people, pushing the planet towards a sixth mass species extinction, and costing more than 10 percent of the annual global gross product through loss of biodiversity and ecosystem services.[3]

An important explanation for these ominous trends is that the real value of ecosystems and their services is poorly understood by policy makers and the wider public. The study on 'The Economics of Ecosystems and Biodiversity' (TEEB), led by Pavan Sukhdev, has published a series of reports, among which was a 2010 report that showed how economic concepts and tools can help equip policy makers with the means to incorporate the values of nature into their decision-making at all levels.[4] The TEEB report also highlighted that degradation of ecosystems could be significantly slowed or reversed if

1 UNEP and MEA (2005), 1.
2 Ibid.
3 IPBES (2018), 2.
4 Kumar (2010), 3.

the full economic value of the services is taken into account in decision-making.[5] For sure, as Baumber states, 'price can never be a substitute for strong policies and legislation regarding the preservation of biodiversity'.[6] He also explains that pricing entails several risks, including that pricing promotes the protection of only those ecosystems that are profitable in economic terms, leaving the rest aside.[7]

An important consideration is that the introduction of new methods and tools to financially quantify ecosystem services carries the risk of distorting how society appreciates the natural environment through a purely economic outlook that displaces legitimate ethical, aesthetic, and cultural considerations. Whilst it is, of course, essential that humankind values the life-sustaining properties of ecosystems, and accepts the necessity to repair damage to them, the 'financialisation' of nature risks entrenching the very economic processes and attitudes that have driven unsustainable development at the outset. Critics of neo-liberalism have called for the *contraction*, not expansion, of capitalist modalities, and thus doubt that economic tools such as pollution trading schemes or biodiversity offsets can ever capture the biosphere's complexity and diversity of values.[8] Financialisation also risks transferring the ownership or control of critical natural resources into private hands, typically business corporations, thereby potentially limiting public oversight of their management. The commodification of environmental properties, even when ostensibly to promote corporate accountability for their conservation and sustainable use, may surreptitiously propagate a values system that impedes ethical respect for the natural world and the development of cultural constraints on its misuse. These are complex issues that this chapter cannot address in detail, or resolve, but they are flagged here as considerations to keep in mind when designing governance systems to fund the costs of ecological restoration.

A general advantage of monetisation is that it could facilitate the integration of ecosystem services' values into decision-making frameworks and promote sustainable development. A 2004 World Bank report states that the monetisation of ecosystem services' values is:

> purely a matter of convenience, in that it uses units that are widely recognized, saves the effort of having to convert values already expressed in monetary terms into some other unit of account, and facilitates comparison with other activities that also contribute to well-being.[9]

The IPBES concluded that

> [i]nvesting in avoiding land degradation and the restoration of degraded land makes sound economic sense; the benefits generally by far exceed the cost.

5 Ibid.
6 Baumber (2017), Ch. 31.
7 Ibid.
8 Kill (2015); Buscher, Dressler and Fletcher (2014).
9 World Bank (2004), 10.

> On average, the benefits of restoration are 10 times higher than the costs, estimated across nine different biomes.[10]

In the context of ecological restoration, monetisation might also be highly useful. To estimate the exact costs and benefits of specific restoration projects, valuations need to be made to rationalise trade-offs, to estimate the size of restoration measures, and to calculate the need for investments and funding from public and private sources.

This chapter explores the concept of green finance as a means to mobilise financial resources for ecological restoration purposes. With much of the ecological degradation being the result of historical and cumulative environmental pressures for which traditional liability rules cannot generally work, new financial instruments and sources must be harnessed to fund restoration projects. This section discusses several offsetting mechanisms, such as biodiversity offsetting, wetland banking, payments for ecosystem services, and cap-and-trade mechanisms. This will be followed by an overview of green financing mechanisms for ecosystem restoration, both those based on the polluter-pays principle and other, more innovative mechanisms. Successful examples and case studies that illustrate the application of these mechanisms in different parts of the world are provided. The third section contains some reflections on the opportunities and bottlenecks entangled in the green financing of ecosystem restoration. Importantly, there is no uniform approach, and actors and institutions are recommended to apply a financing mechanism that is best fit-for-purpose for the planned restoration project, taking into account case-specific circumstances and costs.

### *7.1.2 Financialisation and offsetting mechanisms*

The need for monetisation and financialisation is very evident when large investments are needed to restore large degraded areas. Yet, also in the case of financial conditions attached to development approvals from regulators, such as for biodiversity offsetting, wetland banking, or payment for ecosystem services, financialisation could be useful.

'Payments for ecosystem services' (PES) means 'voluntary transactions where a well-defined ecosystem service is bought by at least one buyer from at least one provider, if – and only if – the provider secures the provision of the ecosystem service'.[11] The overarching principle of PES is to ensure that those who benefit from a particular ecosystem service compensate those who provide it, giving them an incentive to continue doing so. Direct beneficiaries, such as water companies, irrigation authorities, aquaculture operations, or hydropower companies, are often willing to pay to secure the services that underpin their businesses.[12] Conversely,

10 IPBES (2018), 2.
11 Ten Brink (2011), 182.
12 Ibid, 183.

where private actors see PES as unrelated to their business, they may resist making such payments.

One good example of PES is the mechanism for Reducing Emissions from Deforestation and Forest Degradation (REDD+) in developing countries, under the 1992 United Nations Framework Convention on Climate Change (UNFCCC), to help internalise the carbon-related ecosystem services provided by forests.[13] Simply stated, developed countries, as the parties primarily responsible for climate change to date, will pay developing countries to keep their forests standing and hence reduce emissions from forest carbon loss.[14] REDD+ thus incentivises developing countries to reduce their deforestation and, in the event they succeed, to compensate them financially.[15]

Another compensatory mechanism, biodiversity offsetting, has become more popular in the European Union (EU). The EU is committed to halting the loss of biodiversity and the degradation of ecosystem services by 2020. Its Biodiversity Strategy enumerates six targets and 20 specific actions geared towards this overall objective.[16] Action 7 aims to ensure no net loss of biodiversity and ecosystem services. The action calls for the development of a methodology to assess the impact of EU funds on biodiversity, and foresees that the European Commission will develop 'an initiative to ensure there is no net loss of ecosystems and their services' (e.g. through compensation or offsetting schemes).[17]

While there are no universal definitions yet, conceptually no net loss and net positive impact goals can underpin decisions to manage and conserve biodiversity in decision-making about development projects.[18] These goals call for negative biodiversity impacts to be either balanced (for no net loss) or outweighed (for net positive impact, also referred to as net gain) by biodiversity gains through offsetting compensation measures implemented in the project region. The biodiversity gains are evaluated against a baseline (e.g. a reference point or trajectory without the project occurring, or prior to the project occurring) of the relevant biodiversity values being impacted by the project. From a conservation perspective, achieving a no-net-loss (NNL) or net-positive-impact goal (NPI) for a given project ultimately means no net reduction in the 1) diversity within and among species and vegetation types; 2) long-term viability of species and vegetation types; and 3) functioning of species assemblages and ecosystems, including ecological and evolutionary processes.[19] The 'net' in this formula acknowledges that some

13 Ibid, 200.
14 This simplicity, however, masks the myriad complexities of international and domestic law, tropical forest governance, and the science of tropical forests and carbon, all of which underpin the ability of different countries and stakeholders to implement effectively and equitably, and benefit from, REDD+. See Lyster, MacKenzie and McDermott (2013), xi.
15 Haug and Gupta (2013), 77.
16 European Commission (2011).
17 Ibid, Annex paragraph 7b.
18 Aiama et al (2015), 7.
19 Ibid.

biodiversity losses at the development site are inevitable, and that biodiversity gains may not be perfectly balanced in regard to the time, space, or type of biodiversity impaired. This is due to the inherent limitations of scientific information available on the species and ecosystems involved. It is therefore always recommended to overcompensate for residual impacts – meaning that defining and achieving an NPI goal is a precautionary way of ensuring an NNL outcome for biodiversity.[20]

An example of a regularly used offsetting mechanism is wetland mitigation banking, an innovation pioneered in the United States. A wetland mitigation bank is a wetland area that has been restored, created, enhanced, or (in exceptional circumstances) preserved, which is then set aside to compensate for future conversions of wetlands for development activities. The value of a bank is determined by quantifying the wetland functions restored or created in terms of 'credits'. Permittees, upon approval of regulatory agencies, can acquire these credits to meet their requirements for compensatory mitigation.[21]

The use of credits occurs not only in the context of wetland mitigation banking, but also in cap-and-trade mechanisms such as the greenhouse gas emissions-trading regimes. They present a market-based approach to controlling greenhouse gas emissions and mitigating the effects of climate change by limiting the quantity of industrial discharges of greenhouse gases, either through the allocation or purchase of emissions allowances from a central authority or the purchase of emissions credits from market participants.[22]

These cap-and-trade systems and biodiversity offsetting mechanisms can potentially avoid the degradation of biodiversity or ecosystems in larger geographical contexts such as biodiversity in country X or the global climate system. A disadvantage of these offsetting mechanisms is, however, that certain degraded sites or areas will not necessarily be restored. Certain ecosystems and biodiversity values are of very high importance to local communities, including indigenous communities, and their value is difficult to compensate for by restoring or conserving areas elsewhere. Furthermore, certain ecosystems may be degraded to such an extent that they cause significant adverse effects on human well-being, for instance in regard to contaminated soil or sediments or polluted rivers. In these cases, there is an evident need for restoration at that particular area or site. In any event, some of the preconditions for use of such market mechanisms, such as insufficient market liquidity and too few market actors, may be absent.

### *7.1.3 The costs of restoration*

The IPBES calls for acceleration of measures for ecological restoration of degraded land.[23] To attain this, however, significant financial flows should be mobilised.

20 Ibid, 11.
21 INR and CEAD (2007), 5.
22 Hussain, Kerschner and Kawano (2017), 1.
23 IPBES (2018), 23.

Even though offsetting and cap-and-trade mechanisms might succeed through the trade in credits, the implementation of concrete restoration projects sometimes requires huge amounts of financial resources. The costs to be funded may include acquisition of private property, scientific research and environmental monitoring, removal of pollutants, translocation of species, and compensation to local resource users for any reduction in their existing economic rights, such as for forestry and farming. With many states functioning within substantial and protracted budget deficits, and public opposition to higher taxes, the public sector has limited capacity to fund such expenses. Enhancing private parties' involvement is therefore essential for meeting the expectations of the IPBES.

To mobilise financial flows, the total cost of the restoration project should be estimated in advance. As an illustration of the different costs involved in a restoration project, three major phases can be distinguished: planning, construction, and post-construction monitoring. In each of these phases, there are factors that can significantly increase the costs of a project. Such factors in the planning phase can include permitting, feasibility studies, sediment testing, safety analyses, and mapping and ground truthing. During the construction phase, cost-increasing factors might be the construction of roads or other access; presence of contaminated materials; means of excavation; revegetation; control of phragmites or other invasive plans; or aesthetic or other considerations at historical sites. During the post-construction phase, the need for monitoring the performance of a project can greatly increase its cost. Monitoring may also be a legal requirement for some projects to assure that the conditions of the permits have been met. The exact cost of restoring a degraded ecosystem or a specific site thus varies highly and depends on the conditions of the site, the legal requirements, the nature and magnitude of monitoring requirements, and many other technical, scientific, and engineering considerations.[24]

As public budgets tighten, governments around the world are looking at opportunities to attract private capital participation in restoration activities. The Subsidiary Body on Scientific, Technical and Technological Advice, under the Convention on Biological Diversity, also recommended to create a framework for mobilising resources to support ecosystem restoration from national, bilateral, and multilateral sources (such as the Global Environment Facility); leveraging national budgets, donors, and partners, including the private sector; indigenous peoples and local communities; and non-governmental organisations (NGOs), to implement and accelerate restoration activities.[25] In addition, many states have already accepted international legal responsibilities to restore degraded ecosystems, so they can hardly use financial reasons to walk away from their duties.[26]

Ecosystem restoration is indeed important for attaining the objectives of many international legal conventions and other initiatives. The effective implementation

24 Habitat Restoration Portal, The Costs of Environmental Restoration Projects www.edc.uri.edu/restoration/html/tech_sci/socio/costs.htm.

25 CBD-SBSTTA (2016).

26 Telesetsky, Cliquet and Akhtar-Khavari (2016).

of ecosystem restoration helps to achieve not only many of the Aichi Biodiversity Targets,[27] but also several Sustainable Development Goals (SDGs).[28] Furthermore, restoration contributes to ecosystem-based adaptation and mitigation under the UNFCCC; land degradation neutrality under the United Nations Convention to Combat Desertification;[29] the wise use of wetlands under the Ramsar Convention on Wetlands;[30] the four Global Objectives on Forests of the United Nations Forum on Forests; commitments under the Convention on the Conservation of Migratory Species of Wild Animals;[31] the Bonn Challenge of the Global Partnership on Forest and Landscape Restoration; and the objectives of many other initiatives.[32] In 2016, an action plan on ecosystem restoration was developed under the auspices of the Convention on Biological Diversity. The overall aim of this plan is to

> promote restoration of degraded natural and semi-natural ecosystems, including in urban environments, as a contribution to reversing the loss of biodiversity, recovering connectivity, improving ecosystem resilience, enhancing the provision of ecosystem services, mitigating and adapting to the effects of climate change, combating desertification and land degradation, and improving human well-being while reducing environmental risks and scarcities.[33]

Given the imperative to accelerate and upscale activities on ecosystem restoration,[34] mobilising financial resources must be incorporated into policy makers' strategies now rather than deferred for later consideration. To implement restoration activities, a framework for mobilising resources to support ecosystem restoration should be created. Public funds and instruments can be used to leverage private funding through innovative financial approaches. The following section explores the concept of green financing and provides an overview of innovative financial mechanisms with examples from different parts of the world to illustrate their successful application.

## 7.2 Green financing mechanisms

The financing of projects providing environmental benefits is generally known as 'green finance'. It is a relatively novel concept, and a definition was first developed at the 2016 G20 meeting between finance ministers and central bank governors in

27 Aichi Biodiversity Targets 5, 12, 14 and 15.
28 United Nations General Assembly (2015).
29 United Nations Convention to Combat Desertification in those Countries Experiencing Serious Drought and/or Desertification, Particularly in Africa, 1994, 1954 UNTS 3.
30 Convention on Wetlands of International Importance especially as Waterfowl Habitat, 1971, 996 UNTS 245.
31 Convention on the Conservation of Migratory Species of Wild Animals (adopted 23 June 1979, entered into force on 1 November 1983) 1651 UNTS 333.
32 CBD-SBSTTA (2016).
33 Ibid.
34 Aronson and Alexander (2013), 293–296.

China. Here, green finance was defined as the 'financing of investments that provide environmental benefits in the broader context of environmental sustainable development'.[35] Green finance includes both the private and public financing of green investments, as well as efforts to internalise environmental externalities and influence risk perceptions to promote environment-friendly investments.[36]

'Green finance' is often used interchangeably with 'green investment'. Zadek and Flynn clarified that, in practice, green finance is a wider lens beyond financial investment. Most importantly, it includes operational costs of financing environmental projects that are not conventionally included under the notion of green investment. Most obviously, it would include costs such as project preparation and land acquisition costs, both of which are not only significant but can pose distinct financing challenges.[37]

### *7.2.1 Mechanisms based upon the polluter-pays principle*

The polluter-pays principle is a cardinal legal principle in environmental law. Sustainable development and the sustainable use of any natural resource rely on regulatory mechanisms that prevent the user from reducing the ability of natural systems to provide ecosystem services.[38] The polluter-pays principle has been considered as a manner to proportionally allocate the effects of such alterations to those users that are responsible for them, thus producing a regulatory effect on the use of the natural resource or ecosystem service.[39] The principle postulates that the polluter should bear the cost of abating waste and restoring the environment to an acceptable condition. By compelling the polluter to bear the expense, the cost of these clean-up measures is often reflected in the retail cost of the goods and services that generate the pollution through their production or consumption.[40] The principle is also generally used in compensation for liability for environmental damages, and has been widely applied in America and Europe[41] in cases such as oil spills or other hazardous releases that require remediation and habitat protection.[42]

Internationally, the principle is acknowledged in several legal instruments addressing environmental issues. Already in 1992, the United Nations Conference on Environment and Development included the principle in its Rio Declaration on

35 UNEP (2016).
36 Ibid.
37 Zadek and Flynn (2013). For a discussion of the concept of green finance in the context of sustainable development in developing countries, see also UNEP (2016). For a broad definition of green finance, see Lindenberg (2014).
38 García de Jalón et al (2017).
39 Ibid.
40 Larson (2005).
41 In the EU, application of the polluter-pays principle has been encouraged in the EU Water Framework Directive (WFD) and the Environmental Liability Directive. For the application of the polluter-pays principle in the context of the WFD, see García de Jalón et al (2017).
42 Bing Yu and Linyu Xu (2016).

Environment and Development.[43] Principle 16 of the Rio Declaration provides that

> [n]ational authorities should endeavour to promote the internalization of environmental costs and the use of economic instruments, taking into account the approach that the polluter should, in principle, bear the cost of pollution, with due regard to the public interest and without distorting international trade and investment.

The principle's inclusion in one of the most influential international statements of the fundamental principles of environmental law has affected legal developments at the regional level and national levels too.[44] The principle has, for instance, played a significant role in the formation of law and policy in the European Community, such as in its environmental liability directive.[45]

Interestingly, the progressive harmonisation of environmental liability regimes in the EU, USA, and Japan has transpired under the polluter-pays principle. This harmonisation is driven mainly by the prevalence of international treaties, the increasing availability of information concerning the environment, domestic and foreign laws that influence other countries, NGOs that exert pressure on regulatory bodies, bilateral and multilateral development institutions that condition their lending on pro-environmental criteria, and the growing standardisation of environmental standards worldwide through corporate social responsibility.[46]

In the context of restoration and remediation, the polluter-pays principle is applied in financing mechanisms that aim to mobilise financial flows, mainly to remediate unlawful contamination caused by identifiable polluters. The most well-known mechanism in which the polluter-pays principle is crucial is funds, such as the Superfund in the United States, where public anxiety over contaminated sites emerged in the late 1970s following several catastrophic incidents including the Love Canal, Times Beach, and the Valley of the Drums. In response to these scandals, the Comprehensive Environmental Response, Compensation, and Liability Act (CERCLA) was passed in 1980, commonly referred to as the Superfund. It enabled the US Environmental Protection Agency (EPA) to establish a remediation fund which supports the remediation of contaminated sites. Also, CERCLA authorised the government to retroactively charge the polluters for contamination. Superfund is thus a federal government program that funds the remediation of contaminated sites. The EPA may identify the responsible polluters, require them to clean up the sites, or may remediate the site itself using the Superfund. Historically, about 70 percent of the Superfund clean-up activities has been paid for by the polluters.

43 Rio Declaration on Environment and Development (13 June 1992) 31 ILM 874.
44 Sally-Ann Joseph (2014).
45 Council Directive 2004/35/CE of 21 April 2004 on environmental liability with regard to the prevention and remedying of environmental damage [2004] OJ L 143:56–75.
46 Larson (2005).

Prior to a 2005 amendment to the legislation, the courts were interpreting liability as capable of being extended to financial institutions such as banks that had provided funds to the offending companies or had acquired control of land from their insolvent borrowers.[47] The Superfund was originally financed primarily through a tax on the petroleum and chemical industries. Since 2001, the fund has mostly been financed through the federal budget. Between 2000 and 2015, Congress allocated about $1.26 billion of general revenue to the Superfund program annually.[48]

Another example is the BOFAS fund in Belgium, which was set up as a statutory interregional cooperation agreement between the three regions of Flanders, Wallonia, and Brussels, and the federal government of Belgium. The BOFAS fund provides reserve funds for meeting the costs of soil remediation associated with petrol stations in Belgium. It is replenished by levies, based on the polluter-pays principle, on petrol stations and the petroleum industry (50 percent), and an excise duty levied per litre of gasoline and diesel sold to consumers (50 percent). For every litre of gasoline, €0.0030 is levied; and for every litre of diesel, €0.0020 is levied. The annual budget of the BOFAS fund in 2016 was €36 million.[49]

A similar approach has been being applied in Denmark, where the polluters paid the total cost for remediation of contaminated sites. The Danish Oil Industry Association (on behalf of its member firms) proposed a voluntary scheme to overcome the difficulties, leaning on collaboration between the Danish Oil Industry Association and public authorities. By the end of 1992, an agreement on the remediation of former and soon-to-be-closed petrol stations was signed by the Danish Oil Industry Association, and several public authorities and other stakeholders. Thereafter, all nine oil companies in Denmark reached an agreement regulating the funding, and established the Danish Oil Industry's Remediation Fund, financed through a fee levied on the sales of petrol. Distortion of market competition was avoided through the inclusion of all oil companies operating in Denmark. Through this agreement, the oil companies created a cost-effective way to implement the polluter pays principle and avoid litigation.[50]

Funds based upon the polluter-pays principle are mainly being applied in the context of remediation of contaminated sites. In many other contexts, however, the polluter-pays principle might fall short in incentivising the restoration of degraded areas or ecosystems, especially in cases where polluters may not be identifiable or no longer exist. To accelerate and incentivise the restoration of areas under these circumstances, other financing avenues should be explored.

47 *United States of America v Fleet Factors Corp*, 901 F.2d 1550 (11th Cir. 1990).
48 Skumlien Furuseth et al (2018).
49 Perera et al (2018).
50 Skumlien Furuseth et al (2018).

### *7.2.2 Green financing mechanisms for ecosystem restoration*

In addition to the foregoing government-managed industry funds, a range of other financing mechanisms has been increasingly used to finance environmental projects. Examples include green bonds, debt-for-nature swaps, crowd funding or crowd investments, and public–private partnerships.

#### *Green bonds*

Green bonds are one of the most widely used green finance instruments.[51] In China, for instance, total green bond issuance in 2017 reached US$37.1 billion in 2017 – an increase of 4.5 percent over 2016.[52] China is now the second-largest green bond issuer after the United States, spurred by a combination of policy development, growing momentum at local level, and the diversification of issuer types and use of proceeds. Approximately 15 percent of the proceeds from green bonds have been allocated to a variety of water projects, including wastewater treatment, silt removal, water resource management, and ecological restoration.[53] Bonds are issued to finance specific projects and are paid back exclusively by the revenues generated by the same project. Bondholders hence have no recourse to other revenues generated by the project initiator or bond issuer. Project bonds pay a predetermined coupon (interest) during a defined time period while the principal (the borrowed amount) is paid back when the bond reaches maturity. Green bonds are issued in particular to fund projects that have quantifiable environmental benefits.[54] The principal virtue of bond financing is its relative flexibility, and it can be used for the direct financing of remediation and restoration projects. Depending on how they are designed and which institution is issuing the green bond, they can be examples of credit enhancement, export credit guarantees, and in some cases securitisation. When public institutions issue green bonds, their higher credit ratings can be a form of credit enhancement, providing assurance to investors that the issuer will commit to repaying the bond. Approximately 75 percent of green bonds have been issued by government-owned or government-backed entities, including bonds issued by national government institutions, municipalities and cities, and multilateral and bilateral development institutions. The public sector has thus played a seminal role in providing high credit ratings for green bonds.[55]

One good example of a green bond mechanism is the Hong Kong and Shanghai Banking Corporation Limited (HSBC)'s SDG Bond.[56] As a response to the United Nations' SDGs and the Paris Climate Agreement,[57] HSBC issued the

51 UNEP (2016).
52 Dai and Matthews (2018).
53 Ibid.
54 Perera et al (2018).
55 Torvanger et al (2016), 15; Nordic Council of Ministers (2016).
56 Perera et al (2018).
57 United Nations Framework Convention on Climate Change, Paris Agreement (12 December 2015) FCCC/CP/2015/L.9/Rev.1.

world's first corporate sustainable development bond in November 2017. It raised US$1 billion in proceeds, which are assigned to be invested in businesses and projects that contribute to the following SDGs: Good Health and Well-being; Quality Education; Clean Water and Sanitation; Affordable and Clean Energy; Industry, Innovation and Infrastructure; Sustainable Cities and Communities; and Climate Action.[58] As stated in the HSBC SDG Bond Framework, the 'proceeds will be used to finance entirely or partly new and/or re-finance existing businesses and projects that derive 90 percent or more of its revenues from one or several of the seven SDGs that have been selected by HSBC'.[59] To provide transparency on use of the proceeds, HSBC has included both eligibility and exclusion criteria in its SDG Bond Framework.[60]

The design of HSBC's SDG Bond can be replicated to raise financing for restoration projects that can demonstrate stable and predictable revenues post-remediation. Restoration projects might be particularly relevant in relation to SDG 6: Clean Water and Sanitation, if these activities expand public access to safe and affordable drinking water, or if these activities improve water quality.[61] Projects that aim to restore aquatic ecosystems may fulfill these eligibility criteria.

Another seminal example is the green bond issued by the Mass Transit Railway (MTR) Corporation, Hong Kong,[62] which owns and manages the Hong Kong Mass Transit Railway serving more than 5.5 million passengers each day. MTR is also investing in globally operating rapid transit systems in London, Stockholm, Beijing, Shenzhen, and Hangzhou. MTR issued its first green bond in October 2016 as a part of the corporation's green investment strategy and environmental performance goals. The bond was arranged as a part of the MTR US$4 billion debt issuance programme and was entirely guaranteed by the corporation's balance sheet. Prior to the issuance of this green bond, MTR published a green bond framework to define how the proceeds would be invested. The eligible areas include biodiversity and conservation, such as projects aimed at the reforestation of forests and wetlands, and the remediation of contaminated soil.[63]

Though promising, the challenge with bonds is that the coupon and the principal of the bond should be directly backed by the bonds issuer's balance sheet or the revenue of the project. The restoration project therefore should demonstrate that sufficiently stable and predictable revenue streams can be expected after remediation, as otherwise the bond financing will not be feasible. In addition, bond issuers are legally obligated to obtain guarantees to cover default risks, and this increases the costs of debt financing.[64]

Furthermore, the rapid growth in the green bond market has spurred questions about the integrity of the environmental claims of these bonds, both globally and

58 HSBC (2017a).
59 HSBC (2017b), 2.
60 HSBC (2015).
61 HSBC (2017b).
62 Perera et al (2018).
63 Ibid.
64 Ibid.

in China. Without clear and widely accepted standards, many investors have raised concerns about 'greenwashing', a practice in which bond proceeds are allocated to assets that have little or unclear environmental value. This lack of certainty shakes confidence in the market and hampers efforts to finance the transition to a low-carbon economy.[65] To address the scourge of greenwashing within water infrastructure investments, a consortium of NGOs developed the Climate Bonds Standard Water Criteria.[66] It defines low-carbon and climate-resilient water infrastructure by evaluating the impact of water-related investments on climate mitigation and climate adaptation. The criteria provide guidance on the type of water projects that should be included in green bonds, including infrastructure for water capture and collection, water storage, water treatment, flood and drought defense, stormwater management, and ecological restoration and management. The Water Criteria certification scheme requires an independent 'approved verifier' to assess green bonds. The verifier's assessment indicates to investors that the underlying water assets have undergone environmental due diligence, thus enabling investors to more easily identify and invest in water infrastructure with genuine climate benefits.[67]

### *Debt-for-nature swaps*

Another interesting mechanism in the context of restoration is debt-for-nature swaps (DNS).[68] They involve cancellation of debt obligations of a developing country on the agreement that the funds released from debt servicing will instead be invested in the protection of biodiversity. This arrangement is valuable for developing countries, which face pressing conservation challenges. Debt-for-nature swaps can be structured in two ways: 1) the creditor government waives all or part of its credit rights and the debtor government invests the equivalent value in biodiversity conservation; and 2) the creditor government sells all or part of the outstanding debt to an organisation with expertise to carry out biodiversity conservation work in the debtor country.[69]

The proceeds from DNSs are best invested in environmental trust funds in the debtor country. These funds can then be used to provide grants and loans for biodiversity conservation and the conservation of protected areas. Proceeds from DNS can also be invested in endowment funds that ensure long-term annual budgets for earmarked projects. In either case, the establishment of a fund is important to secure the long-term commitment for conservation from the debtor government and adequate follow-up from the creditor government.[70] Despite their obvious value in promoting sustainability in the host countries, DNSs have been controversial for some observers as potential means of overriding developing

65 Dai and Matthews (2018).
66 Ibid.
67 Ibid.
68 Sheikh (2010).
69 Perera et al (2018).
70 Ibid.

countries' sovereignty over their natural resources and entrenching new forms of control by institutions in the Global North.[71]

Debt-for-nature swaps are a financing mechanism especially relevant for the restoration and conservation of biodiversity in developing countries, and can help finance the restoration of areas where the value capture is not so lucrative and hence cannot be easily co-financed by private investors. The DNS is also an instrument that can attract philanthropic capital from charitable foundations and high-network individuals. These funds can be combined with the proceeds from debt forgiveness or, indeed, used to set-up a 'debt-for-remediation swap preparation fund'.[72] Challenges could, however, arise in the arrangement of DNSs. This needs to be undertaken by an agency specialising in restoration along with public sector financiers and appropriate experts on investment law. In addition, for the swap to generate the desired outputs, there should be baseline knowledge on remediation or restoration in the recipient country, otherwise the conditions of the swap might not be fulfilled.[73]

One example is the Guatemala Forest Conservation Debt-for-Nature Swap. The US and Canada have been the leading creditor countries in the structuring of DNS. Impetus and legal security for these initiatives were also provided by the 1998 US Tropical Forest Conservation Act,[74] which provided for the allocation of public and private capital towards forest conservation. In 2006, the US government agreed to waive US$24 million of debt owed by the Government of Guatemala, amounting to 20 percent of the country's entire debt, in a DNS arrangement that was facilitated by the Nature Conservancy and Conservation International. Four earmarked conservation projects were chosen to benefit from the DNS arrangement, comprising the Maya Biosphere Reserve which makes up 10 percent of Guatemala's land area; the Sierra del Lacandón National Park and Tikal National Park that lies within the Maya Biosphere Reserve; the Motagua/Polochic System, which is home to 80 percent of Guatemala's biodiversity; and the Sierra Madre Volcanic Chain including Lake Atitlán. Currently, Panama, Jamaica, Colombia, and Peru have also entered into DNSs under the Tropical Forest Conservation Act of 1988.[75]

### *Crowd funding/crowd investments*

An increasingly used financing mechanism to support environmental projects is crowd funding or crowd investments. Crowd-investment platforms connect private investors with projects and entrepreneurs that need capital to start or scale-up their impact-driven venture. The value of such platforms lies in their function to aggregate capital from multiple investors and channel it to multiple projects and

71 Eshbach (1990).
72 Perera et al (2018).
73 Ibid.
74 Tropical Forest Conservation Act of 1998 (P.L. 105–214) (22 U.S.C. 2431).
75 Perera et al (2018).

businesses seeking investment. The value added of crowd-investment platforms for borrowers includes the acquisition of affordable capital, the promotion of business plans, assistance with risk analysis, the aggregation of projects, and the reduction of transaction costs. The value added for investors includes access to financially viable ventures, reliable due diligence on the business case, and, again, the reduction of transaction costs.[76]

A successful example of crowd funding is TRINE's Solar for Local Businesses project in Zambia.[77] In 2016 the Swedish company TRINE launched a crowd-investment platform to channel investments from large and small private impact investors to solar energy entrepreneurs in East Africa. The objective was to reduce energy poverty and introduce solar technology to rural and off-grid areas. The TRINE platform offers investors a portfolio of investment opportunities and transparency on project risks, revenues, and rates of return. First, funds provided by private investors are channeled into an e-wallet under the investor's name. Funds are maintained in the e-wallet until the total amount required for a project is raised. TRINE then enters into a loan agreement under commercial terms with the selected borrower. The borrower is required to pay a 5 percent administration fee for setting up the funding scheme and an interest payment between 10 and 16 percent on a declining balance of their loan.[78]

An important feature of TRINE is that it combines matching funding from international donors and larger investment funds. For example, Gullspång Invest (one of the founders of TRINE) and UK Aid have provided matching funding for given portfolios. Similarly, the solar technology provider Omnivoltaic collaborated with TRINE to provide 25 percent first-loss protection for loans provided to a solar lighting venture in Tanzania.[79]

The use of crowd-investment platforms for impact-driven projects such as remediation or restoration is very attractive, especially when projects have stable revenue streams post-restoration. The value added of crowd-investment platforms is the improved transparency on risks and returns, and the substantial reduction of transaction costs associated with raising funds from a multitude of investors. In addition, impact investors may be willing to accept lower returns for higher social and environmental benefits, and hence take on the risks of soil remediation ventures where post-remediation revenues are not so robust.[80] Crowd-investment platforms can also attract small-scale investors (for instance, retail investors) and enable them to invest in restoration projects in their neighborhoods. Moreover, platforms can enable small investors to combine their capital with that of public donors and larger institutional investors to give rise to 'community financing' for projects that are a strong priority for local communities.[81] Currently, the biggest challenge in using crowd funding for remediation and restoration purposes is the

76 Skumlien Furuseth et al (2018); Perera et al (2018).
77 Perera et al (2018).
78 Ibid.
79 Ibid.
80 Ibid.
81 Ibid.

significant uncertainty of how investment returns could be monetised and generate returns for investors. Many so-called crowd-funding remediation projects are essentially a form of voluntary donation funding.[82]

*Public-private partnerships*

Many restoration and remediation projects are examples of public-private partnerships (PPPs), which are cooperative institutional arrangements between public and private sector actors. PPPs are generally recognised as long-term cooperative institutional arrangements between public and private actors to achieve various purposes.[83] The growing use of PPPs reflects broader pressures on government resources and ideological shifts in governance towards greater delegation of or sharing of responsibilities with non-state actors.[84]

The Dolores River Restoration Partnership (DRRP) illustrates a successful PPP. The Dolores River is a major waterway in south-western Colorado and eastern Utah, and has important scenic, recreational, and natural value for the region. The riparian habitat has been degraded by several factors, including invasion of tamarisk and other exotic, invasive plants as well as loss of native vegetation in the riparian corridor. The invasive species and ecologically depleted riparian corridor are causing harmful impacts to wildlife habitat, water resources, livestock forage, and recreational use along the Dolores River and its tributaries within Dolores, San Miguel, Montrose, and Mesa counties in Colorado, and Grand County in Utah.[85]

Since 2009, the DRRP, a public–private collaborative, has been working to address these impacts and to restore the river through the Dolores River Riparian Action Plan.[86] The partnership consists of local communities, state and federal agencies, non-profit and private companies, and private landowners that aimed initially for a period of five years of collaboration to protect the shared investment in restoring the riparian corridor of the Dolores River. To mobilise funding, a funding subcommittee has been established to strategically pursue funding options to enhance and leverage resources for supporting partnership activities. Fundraising will include applying for grants, renewing the Bureau of Land Management Assistance Agreement, and working with the core team to develop long-term and multiple funding sources.[87] Total costs for the restoration and management of the Dolores River watershed for the first five years were estimated at $5,191,172[88], including costs related to total tamarisk control, biomass reduction, revegetation, secondary weed control, and short-term monitoring and maintenance for each

82 Skumlien Furuseth et al (2018).
83 Akhter Khanom (2010), 1.
84 Mert (2015).
85 Dolores River Restoration Partnership (2015),1; see also Ratcliff (2015); Oppenheimer et al (2015).
86 Dolores River Restoration Partnership (2015), 1.
87 Ibid, 5.
88 Ibid, 34.

site.[89] Over the past nine years, the economic cost has been US$8,500,000. Half of this cost has been funded by public resources (mainly federal and state governments and some local government funding); the remaining 50 percent has been financed through private resources (private foundations, corporate initiatives, and individual philanthropic funds).[90]

A survey among the main partners in this PPP identified four key areas critical to the partnership's success: 1) effective planning, 2) responsive governance, 3) monitoring progress, and 4) collective learning. Implementing these four components has been critical to the partnership's restoration achievements, which to date include enhancing resilience and diversity on over 1,200 riparian acres of the Dolores River.[91]

Public-private partnerships are a promising mechanism for the financing of ecosystem restoration as they facilitate the funding of restoration projects in areas of ecological importance but minor commercial value. An advantage of public-private partnerships is the risk-sharing element, lowering the risk and threshold for partners to collaborate and invest. Such partnerships can help overcome the limitations of purely market-based mechanisms that may fail to recognise the full values of natural capital and lack the means to engage non-market stakeholders in their restoration.

## 7.3 Reflections

The scale of ecological degradation in the Anthropocene requires ever-greater natural resources to be repaired. With much of the degradation being the result of historical and cumulative environmental pressures, for which traditional liability rules cannot generally work, new financial instruments and sources must be harnessed to fund ecological restoration. This chapter has presented several financing instruments that have been applied in the context of remediation and ecosystem restoration, and provided case study examples from different parts of the world where these instruments are used successfully.

Two main lessons can be learned from this chapter. First, in addition to the full exploitation of compensation or offsetting mechanisms such as wetland banking or cap-and-trade systems, the restoration of already degraded ecosystems is at least as important as to halt the further loss of biodiversity and to conserve key ecosystem functions. In regard to negative trends impacting the functioning of the biosphere, mitigation or offsetting compensation activities can be useful. This is particularly so with climate change, where an increase in greenhouse gas emissions in one part of the world could be compensated for by a reduction in such emissions elsewhere. Even though this might engender ethical and equity-related issues in how the burden of shifting to a low-carbon world is managed, this approach could facilitate the implementation of international legal obligations and commitments

89 Dolores River Restoration Partnership (2010), 33.
90 Personal communication with the Funding Subcommittee, 26 April 2018.
91 Oppenheimer et al (2015), 144.

to halt or at least slow global warming.[92] Often, however, environmental degradation will also have direct, negative impacts on human well-being, most evidently on local communities living near affected areas. When degraded ecosystems pose an immediate threat to people's livelihoods and health, such as in the case of contaminated soil or polluted rivers, the political imperative to undertake restoration will be particularly strong.

Second, though the chapter's case studies are examples of successful remediation or restoration projects, it must be emphasised that there is no uniform approach to financing restoration. Both the costs of restoration and the financing instrument(s) to be applied may be highly case-specific. In fact, as De Groot correctly argues: '[a]ny given level of degradation in a given biome can be addressed in many different ways, ranging from natural regeneration of abandoned landscapes to very active, hands-on restoration interventions over many years'.[93] In the latter case, explains De Groot, 'the work is often labor intensive, capital intensive, or both, and among these possibilities there are myriad options and cost structures'.[94]

More specifically, a restoration strategy intended to ameliorate degradation in a specific ecosystem can have different costs and different results when applied to the same ecosystem type in a different area, due to variations in regional or country-specific cost structures such as workers' salaries and fuel costs. The size of restoration projects is also a material factor. Small-scale restoration, often done for experimental purposes to develop scientific knowledge, tends to be much more expensive than large-scale operations. Furthermore, when restoration activities entail enhancing ecological resilience, the costs of maintenance and of further restoration should gradually become lower. For that reason, there could be substantial differences in restoration costs and needs for funding and financialisation.[95]

Despite the lack of a uniform approach to finance ecosystem restoration, a number of general reflections can be offered. When ecosystem degradation is partly caused through contamination from industrial activities, the polluter-pays principle should be applied and the polluters should be held liable for the contamination. Rather than awaiting expensive litigation processes, however, industry often prefers an approach based on the idea of shared corporate responsibility, where the different polluting actors together finance the remediation of contaminated sites. The US Superfund, the Belgian BOFAS fund, and the Danish Oil Industry's Remediation Fund illustrate this approach.

Where restoration is expected to provide stable revenue streams post-restoration, private investors might be interested in investing in the restoration or remediation of the area through bonds. Bonds have become a popular green finance instrument and are relatively low-risk for investors, but only if the restored area

92 Platjouw (2009).
93 De Groot et al (2013), 1291.
94 Ibid.
95 Ibid.

will eventually result in revenue streams. HSBC's SDG Bond and the green bond issued by Hong Kong's MTR Corporation are examples of the successful use of bonds for environmental purposes.

In the case of ecosystem restoration, many projects do not offer any clear commercial opportunities and will not necessarily result in any revenue streams post-restoration. This might be the case for landscape restoration or river restoration. The restoration will undoubtedly enhance the integrity and functioning of the ecosystem, and thus its capacity to provide ecosystem services to people. Investing in the restoration will thus increase the economic value of the restored natural capital; however, the restoration in itself will not provide any revenue-increasing opportunities for investors. These restoration projects therefore require more public funding and philanthropic contributions from local communities and NGOs. Both public–private partnerships and crowd funding may be effective financing avenues in these circumstances. One example is the DRRP, which up to 2015 has accomplished the restoration of 1,200 riparian acres of the Dolores River.

Finally, to conserve biodiversity and restore ecosystems in developing countries in particular, DNSs have proven to be an effective means to incentivise host countries to prioritise restoration or nature conservation activities. This method could be utilised in a broad array of projects, ranging from the remediation of contaminated soil and sediments to the reforestation of rainforests and the restoration of wetland and river ecosystems. However, as many nations in the Global South have achieved economic success in recent decades, especially in East Asia, the number of states willing to embrace DNSs is declining. Even for the most heavily indebted and poorest countries, more intensive economic utilisation of their natural resources, as is occurring in Brazil and Indonesia, for instance, can be more attractive politically and economically than entering into DNSs.

In sum, there is no uniform approach, and actors and institutions are recommended to apply a financing mechanism that is best fit for purpose for the planned restoration project, taking into account case-specific circumstances and costs. We need both global and national-level funding strategies, as well as multi-actor regimes that combine state and non-state actors. Green financing is also not without risks and limitations if we fail to address the underlying economic drivers associated with a financial system that degrades and overconsumes natural resources. Scholarship on socially responsible investing has argued for a paradigm shift away from market capitalism towards new economic models that are compatible with maintenance of ecological integrity.[96] Valuation of natural capital must be embedded in the heart of any economic system to achieve this goal. From this perspective, all restoration processes should be appreciated from a different perspective, namely that '[m]oney spent on ecosystem restoration is not simply a cost; rather, it is a worthwhile investment that brings multiple benefits and can help achieve policy goals'.[97]

96 Richardson (2008); Ramiah and Gregoriou (2015).
97 De Groot et al (2013), 1292.

## Acknowledgements

The case study descriptions in section 7.2 (except that of the DRRP) are based in part on material belonging to the Norwegian Institute for Water Research (Norsk institutt for vannforskning, NIVA) and the International Institute for Sustainable Development (IISD), and are not necessarily a full or accurate reflection of the material. NIVA and IISD have granted permission to the author to reproduce and adapt the material for academic and non-commercial purposes.

## References

### *Primary resources (law materials)*

CBD-SBSTTA (2016), 'Recommendation Adopted by the Subsidiary body on Scientific, Technical and Technological Advice. XX/12. Ecosystem Restoration', UNEP/CBD/SBSTTA/REC/XX/12.

Convention on the Conservation of Migratory Species of Wild Animals, 1979, 1651 UNTS 333.

Convention on *Wetlands* of International Importance especially as Waterfowl Habitat, 1971, 996 UNTS 245.

European Council Directive 2000/60/EC of 22 December 2000 establishing a framework for Community action in the field of water policy [2000] OJ L327/22.

European Council Directive 2004/35/CE of 21 April 2004 on environmental liability with regard to the prevention and remedying of environmental damage [2004] OJ L 143:56–75.

Rio Declaration on Environment and Development (13 June 1992) 31 ILM 874.

Tropical Forest Conservation Act of 1998 (P.L. 105–214) (22 U.S.C. 2431).

United Nations Convention to Combat Desertification in those Countries Experiencing Serious Drought and/or Desertification, Particularly in Africa, 1994, 1954 UNTS 3.

United Nations Framework Convention on Climate Change, 1992, 1771 UNTS 107.

United Nations Framework Convention on Climate Change, Paris Agreement, 2015, FCCC/CP/2015/L.9/Rev.1.

United Nations General Assembly, Resolution 70/1 'Transforming our World: the 2030 Agenda for Sustainable Development', 21 October 2015, A/70/L.1

*United States of America v Fleet Factors Corp*, 901 F.2d 1550 (11th Cir. 1990).

### *Secondary resources*

Deviah Aiama et al (2015) *No Net Loss and Net Positive Impact Approaches for Biodiversity: Exploring the Potential Application of these Approaches in the Commercial Agriculture and Forestry Sectors*, IUCN.

Nilufa Akhter Khanom (2010) 'Conceptual Issues in Defining Public Private Partnerships (PPPs)' 6(2) *International Review of Business Research Papers* 150.

James Aronson and Sasha Alexander (2013) 'Ecosystem Restoration is Now a Global Priority: Time to Roll up our Sleeves' 21(3) *Restoration Ecology* 293.

Alex Baumber (2017) 'Restoration and Market-based Instruments' in Stuart K. Allison and Stephen D. Murphy (eds) *Routledge Handbook of Ecological and Environmental Restoration*, Routledge.

Bram Buscher, Wolfram Dressler and Robert Fletcher (eds) (2014) *Nature Inc: Environmental Conservation in the Neoliberal Age*, University of Arizona Press.
CCDC (2018) 'China Green Bond Market 2018', Climate Bonds Initiative and China Central Depository and Clearing Company, www.climatebonds.net/resources/reports/china-green-bond-market-2017.
Climate Bonds (2018) 'The Water Infrastructure Criteria', www.climatebonds.net/standard/water.
Lily Dai and John Matthews (2018) *China's Green Bonds Finance Climate Resilience*, China Environment Forum.
Rudolf De Groot, et al (2013) 'Benefits of Investing in Ecosystem Restoration' 27(6) *Conservation Biology* 1286.
Dolores River Restoration Partnership (2010) *Dolores River Riparian Action Plan (DR-RAP), Recommendations for Implementing, Tamarisk Control & Restoration Efforts*, https://drrpartnership.org/pdf/2010_Dolores_River_Riparian_Action_Plan.pdf.
Dolores River Restoration Partnership (2015) *Memorandum of Understanding*, 23 December 2015, https://drrpartnership.org/pdf/DRRP_MOU_12_23_2015.pdf.
Rossean Eshbach (1990) 'A Global Approach to the Protection of the Environment: Balancing State Sovereignty and Global Interests' 4 *Temple International and Comparative Law Journal* 271.
European Commission (2011) *Our Life Insurance, Our Natural Capital: An EU Biodiversity Strategy to 2020*, EU COM/2011/0244 final.
G20 Green Finance Study Group (2016) *G20 Green Finance Synthesis Report* UNEP.
Silvestre García de Jalón, Marta González del Tánago, Carlos Alonso and Diego García de Jalón (2017) 'The Environmental Costs of Water Flow Regulation: an Innovative Approach Based on the 'Polluter Pays' Principle' 31 *Water Resource Manage* 2809–2822.
Constanze Haug and Joyeeta Gupta (2013) 'The Emergence of REDD on the Global Policy Agenda' in J. Gupta, N. van der Grijp and O. Kuik (eds) *Climate Change, Forests and REDD. Lessons for Institutional Design*, Routledge.
HSBC (2015) 'HSBC Green Bond Framework', 6 November.
HSBC (2017a) 'HSBC Issues World's First Corporate Sustainable Development Bond', 23 November, media release.
HSBC (2017b) 'HSBC Sustainable Development Goal (SDG) Bond Framework', 17 November.
Talat Hussain, Seth Kerschner and Ayako Kawano (2017) *Greenhouse Gas Emissions Trading Schemes: A Global Perspective. An Overview of Rules and Developments in Major Jurisdictions Globally, Including the US, Canada, Mexico, Japan, the UK and the EU*, White and Case LLP.
INR and CEAD (2007), *Wetland Mitigation Banking: Assessing the Appropriateness of Wetland Mitigation Banking for Securing Aquatic Biodiversity in the Grassland Biome of South Africa*, Institute of Natural Resources and Centre for Environment, Agriculture and Development.
IPBES (2018) *Summary for Policymakers of the Thematic Assessment Report on Land Degradation and Restoration of the Intergovernmental Science-Policy Platform on Biodiversity and Ecosystem Services*, Intergovernmental Science-Policy Platform on Biodiversity and Ecosystem Services.
Sally-Ann Joseph (2014) 'The Polluter Pays Principle and Land Remediation: A Comparison of the United Kingdom and Australian Approaches' 1(1) *Australian Journal of Environmental Law* 24.

Jutta Kill (2015) *The Financialization of Nature: Creating a New Definition of Nature*, Friends of the Earth.

Pushpam Kumar (ed.) (2010) *The Economics of Ecosystems and Biodiversity: Ecological and Economic Foundations*, Earthscan.

Eric Thomas Larson (2005) 'Why Environmental Liability Regimes in the United States, the European Community, and Japan Have Grown Synonymous with the Polluter Pays Principle' 38 *Vanderbilt Journal of Transnational Law* 541.

Nannette Lindenberg (2014) *Definition of Green Finance*, Deutsches Institut für Entwicklungspolitik, www.cbd.int/financial/gcf/definition-greenfinance.pdf.

Rosemary Lyster, Catherine MacKenzie and Constance McDermott (eds) (2013) *Law, Tropical Forests and Carbon. The Case of REDD+*, Cambridge University Press.

Aysem Mert (2015) *Environmental Governance through Partnerships*, Edward Elgar.

Nordic Council of Ministers (2016) *Green Financing – The Nordic Way*.

J. Daniel Oppenheimer et al (2015) 'A Collaborative Model for Large-scale Riparian Restoration in the Western United States' 3(2) *Restoration Ecology* 143.

Oshani Perera, Laurin Wuennenberg, David Uzsoki and Andres Cuellar (2018) *Financing Soil Remediation: Exploring the Use of Financing Instruments to Blend Public and Private Capital*, International Institute for Sustainable Development.

Froukje Maria Platjouw (2009) 'Reducing Greenhouse Gas Emissions at Home or Abroad? The Implications of Kyoto's Supplementarity Requirement for the Present and Future Climate Change Regime' 18(3) *Review of European Community and International Law* 244.

Vikrash Ramiah and Greg Gregoriou (eds) (2015) *Handbook of Environmental and Sustainable Finance*, Elsevier.

Lindsey Ratcliff (2015) 'Lessons Learned from Large-scale River Restoration' *Yale Environment Review*, 2 November, https://environment-review.yale.edu/lessons-learned-large-scale-river-restoration-0.

Benjamin J. Richardson (2008) *Socially Responsible Investment Law: Regulating the Unseen Polluters*, Oxford University Press.

Pervaze A. Sheikh (2010) *Debt-for-Nature Initiatives and the Tropical Forest Conservation Act: Status and Implementation*, CRS Report for Congress 7-5700, Congressional Research Service.

Ingvild Skumlien Furuseth, Karl Jakob Kammler, Wenting Chen, Froukje Maria Platjouw, Yan Lin, Morten Jartun, Thorjorn Larssen and Mathias Lund Larsen (2018) *Green Finance Approaches to Soil Remediation: International Examples, Norwegian Institute for Water Research.*

Anastasia Telesetsky, An Cliquet and Afshin Akhtar-Khavari (2016) *Ecological Restoration in International Environmental Law*, Routledge.

Patrick Ten Brink (ed.) (2011) *The Economics of Ecosystems and Biodiversity in National and International Policy Making*, Earthscan.

Asbjørn Torvanger, Patrick Narbel, Kamleshan Pillay and Christa Clapp (2016) *Instruments to Incentivize Private Climate Finance for Developing Countries*, CICERO.

UNEP (2016) *Green Finance for Developing Countries: Needs, concerns and innovations. Inquiry: Design of a Sustainable Financial System*, United Nations Environment Programme.

UNEP and MEA (2005) *Ecosystems and Human Well-Being: Our Human Planet, Summary for Policy Makers*, United Nations Environment Programme (UNEP) and Millennium Ecosystem Assessment (MEA), Island Press.

World Bank (2004) *How Much is an Ecosystem Worth? Assessing the Economic Value of Conservation*.

Bing Yu and Linyu Xu (2016) 'Review of Ecological Compensation in Hydropower Development' 55 *Renewable and Sustainable Energy Reviews* 729.

Simon Zadek and Cassie Flynn (2013) *South-Originating Green Finance: Exploring the Potential, The Geneva International Finance Dialogues*, IISD.

# Part 2

# Case studies of ecological restoration law

# 8 Legal considerations in operationalizing eco-restoration in the European Union

## A Sisyphean task or unlocking existing potential?

*Hendrik Schoukens*

### 8.1 Introduction

Against the backdrop of the current extinction crisis – dubbed the 'sixth mass extinction event' by an increasing body of scientists[1] – the concept of ecological restoration (hereafter 'eco-restoration') has emerged as a novel conservation narrative.[2] This new, more proactive 'conservation' narrative[3] saw its first explicit acknowledgement at policy level with the adoption of the 2010 Aichi Targets[4] within the legal framework of the 1992 Convention on Biological Diversity.[5] In this context, the ambitious goal of restoring at least 15 percent of the degraded ecosystems by 2020 hasbeen put forward, which not only constitutes a target and objective but also serves as a tool to achieve the overall objective to halt biodiversity loss.[6] In 2011, the European Union (EU) followed suit, resulting in eco-restoration being explicitly recognized as a policy target within the EU Biodiversity Strategy to 2020.[7]

This chapter explores in depth the legal difficulties and opportunities in translating the eco-restoration agenda into workable governance regimes. As one of the most important jurisdictions in the world for environmental law practice and reform, the EU provides valuable insights into this subject matter. The chapter considers the myriad legal issues to be addressed when implementing the shift towards eco-restoration in the EU.[8] Most importantly, the question arises whether EU environmental law is currently adequate for achieving the applicable restoration pledges. The Mid-Term Review of the EU Biodiversity Strategy has revealed that the degradation of European ecosystems has not been stopped, while the

1 Barnosky et al (2011).
2 Aronson and Alexander (2013).
3 For a more critical appraisal, see Katz (1992).
4 Aichi Target 15, CBD (2010) COP 10 Decision X/2, Strategic Plan for Biodiversity 2011–2020.
5 Convention on Biological Diversity 31 ILM 818 (1992).
6 Jørgensen (2013).
7 European Commission (2011a).
8 On the role of legal instruments in fuelling restoration actions: Aronson et al (2011).

EU's restoration targets are still far from being met.[9] On a more fundamental level, several authors have underlined that many EU environmental directives, at least when effectively applied on the ground, do not always reflect a clear-cut restoration rationale.[10] At the same time, the genuine lack of concerted action on the side of the European Commission (EC), for instance the lack of comprehensive guidance and priority frameworks, is being singled out by some environmental nongovernmental organizations (NGOs) as one of the main reasons why the EU's targets are not being achieved.[11]

In light of this arguably unsatisfactory state of affairs, this chapter examines the interference between the operational building blocks of eco-restoration strategies – including instrumental concepts such as environmental baselines, restoration targets, legal restoration duties and prioritization schemes – and EU environmental law. This discussion is particularly relevant in an age when cooperation between lawyers and restoration ecologists is being singled out by authors, such as Telesetsky and others, as necessary for the emergence of effective eco-restoration law.[12] In section 8.2 the broad legal and policy context surrounding the concept of eco-restoration within the EU is outlined. Section 8.3 provides in depth-analysis of the major operational governance elements that have to be put in place to ensure a more strategic approach to eco-restoration. The concluding section offers brief remarks on possible future directions to bolster eco-restoration governance in the EU.

## 8.2 Ambivalent policy and legal context: eco-restoration as an explicit EU policy target

### *8.2.1 The wider legal and policy context*

Under the terms of the Treaty on the Functioning of the European Union (TFEU) the EU environment policy is a shared competence, which entails that both the EU and its Member States have jurisdiction to legislate on the matter.[13] However, pursuant to EU law, Member States lose their competence to enact rules whenever the EU within its mandate legislates on a certain topic. Concerning eco-restoration, it is a shared competence, leaving considerable leeway for the national authorities in the domains that have not been exhaustively tackled at the EU level. Only when the purported actions intrude on a so-called exclusive competence under EU law, which might be relevant for measures on conservation of marine biological resources in the context of the Common Fisheries Policy (CFP), might Member States not be entitled to unilaterally adopt further measures regarding eco-restoration. Furthermore, Member States should also refrain from

9 European Commission (2015).
10 See e.g. Cliquet et al (2009).
11 Langhout (2014).
12 Telesetsky, Cliquet and Akhtar-Khavari (2017), 287.
13 Treaty on the Functioning of the European Union, art. 4.

defeating the objective of the existing EU rules on nature protection. While no explicit reference is made to the term 'ecological restoration' in the TFEU, it is clear that eco-restoration is at least implicitly covered by the over-arching principles mentioned in the environmental provisions of the Treaty.[14] To give but one indication, robust restoration and recovery policies are crucial to avoid further biodiversity losses in the EU and therefore are certainly encompassed under the umbrella of the TFEU's prevention and precaution principles.

An implicit reference to the concept of eco-restoration can even be found in the Treaty on European Union. Article 3(3) explicitly recognizes that the internal market, which still constitutes the bedrock of the European legal system,

> shall work for the sustainable development of Europe based on balanced economic growth and price stability, a highly competitive social market economy, aiming at full employment and social progress, and a high level of protection and improvement of the quality of the environment [...].[15]

Of course, the precise content of the term 'sustainable development' will be determinative for the exact ambition level to be adhered to within the EU's environmental policy.[16] Nonetheless, the explicit references to the principle of 'a high level of protection and improvement of the quality of the environment' indirectly underpin the EU's eco-restoration policy.

### *8.2.2 The applicable EU restoration targets*

The more concrete targets regarding eco-restoration are in the EU's comprehensive Biodiversity Strategy to 2020.[17] Its overarching objective is to halt and reverse biodiversity loss within and outside the EU. While this general objective already leaves ample room for eco-restoration actions, the EC opted to issue a more specific target regarding eco-restoration; Target 2 of the EU Biodiversity Strategy stipulates that '[b]y 2020, ecosystems and their services are maintained and enhanced by establishing green infrastructure and restoring at least 15% of degraded ecosystems',[18] thereby incorporating the target set at international level.

A multitude of more concrete actions are associated with the EU's biodiversity targets. Target 2 (in action 6a) expects Member States, with the assistance of the EC, to develop a strategic framework to set priorities for eco-restoration at subnational, national and EU levels by 2014. This prioritization framework has to define the scale of the restoration target and the criteria on which prioritization should be based. Likewise, the EC pledged to developing a Green Infrastructure

14 Ibid, art. 191(1).
15 Ibid, art. 3(3).
16 See on the concept of sustainable development within the EU: Douma (1998). See more generally on the precise meaning to be given to the concept of sustainable development: Redclifft (2006); Owens (2003).
17 European Commission (2011b).
18 European Commission (2011a).

Strategy by 2012 to promote the deployment of Green Infrastructure in the EU in urban and rural areas (action 6b). Finally, the EC vowed to carry out further work in order to propose by 2015 an initiative to ensure no net loss of ecosystems and their services (e.g. through compensation or offsetting schemes) (action 7b). Of course, the EU restoration pledges cannot be treated separately from the legally binding conservation and protection duties within the specific context of the Nature Directives and, more generally speaking, the Water Framework Directive and the Marine Strategy Framework Directive,[19] which are treated in more detail below.

### *8.2.3 Further implementation challenges*

The policy pledges presented above, and the catalogue of concrete follow-up action, might come across as impressive at first glance. To date, however, the EC has failed to produce a comprehensive prioritization template for the Member States to establish national and/or regional eco-restoration strategies. Thus several practical questions relating to the 15 percent target remain unanswered, such as its territorial scope and the environmental baseline from which to work. Admittedly, it would be incorrect to accuse the EC of total inaction; it has tried to propose substantive guidance. However, it has only partly succeeded in doing so. In a relatively obscure 2014 *Note to the Nature Directors*,[20] the EC underscored the relevance of the conclusions of a study entitled 'Priorities for Restoration of Ecosystems and their Services in the EU', which it had commissioned to help Member States with the development of prioritization frameworks for the restoration of their ecosystems.[21] Opting for a rather pragmatic approach to eco-restoration, it is assumed here that for any ecosystem type, a variety of ecological conditions can be described along a continuum from poor to excellent. A so-called 'four level model for ecosystem restoration' is derived from these principles, which divides the continuum of ecosystem conditions from poor (e.g. urban areas) to excellent (e.g. wilderness areas). In the Commission's pragmatic view, any significant improvement that moves an area to a better condition should be regarded as a contribution to the EU's 15 percent restoration target.

It is interesting to note that the EC, with the issuance of the 2014 Note, also tried to provide comprehensive answers to some of the persisting uncertainties surrounding the exact scope of the 15 percent restoration target. The EC assumes that the entire EU territory is included in the scope of the quantified restoration target. Moreover, the 2014 Note suggests that the 15 percent target applies to both the marine and the terrestrial area. It remains debatable whether it makes sense to include rather far-reaching substantive guidelines on how to apply the

19 Directive 2008/56/EC of the European Parliament and of the Council of 17 June 2008 establishing a framework of community action in the field of marine environmental policy [2008] OJ L 164/136.

20 European Commission (2014).

21 Lammerant et al (2013).

eco-restoration target on the ground in a document without binding legal status. Evidently, this entails that Member States cannot be directly sued before the EU courts for flouting the 15 percent target on their territories. Furthermore, there are also major substantive problems with the restoration template presented. For instance, the EU's approach negates the fact that environmental degradation has two components: the extent of the area that has become degraded or restored; and the magnitude of the degradation or its counterpart improvement at any location.[22] Some scholars argue that the focus should be on EU protected habitats that currently have an unfavourable conservation status.[23] Either way, the lack of consensus among academic commentators aptly underlines the unstable substantive and legal foundations of the EU's current approach towards eco-restoration.

## 8.3 Implementing eco-restoration within EU environmental law: key considerations

The limited guidance available at the EU level, and the lack of an overarching binding instrument regarding eco-restoration, partly explain why, as of 2018, only two Member States have come forward with an explicit national restoration strategy.[24] The question as to whether existing EU environmental legislation already contains the necessary mandates and instruments to prompt a shift towards eco-restoration therefore turns out to be all the more relevant.

### *8.3.1 Defining eco-restoration in EU environmental law: squaring the circle?*

Surprisingly, the EU Biodiversity Strategy lacks a specific definition of eco-restoration. In its 2014 Note, the EC confined itself to underlining that restoration is to be regarded primarily as a process and not necessarily as a final destination, which reflects a rather pragmatic but not necessarily unworkable approach to the difficulties of defining eco-restoration and its purpose. Yet the Commission's working paper accompanying the EU Biodiversity Strategy defined eco-restoration as follows:

> The restoration of ecosystems and their services is understood as actively assisting the recovery of an ecosystem that has been degraded, damaged, or destroyed, although natural regeneration may suffice in cases of low degradation. The objective should be the return of an ecosystem to its original community structure, natural complement of species, and natural functions to ensure the continued provision of services in the long term, although in cases of extreme degradation, the focus on specific services may be justified.[25]

22 Kotiaho and Moilanen (2015).
23 Egoh et al (2014).
24 Cortina-Segarra et al (2016).
25 European Commission (2011b).

This definition closely resembles the definition of eco-restoration set forward by the Society for Ecological Restoration.[26] Both definitions highlight the intentional character of restoration actions, which are thereby distinguished from the process of natural ecological succession. Still, the EC's definition contains some weaknesses. Amongst others, it appears to imply that the recovery of an ecosystem can be exclusively human-induced, while in reality eco-restoration actions can merely facilitate the recovery process.[27] Second, and perhaps more fundamentally, it ignores that in many instances complete restoration will not be technically feasible.[28]

Presently, several EU environmental directives include implicit definitions of the notion of 'significant damage', yet only one defines eco-restoration. The Environmental Damage Directive[29], dating from 2004, includes explicit definitions and standards to be observed when implementing restoration measures in the specific context of incidents and accidents giving rise to ecological damage. The Court of Justice of the European Union (CJEU), when asked to interpret the concept of environmental damage in the context of water bodies, has moreover reaffirmed that the latter covers damage that has a significant adverse effect on the ecological, chemical or quantitative status or ecological potential of the water in question, irrespective of the fact that it is covered by an explicit authorization.[30] Interestingly, the Environmental Damage Directive also contains a definition of 'recovery', in Article 2(15), which includes natural recovery, and implies the return of damaged natural resources and/or impaired ecosystem services to baseline condition. This definition thus seems to encompass both active and passive restoration. When defining the notion of remediation measures, the Environmental Damage Directive refers to 'any action, or combination of actions, including mitigating or interim measures to restore, rehabilitate, or replace damaged natural resources and/or impaired services, or to provide an equivalent alternative to those resources or services (...)'. In addition, Annex II implicitly seems to further underline the distinction between restoration and rehabilitation measures by distinguishing between 'primary', 'complementary' and 'compensatory remediation'. Primary remediation aims to restore the damaged natural resources and/or services to, or towards, their original pre-damage baseline condition. If the damaged natural resources and/or services do not return to baseline condition, 'complementary remediation' will be undertaken. Its purpose is to provide a similar level of natural resources and/or services as would have been provided if the damaged site had been returned to its baseline condition. Even when the usage of 'baseline' within the Environmental Damage Directive merely refers to the pre-existing ecological situation prior to the damage, it is clear that the concept of 'complementary

26 SER (2004).
27 Kotiaho et al (2016a).
28 Ibid.
29 Directive 2004/35/CE of the European Parliament and of the Council of 21 April 2004 on environmental liability with regard to the prevention and remedying of Environmental Liability [2004] OJ L 143/56.
30 Case C-529/15, *Folk* [2017], para. 33–34.

measures' relates more to 'rehabilitation' and 'replacement' than eco-restoration measures associated with landscape or ecosystem scales. Yet, if one were to consider the adoption of a new comprehensive legislative instrument on eco-restoration in the EU, it would be desirable not to start from scratch but to take into account existing useful elements in the Environmental Damage Directive.

### *8.3.2 Setting up benchmarks: how to avoid degrading environmental baselines?*

It is not hard to grasp that setting objectively determined, historical baseline conditions remains an extremely difficult goal for eco-restoration, especially in the European context. In the Americas and Australia, the arrival of Europeans gave rise to significant changes in the environment, making it a more expedient reference point to which to direct restoration efforts.[31] Such pre-degradation reference is evidently lacking in Europe, which has seen more gradual land-use changes from Roman times. Whereas the EU Biodiversity Strategy in itself lacks a well-defined baseline scenario against which progress is to be measured, it is generally accepted that the progress of the restoration actions should be measured against the 2010 EU Biodiversity baseline report. Also, efforts have been made to map the pressure on ecosystems and to assess the current condition of ecosystems, for instance in the context of the European Environment Agency's Mapping and Assessment of Ecosystems and their Services initiative. In this regard, I want to highlight three major elements which are crucial when considering the establishment of environmental baselines.

First, I contend that it will be crucial to maintain a categorical distinction between two distinct approaches to the concept of environmental baseline. On the one hand, there is the use of the term 'baseline' as a 'natural state' or 'reference point', which is of major importance to assess the difference between the current, often degraded condition of an ecosystem and the pre-degradation condition (and thus to define the concept of 'degradation').[32] In this context, the notion of a baseline serves as an ecological yardstick against which the competent authorities are to specify their more concrete restoration objectives, which do not necessarily correspond to a pre-degradation state. On the other hand, one might refer to the concept of 'baseline' as a benchmark against which to measure progress toward the 15 percent restoration target. Only in respect of the latter context does the choice of 2010 as a temporal benchmark appear justifiable. This date unequivocally represents the policy shift to eco-restoration at international and EU level. However, this should not lead us to negate the fact that some of the existing restoration duties included in the EU environmental directives already applied prior to that date (see *infra*). Also, past damage, sometimes referred to as 'extinction debt', is to be taken into account when enforcing compliance with existing conservation duties under EU law.[33]

31 Allison (2004).
32 Kotiaho et al (2016b).
33 Cliquet and Decleer (2017).

Second, I recommend a more fundamental reflection on the usage of a pre-human reference state as a baseline scenario, as it might in some instances stand at odds with the content of some applicable environmental rules. Additional caution is thus warranted. The latter is aptly illustrated by the content of the 1992 Habitats Directive,[34] which constitutes the cornerstone of EU nature conservation law. Its definition of 'natural habitat' includes both 'entirely natural' and 'semi-natural' habitats, which entails that the conservation and even restoration of secondary, anthropogenic habitats in extensive cultural landscapes needs, at least in some cases, also to be treated as a so-called baseline scenario. This was aptly demonstrated by the outcome of the notable French hamster case, in which France was ultimately condemned by the CJEU for not having put in place sufficient anthropogenic habitat – croplands, amongst others – to guarantee the survival of this endangered rodent species.[35] While in many instances the implementation of the Habitats Directive can be aligned with the restoration of a pre-human reference state, it is clear that establishing a clear-cut baselines can in some instances lead to a difficult balancing exercise between conflicting biodiversity goals, especially when new concepts and narratives, such as rewilding, come into the mix.[36]

In sharp contrast to the Habitats Directive, the 2000 Water Framework Directive[37] seems to opt for a more stringent understanding of the concept of environmental baseline. Although no explicit temporal and/or spatial benchmark is put forward by the Water Framework Directive to determine the reference state to be used when Member States are to establish their national environmental quality standards, it still appears to lay down reference conditions prior to the intensification of agriculture 100–150 years ago, which was appraised by several authors as revolutionary and ground-breaking in itself.[38] However, most remarkably, the rationale of the EU Water Framework Directive is dismissed by some as a so-called 'paradise-lost' approach, which fails to take into account more realistically attainable human benefits.[39] In my view this criticism lacks persuasiveness in view of the often lenient and unambitious national implementation policies and the lack of strict enforcement in the field, which evidently renders the use of such baselines very challenging. Moreover, concepts like 'heavily modified water body' were explicitly introduced into the Water Framework Directive in recognition of the very fact that water bodies in Europe have been subject to major physical alterations (see also *infra*).

34 Council Directive 92/43/EEC on the conservation of natural habitats and of wild fauna and flora [1992] OJ L 206/7.
35 Case C-383/09 *Commission v France* [2011]. Schoukens (2014).
36 See more extensively on rewilding: Navarro and Pereira (2012); Nogués-Bravo et al (2016).
37 Directive 2000/60/EC of the European Parliament and of the Council of 23 October 2000 establishing a framework for Community action in the field of water policy [2000] OJ L 371/1.
38 Moss (2008).
39 Dufour and Piegay (2009).

Lastly, it is very important to have in place effective legal instruments to address unsustainable 'lock-in' and 'death by a thousand cuts' scenarios, especially in the context of eco-restoration. As such, enforceable standstill obligations are paramount to avoid degrading baselines. In this respect, a more detailed analysis of the most prominent EU environmental protection instruments reveals a mixed picture. For a considerable time, at least a relevant part of the national permitting agencies in many countries assumed the non-deterioration obligation present in Article 4(1) of the Water Framework Directive was not to be regarded as an obstacle for the authorization of projects prone to impact on the quality status of the water body. Yet in a landmark ruling in 2015, the CJEU overturned this approach.[40] A similar rigidity prevailed in respect of the non-regression provisions present in the Nature Directives, such as Article 6(2) of the Habitats Directive, which allows Member States to pursue more ambitious recovery policies in the context of EU protected (Natura 2000) sites.[41] The EU judges have underlined that even lawful ongoing activities, such as unsustainable logging or agricultural practices, that were initiated before the designation of an EU protected site, need to be reassessed.[42] In cases of persisting non-compliance, the application of the non-regression clauses can even lead to restoration duties, entailing that Member States need to rectify the loss incurred by biodiversity since the entry into force of the relevant EU environmental protection duties.[43]

### *8.3.3 Explicit restoration targets: what ambition level should we aim for in the EU?*

While defining an ecological reference point that differs only slightly from a pristine condition might be sensible for assessing the scope of the degradation to be healed, it does not necessarily mean that the restoration target should reflect that utopian historical condition in its entirety. Rather, restoration targets are the result of a political decision-making process that also takes into account social and economic interests.[44] The concept of 'favourable conservation status' (FCS), which is central to the Habitats Directive, will likely be at the forefront of many national restoration strategies in the EU over the coming decades.[45] It is thus worthwhile to look in more depth at what precisely this concept, which is defined in Article 1 € of the Habitats Directive, entails.

Not unsurprisingly, a plethora of additional interpretation questions arise when applying the notion of FCS in the field. For instance: at what geographical level does the FCS need to be achieved? What is to be regarded as 'a population maintaining on a long-term basis as a viable component of its natural habit'?[46]

40 Case C-461/13, *Bund für Umwelt und Naturschutz Deutschland eV* [2015], para. 43.
41 For an overview, see Schoukens (2017a).
42 See amongst others: Case C-226/08 *Stadt Papenburg* [2010] para. 49; Case C-399/14 *Grüne Liga Sachsen* [2016], paras 44–46.
43 See more elaborately, Schoukens (2017c).
44 Kotiaho et al (2016b).
45 See more extensively, Epstein et al (2015); Epstein (2016).
46 Article 1(e) Habitats Directive.

How many individuals of an endangered species are needed for it to be considered as 'recovered'. As is obvious from the text of the Habitats Directive itself, the specific ambition level to be adhered to when setting up national, regional and site-specific conservation targets will also hinge upon the environmental science available as to the existing population levels, historical data and future developments.[47] One of the major legal issues to address is the geographical benchmark to be taken into account. For instance, there still remains considerable uncertainty as to whether the FCS is to be reached principally at the European and/or biogeographical level or, alternatively, at the national and/or local (site) level. In its case law on assessing compliance with the protection mandates regarding strictly protected species, such as wild hamsters and gray wolves, the CJEU suggests that the FCS exists exclusively at the national level, which indirectly precludes Member States from taking advantage of conservation efforts in neighbouring states or regions as an excuse for lack of action in their own territories.[48] However, in none of these cases was the issue explicitly raised as a prime legal argument. In a recent decision regarding the reduction of the size of a Dutch Natura 2000 site, though, the CJEU opted for a middle ground by holding that the FCS of a natural habitat and/or species is to be assessed in relation to the entire European territory of the Member States to which the treaty applies, thereby granting Member States some more leeway.[49] This entails that, when setting up national and site-specific restoration targets, Member States are required to ensure that their restoration and conservation actions contribute considerably to the achievement of the FCS at the EU level.

The definition of FCS seems to indicate that the targets do not necessarily need to be set at a population that reflects historical levels.[50] Member States are thus not required to revive extirpated species that disappeared several centuries ago. This is reasonable since some parts of the territory of Member States might have been irreversibly damaged – for instance due to urbanization or industrialization – and are therefore no longer suitable as potential habitats for strictly protected species, such as European otters or wild hamsters. Even so, recent literature has aptly underscored that further improvement of the conservation status of many European habitats and species currently in an unfavourable conservation status is necessary.[51] This stance finds support in the case law of the CJEU, which indirectly seems to indicate that a recovery-based approach is imperative for threatened species.[52] Some authors claim that Member States, when setting up recovery strategies, need to go beyond the so-called 'extinction baseline'. The latter interpretation is endorsed by

47 Schoukens (2017b).
48 See Case C-383/09 *Commission v France* [2011] (Wild hamsters); Case C-342/05 *Commission v Finland* [2007], para. 37–28 (Gray Wolves).
49 Case C-281/16, *Vereniging Hoekschewaards Landschap* [2017], para. 32. This stance was already put forward by the CJEU in 2000: Case C-371/98, *First Corporate Shipping* [2000], para. 23.
50 Epstein (2016), 238–239.
51 Schoukens (2017b).
52 Case C-383/09 *Commission v France* [2011].

reference to the so-called 'carrying-capacity' approach, which puts more emphasis on the ecological role to be played in their ecosystems by, for instance top predators, grazers and large rodents.[53] In turn, this might lead to more synergies between the traditional conservation movement and the recently emerged rewilding narrative (as discussed in the conclusion of this article).

Concerning 'good ecological status' (GES), which was principally to be achieved by the Member States for their waters, lakes and estuaries by 2015, a more holistic, ecosystem-based approach seems to prevail.[54] The concept of GES, laid down by the Water Framework Directive, clearly serves as a counterpart to the notion of FCS for EU protected biodiversity. However, it puts forward an ecosystem-based approach, moving away from protecting specific species and habitats. And while, as mentioned above, the GES is to be equated to a condition that shows a low level of distortion resulting from human activity, it deviates only slightly from the values associated with the water body under so-called undisturbed conditions.[55] Even so, the EC's own guidance documents seem more preoccupied with negative protection duties than with enforcing the restoration rationale underpinning Article 4(1)(ii) of the EU Water Framework Directive.[56] The progressive timeframe that is included in the Water Framework Directive seems largely utopic in view of the degraded state of many water bodies due to the cumulative impacts of industrialization and intensive agriculture over the past centuries, and the extensive application that is made of the possibility to designate water bodies as heavily modified water bodies, which allows for less stringency.[57] The potential success in terms of restoration of the Water Framework Directive thus hinges upon a fundamental rethink of some intensive agricultural practices in several Member States, which might allow the EU water bodies to recover swiftly, whereas more robust restoration measures, which might include the removal of dams and other constructions, are necessary to restore severely degraded streams and rivers.

### *8.3.4 Restoration mandates in the EU: mandatory or voluntary?*

In recent literature, it has been noted that legal mandates for eco-restoration exist in different contexts.[58] Evidently, having put in place clear-cut legal restoration duties is one technique to bolster the implementation of restoration schemes on the ground. Within the context of EU environmental law, three major restoration duties exist: autonomous restoration duties aimed at the recovery of degraded habitats and species; specific remediation duties in the context of environmental

53 Epstein et al (2015).
54 For the exact definition of this concept, Directive 2000/60/EC of the European Parliament and of the Council of 23 October 2000 establishing a framework for Community action in the field of water policy [2000] OJ L 371/1, art. 2(22).
55 European Commission (2003).
56 Moss (2008).
57 Hering et al (2010).
58 Richardson (2016).

accidents; and more generally applicable offsetting schemes, which might include restoration actions.

On generic positive restoration duties, Article 6(1) of the Habitats Directive, even while being defined in a conspicuously vague manner,[59] should be regarded as one of the most powerful restoration mandates within EU environmental law. Admittedly, this article does not explicitly refer to 'restoration' or 'recovery actions'. However, since the Habitats Directive explicitly includes 'restoration' within its relatively broad definition of 'conservation', laid out by Article 1(a), this omission does not limit its potential to encompass restoration. Moreover, the CJEU has indirectly underscored the restoration rationale when interpreting the designation duties imposed on Member States in the context of degraded sites that still feature important restoration possibilities.[60] The Court reasoned that Member States also must restore degraded sites with important restoration potentialities. The restoration imperative present in this provision is also reasserted by the guidance documents of the EC.[61] In order to prompt the Member States to more explicit restoration schemes, Article 6(1) of the Habitats Directive should become the object of strategic litigation, at both EU and national levels, which might further underline its important restoration potential, for instance when assessed in light of the applicable site-specific conservation objectives for an EU protected Natura 2000 site.

Article 12(1) of the Habitats Directive, which sets protection duties for strictly endangered species, appears to leave less room for a restoration-friendly approach. And while the CJEU has indirectly revealed the surprising restoration potential of the latter provision, especially in a case of persisting non-application of the protection rules regarding wild hamsters,[62] it probably cannot be used to compel Member States to consider active restoration measures in areas where the protected species are no longer present. This view was also reinforced by the EC in its 2007 guidance document on strict species protection.[63] However, a recent ruling of a French administrative court, in which France was condemned for not having put forward sufficient recovery policies for its dwindling population of brown bears in the Pyrenees, reveals that when faced with decades of declining population numbers and continuous political unwillingness to initiate robust recovery policies for species threatened with immediate extinction, national courts might still be willing to interpret a clear-cut reintroduction duty in the latter provision.[64] In 2011, a Dutch court even compelled the Dutch government to come forward with additional connection corridors for threatened wild hamsters.[65] Yet the same court subsequently declined to order the Dutch government to actively consider

59 Garcia-Ureta and Lazkano (2015).
60 Case C-281/16, *Vereniging Hoekschewaards Landschap* [2017], para. 37.
61 European Commission (2000).
62 Schoukens (2017a).
63 European Commission (2007).
64 Administrative court of Toulouse, case Nos 150.1887 & 1502320, Association Pays de l'Ours – ADET et al (2018).
65 Court of First Instance of The Hague (2011).

active habitat restoration measures in the context of the ongoing otter reintroduction program.[66]

Both the Water Framework Directive's and the Marine Strategy Framework Directive's provisions on management and recovery are arguably more detailed compared to the Nature Directives' provision. Even so, there exists a remarkable lack of clear-cut restoration mandates. For example, explicit restoration measures are not at the forefront of the catalogue with mandatory measures included in the Water Framework Directive, which might again render them less legally enforceable. The Water Framework Directive seems to treat 'habitat restoration measures' as a mere 'fall-back' option in this context.[67] A similar obfuscation is also noticeable within the Marine Strategy Framework Directive.[68] Here, the available science will thus determine to what extent Member States are actively required to ponder robust restoration strategies regarding the water-related ecosystems. For now, no national case law exists in which agencies were forced by court action to consider more robust river or marine restoration programs.

Second, as far as incident-linked restoration duties are concerned, again reference is to be made to the Environmental Damage Directive, which ostensibly has a clear duty to remedy ecological damage, a duty on the operators of certain types of occupational activities, such as large-scale industrial activities, as listed in Annex III. However, the remediation scheme is more complex than it appears at first sight, with several exemptions and applicable significant thresholds. Admittedly, the scope of the provisions on damage to EU protected nature is relatively wide, also covering operational activities which are not listed in Annex III, such as ordinary construction works, but are still carried out in a negligent or faulty manner.[69] By contrast, the very fact that only 'significant damage' to natural habitats and protected species triggers the administrative remediation scheme serves as a major obstacle for its more widespread application.[70] Nonetheless, the national authority overseeing the application of this remediation scheme has the obligation to require that responsible operators take preventive or remedial action.

Third, we must consider the mitigation and offset duties that exist in EU environmental law, which might also, when strictly enforced and applied with the necessary caution, stimulate the implementation of more ambitious restoration schemes.[71] In this respect, it is to be reiterated that the EU Biodiversity Strategy foresees that the Commission proposes 'an initiative to ensure there is no net loss

66 Court of First Instance of The Hague (2013).

67 Directive 2000/60/EC of the European Parliament and of the Council of 23 October 2000 establishing a framework for Community action in the field of water policy [2000] OJ L 371/1, art. 11(3) and Part B of Annex VI.

68 Directive 2008/56/EC of the European Parliament and of the Council of 17 June 2008 establishing a framework of community action in the field of marine environmental policy [2008] OJ L 164/136, art. 1(2)(a).

69 Directive 2004/35/CE of the European Parliament and of the Council of 21 April 2004 on environmental liability with regard to the prevention and remedying of environmental liability [2004] OJ L 143/56, art. 3(1)(b).

70 Fogleman (2015); European Commission (2016).

71 Lapeyre et al (2015).

of ecosystems and their services (e.g. through compensation or offsetting schemes)'. In spite of the many studies commissioned by the EC, investigating, among other things, the potential for more economic approaches to biodiversity offsetting such as habitat banking, at present no legal instrument has been adopted which is apt to address generic biodiversity loss.

However, some existing mitigation schemes might be used to spur restoration actions in the context of development projects. At a principal level, one should evidently denounce an approach under which restoration is almost exclusively based upon development-led restoration. These reservations seem to be shared by EU judges, at least when it concerns offset measures in the context of EU protected sites. In a 2016 landmark ruling on a proactive restoration program tied to future harbor expansion in the Port of Antwerp, the EU judges held that future recovery actions cannot be taken into account as general mitigation measures in the context of an appropriate assessment for a port expansion plan encroaching upon EU protected sites.[72] This restrictive jurisprudential approach to no-net-loss clauses regarding EU protected nature, while remarkably rigid, ensures that compensation measures are effectively aimed at creating additional ecological benefits, which go beyond the existing commitments. It also avoids restoration actions being abused as a license to hinder unsustainable project developments. The impact of the latter case law developments cannot be underestimated in the specific domain of biodiversity offsetting and EU protected nature, especially since it guarantees that restoration actions are not exclusively development-led.[73] Scientific research has, moreover, sufficiently illustrated that, due to the major time lags and uncertainties to be taken into account when recreating nature, offset measures only rarely avoid net losses.[74] As such, the derogation scheme included in Article 6 (4) of the Habitats Directive grants national permitting authorities the prerogative to take into account restoration actions, whenever they go beyond the existing conservation duties,[75] yet only when it has been established that the project development relates to an 'imperative reason of overriding public interest' and no other, more environmental friendly alternatives are available.

Recently, restoration actions have become increasingly more prevalent when carrying out an Environmental Impact Assessment (EIA) for potentially harmful projects and activities according to EU environmental law.[76] Even assuming that EIA serves primarily as a mere procedural instrument for information gathering and due diligence, the 2014 amended EIA Directive more directly appears to incentivize offsetting actions in an EIA context. The simple fact that the amended directive now explicitly requires the permitting authority to consider measures to 'avoid, prevent, reduce, and, if possible, offset significant effects on the

72 Case C-387/15 *Orleans* [2016], para. 36–38.
73 Schoukens (2017b).
74 Curran et al (2014); Moreno-Mateos (2015).
75 European Commission (2007).
76 Directive 2011/92/EU of the European Parliament and of the Council of 13 December 2011 on the assessment of the effects of certain public and private projects on the environment [2011] OJ L 26/1.

environment' can be heralded as a step in a good direction for landscape-wide restoration linked to harmful project developments.[77] Seeing that in many Member States 'ordinary' nature is left unprotected, the inclusion of more explicit hints to offsetting in the context of environmental evaluation clauses is to be regarded as a substantial improvement, even when the provision does not explicitly compel planning authorities to require offsets in any single instance of 'ordinary' biodiversity loss.

### *8.3.5 Territorial and material issues: how to move beyond paper parks to ecosystem and landscape-based restoration?*

The foregoing discussion established that several distinct restoration mandates exist within EU environmental law. Still, the question as to the precise material and territorial scope of these duties merits closer examination. Given the wide focus of the EU policy targets on eco-restoration – principally targeting landscape-wide and ecosystem-based restoration efforts – it needs to be analysed whether their scope matches with the ambitious rationale of the EU Biodiversity Strategy. I believe there exists an important mismatch in this respect.

A first remark concerns the allegedly limited material scope of the Nature Directives, especially when viewed alongside the EU Water Framework Directive and the EU Marine Strategy Directive, which both put forward ecosystem-based conservation. Ostensibly, EU nature conservation law is not preoccupied with conserving entire 'ecosystems'; rather, it has the more specific aim of ensuring the conservation of a wide range of rare, threatened or endemic animal and plant species. The concept of 'ecosystem' is not even explicitly mentioned in the Habitats Directive. While the Birds Directive grants protection to all native bird species, the Habitats Directive covers a relatively low number of threatened invertebrates.[78] As such, the species-based approach might block a more ambitious approach to eco-restoration. Even so, the relatively large number of species mentioned in the Annexes to the Nature Directives underlines their great potential in terms of ecological coverage. Most importantly, the objective of creating an ecologically coherent network of Natura 2000 sites, as explicitly put forward by Article 3(1) of the Habitats Directive, serves as an important indication that the Nature Directives go beyond traditional species protection. By 2017, a total of 27,522 Natura 2000 sites were designated, covering 1,184,609 $km^2$, making it the most extensive ecological network on the planet.[79] It should also be recalled that the Natura 2000 sites are to be selected according to biogeographical regions, each with their own ecological coherence.

77 Directive 2011/92/EU of the European Parliament and of the Council of 13 December 2011 on the assessment of the effects of certain public and private projects on the environment [2011] OJ L 26/1, art. 8a 1(b).

78 Cardoso (2012).

79 European Commission, Natura 2000 barometer http://ec.europa.eu/environment/nature/natura2000/barometer/index_en.htm.

Moreover, in line with the ecological selection and identification criteria in Annex III of the Habitats Directive, Natura 2000 sites are to be selected on the basis of the conditions in each biogeographical region, which should guarantee that the selected sites represent species and habitat types under similar natural conditions across a suite of countries. Ideally, the Natura 2000 sites can serve as the backbone of progressive national networks of protected sites and corridors. In addition, economic considerations cannot play a prominent role during this selection process.[80] In recent case law, which entails that Member States must re-evaluate their designation policies in view of new monitoring results and emerging scientific studies, the CJEU has underscored that this designation obligation is dynamic by nature.[81] Furthermore, several natural habitats, listed on Annex I, could also be explicitly regarded as ecosystems themselves. This is especially the case for marine habitats, such as lagoons, coastal areas and estuaries.

It is one thing to put forward a relatively straightforward restoration imperative regarding protected sites – such as the Natura 2000 Network – and advocate for a more liberal interpretation thereof; it is quite another to contend that 'ordinary' nature will be saved through area protection rules alone. Even in the unlikely event that all Member States effectively adhere to a more progressive and broad interpretation of restoration commitments in this regard, and also include buffer zones in Natura 2000 sites, such understanding still leaves the majority of the EU territory unaffected by restoration pledges. In fact, the Natura 2000 Network – while ambitious in itself – leaves more than 80 percent of the EU territory 'unprotected'. This underlines the importance of having sufficiently broadly formulated restoration duties, which go beyond the boundaries of designated protected sites. This constitutes one of the major flaws of EU environmental law in view of the applicable restoration objectives.

The most relevant provision in terms of landscape-wide restoration actions is Article 10 of the Habitats Directive, which stipulates that Member States shall endeavour, where they consider it necessary, in their land-use planning and development policies and, in particular, with a view to improving the ecological coherence of Natura 2000 Network, to encourage the management of features of the landscape which are of major importance for wild fauna and flora. This provision could prompt Member States to place additional emphasis on nationally protected sites and ecological corridors between such areas and Natura 2000 sites. However, its flexible wording makes it unlikely to qualify as a legally binding norm.[82] Furthermore, a more progressive, eco-restoration-friendly reading of the latter provision, especially in regions where the degraded state of many Natura 2000 sites is caused by severe fragmentation,[83] is not plausible given the paucity of supportive CJEU case law.[84] Therefore, to achieve a more enduring shift towards

80 Case C-44/95 *Regina* [1996], para. 27.
81 Case C-209/04 *Commission v Austria* [2006], para. 37.
82 Borgström and Kistenkas (2014).
83 Cliquet et al (2009). For a more reserved approach, see Squitani (2012).
84 Case C-418/04 *Commission v Ireland* [2007].

the 15 percent restoration target, more compulsory and broadly formulated provisions regarding restoration of degraded ecosystems in the wider landscape are in order. With the adoption of the Strategy on Green Infrastructure in 2013, the EC highlighted the potential of green infrastructure to contribute to various EU policy objectives, such as increased human health and well-being, and climate change adaptation.[85] Nonetheless, it remains to be seen whether this policy instrument, which has no legally binding effect, will really be able to have a tangible impact on the ground.

When analysing the territorial scope of restoration policies within the context of EU law, it should not be forgotten that ecological considerations are also increasingly integrated in the EU's Common Agricultural Policy (CAP) and in the CFP, two prominent EU policy domains. With regard to the former, reference is to be made to the greening measures that have been further strengthened in the CAP through the most recent 2013 Reform.[86] One of the key measures to enhance biodiversity on arable lands, Ecological Focus Areas (EFA), aptly underscores the weaknesses of the new shift in the EU's agricultural policy towards greening under Pillar 1. However, in light of the EU's restoration ambitions, the EFA as a flagship of the 2013 CAP Reform must still be judged unsatisfactory at many levels.[87] As a mere 5 percent of the arable land needs to be maintained as EFA, additional doubts remain as to its alleged net beneficial impact on biodiversity.[88] Also, the additional funding opportunities for eco-restoration measures, for instance in the context of agri-environment actions, are too loosely formulated to lead to a long-term investment in eco-restoration measures.

A similar, sobering conclusion arises with the intersection between eco-restoration and the CFP. The exclusive nature of the EU's competence in the field of CFP stands in the way of more progressive eco-restoration policies.[89] This has recently been reinforced by a ruling of the CJEU, in which it was clarified that Member States are still precluded from adopting a ban on commercial fishing activities, touching upon the seabed, inside their own Natura 2000 sites whenever they affect the fishing vessels of other Member States.[90] More effective integration of the eco-restoration rationale in the CAP and CFP must be one of the major focal points for future eco-restoration policies at EU level. Achieving ambitious restoration targets at EU level will remain a mirage unless binding restoration commitments can be established in these policy areas.

85 European Commission (2013).
86 For an overview and critical analysis of the 2013 reform, see Doussan and Schoukens (2015).
87 Pe'er et al (2017).
88 Ibid.
89 Schoukens and Dotinga (2015).
90 Case C-683/16 *Deutscher Naturschutzring* [2018], para. 55–57.

### *8.3.6 Prioritization challenges and difficulties: applicable timeframes and exemptions*

A final element to be addressed concerns the subtle balancing act when establishing restoration priorities. In several circumstances, achieving the restoration targets contained in the EU environmental directives will not be realistic in the short term and might exceed the timeframe determined by law. Given the limited financial resources available, the budgetary restraints and the need for societal support, it is necessary to contemplate how to prioritize restoration actions.[91]

As is widely known, the Nature Directives do not put forward specific, legally binding deadlines for the achievement of the FCS. In contrast, both the Water Framework Directive and the Marine Strategy Directive prescribe when Member States must achieve the environmental objectives set by their core provisions. The absence of a time-bound quantified target for the achievement of the FCS seems to leave significant leeway to the Member States. Still, the effect of this lack of an explicit timeframe is mitigated in several ways. Besides the timeframe present in the EU's Biodiversity Strategy, it should be noted that pursuant to Article 4(4) of the Habitats Directive, Member States still have the obligation to designate their pre-selected protected sites as Natura 2000 sites and implement the necessary conservation measures required under Article 6(1) of the Habitats Directive within six years after the sites have been adopted by the EC.[92] Yet the fact that more than half of the EU's designated Natura 2000 sites lack management plans and restoration measures as of 2017[93] suggests that this 'deadline' is not consistently being enforced by authorities.

One of the defining characteristics of the Water Framework Directive, especially when compared to the Nature Directives, is that it explicitly introduces a relatively large number of exemptions and time extensions, which are laid down in Articles 4 (4) to 4(7) and, at least indirectly, pave the way for restoration prioritization frameworks within river management plans. For instance, Article 4(5) grants Member States the leeway to set less stringent environmental objectives for heavily modified bodies of water when their natural condition is such that the achievement of these objectives would be unfeasible and disproportionally expensive. This can therefore be seen as an explicit acknowledgment that full restoration of severely degraded aquatic ecosystems, such as reservoirs, canals or canalized rivers, is not possible without interfering substantially with their economic use. Derogation clauses, such as Article 4(7) of the Water Framework Directive, represent a more classic exemption clause, which allows Member States to derogate from the binding effect of the restoration imperative and the applicable timeframes associated to it for reasons inter alia of overriding 'public interest'. Given the fact that by 2018 the vast majority of EU surface waters were below GES,[94] the presence of such exemption clauses seems to be very reasonable. Even so, expecting that a

91 Kotiaho et al (2016a).
92 Case C-90/10 *Commission v Spain* [2011], para. 64.
93 European Commission (2017).
94 European Environment Agency (2018).

significant portion of the EU's surface water will end up being 'in a state of exception', the EC rightly underscored in its Guidance document that the application of the exemption clauses should be the exception rather than the rule.[95]

The Nature Directives provide relatively little additional guidance regarding eco-restoration prioritization. Although this does not necessarily transform the obligation to reach a FCS into an 'obligation of best efforts', it does somehow limit its legal teeth, especially in combination with the relatively large freedom Member States enjoy when translating the concept into national recovery strategies. With the exception of cases of non-observance of protection schemes after a site has been included in the Natura 2000,[96] I know of no specific restoration-based infringement proceedings launched by the EC against Member States which failed to implement effective restoration actions for degraded protected sites. Precisely this absence of explicit deadlines might help to explain the EC's reluctance in this respect.[97]

However, the Nature Directives should be interpreted as containing an implicit policy prioritization exercise since they focus on the EU's most endangered species and habitats, as listed in their Annexes. By pre-selecting the most endangered species and habitats at EU level, the European legislator has already determined what the priorities should be at the national level. For example, Member States are not obliged to designate all sites where EU protected habitats and species occur as long as they can frame the non-designation of certain of these areas in a sound science-based approach. While Member States primarily need to take into account ecological criteria and restoration possibilities, the EC still allows them to take into consideration the economic costs of the purported restoration actions.[98] They might thus prioritize the designation of the core areas of natural habitats, and leave some relevant yet hard-to-restore habitats outside their designated sites. This is legally acceptable as long as it does not hamper the achievement of the overall FCS. In a 2014 ruling, the CJEU stated that if a designated Natura 2000 site is definitively incapable of contributing to the achievement of the objectives underpinning the Habitats Directive, a Member State is required to propose the declassification of the site.[99] Admittedly, the CJEU has not specified under what precise conditions, for instance with respect to cost-effectiveness, a Member State might still choose to prioritize the designation of a new site instead of opting for the very costly recovery of a severely degraded site. Yet the margin to do so is severely limited, especially when linked to any prior non-observance of species and habitat protection duties.

Lastly, Article 2(3) of the Habitats Directive, which states that measures implemented pursuant to the Habitats Directive are to take into account economic, social and cultural requirements as well as local characteristics, seems to

95 European Commission (2009).
96 Schoukens (2017c).
97 See also Squitani (2012).
98 European Commission (2011).
99 Case C-301/12 *Cascina Tre Pini ss* [2014], paras 31–34.

grant some additional flexibility when formulating eco-restoration strategies. Still, it does not allow for a complete abandonment of recovery actions in the face of degraded environmental quality.[100] Even the application of explicit derogatory provisions, such as Article 6(4), is to remain exceptional and requires, in any event, a balancing between the invoked reasons of overriding interest and the ecological interests linked to the conservation of the said site.[101] Taking account of the strict interpretation of Article 6(4) of the Habitats Directive in recent jurisprudence, it cannot be accepted as a general rule that the economic interests will automatically prevail over the preservation of EU protected sites.[102]

## 8.4 Conclusions and recommendations

This chapter has entertained the proposition that, while the EU is currently lacking an overarching, binding instrument to ensure a proper implementation of its restoration pledges, the existing EU environmental directives already require Member States to develop robust eco-restoration schemes in the context of EU protected nature and, to a large extent, in their water and marine policies. For now, several promising restoration projects are under way at the national level. Yet the analysis has revealed a glaring lack of overarching coordination and binding mandates regarding ordinary biodiversity. Having outlined the main shortcomings in EU law, it might appear tempting to make a case for the adoption of a new EU Directive on Ecosystem Restoration, which might include clear-cut prioritization mandates and prioritization frameworks. Such a new regulatory instrument might also incentivize Member States to designate restoration zones and consider transboundary recovery efforts more intensively. While opting for a new legislative instrument on landscape-wide restoration could very well be presented as a sensible policy option within the EU treaty framework, it appears unlikely that such move would muster the necessary political support any time soon. Moreover, even if such an instrument is adopted one day, it would still render the achievement of the 15 percent target by 2020 unlikely. More fundamentally, one might also contemplate whether additional regulatory instruments, which would further hinge upon quantified targets and would inevitably include generally phrased descriptors and restoration criteria, might not inevitably end up as textbook examples of 'technocratic management', yielding limited results on the ground.

Against this pragmatic backdrop, and in view of the current unfavourable conservation status of many natural habitat and species,[103] I conclude that the more pressing issue is not the adoption of a new legal instrument in the short run, but more focus on conserving the remaining biodiversity and passive recovery. This is in line with the Action Plan for Nature, People and the Economy, adopted by the

100 Opinion Advocate General Kokott in Case C-383/09 *Commission v France* [2014], para. 8.
101 European Commission (2007).
102 Case C-399/14 *Grüne Liga Sachsen* [2016], para. 77.
103 European Environment Agency (2015).

EC in 2017, which urges Member States to step up their implementation of the EU Nature Directives.[104] In view of the available science,[105] active recovery should additionally be considered as an innovative approach where other, more traditional recovery modes have failed. Equally, more should be done about properly integrating effective restoration commitments into the EU's agricultural and fishery policies, which often function more as an obstacle than as a precursor of long-term eco-restoration actions. I also recommend finding an agreement at the EU level on the types of biodiversity worth reviving on the European subcontinent. While one should not abandon the dream of adopting more holistic and effective legislation on eco-restoration, the main recommendation is to focus on better coordinating and earnestly implementing the existing conservation and restoration mandates and, where expedient, ensuring an effective application of the restoration narrative with a strong focus on collaboration and stakeholder involvement. With the rewilding movement gaining momentum, a more fundamental rethink of the EU's biodiversity conservation narrative appears to be in order by 2020. Rewilding, while certainly not uncontested as an ideal,[106] envisions a wilder Europe in which the goal is not further human-managed landscapes so characteristic of the European continent, but rather restoring the self-regulating capacity of ecosystems with a strong emphasis on the presence of large carnivores as agents of top-down control of ecosystems.[107] Thus, when drafting a post-2020 Biodiversity Strategy, the question as to what role humans will play in the restoration of the European landscape will be ever more crucial. This question requires extensive debate with relevant stakeholders and the wider public, taking into account the EU's climate mitigation and adaptation policy. Embedding these diverging approaches in a coherent and participatory policy approach should lead to more instructive guidance on how to advance coherent restoration approaches at the national level. And, who knows, the upcoming discussions might ultimately also persuade the Member States to translate one of the EU's most overarching policy goals into binding legislation. Sooner or later, crossing the Rubicon might be inevitable in order to achieve the necessary paradigm shift for achieving the EU's restoration objectives.

## References

### *Primary sources (law materials)*

Administrative Court of Toulouse, case Nos 150.1887 & 1502320, *Association Pays de l'Ours – ADET et al* (2018).

Case C-44/95*Regina* [1996] ECLI:EU:C:1996:297.

Case C-90/10*Commission v Spain* [2011] ECR I-00134,

Case C-209/04*Commission v Austria* [2006] ECLI:EU:C:2006:195.

104 European Commission (2017).

105 In recent reports, it is highlighted that the benefits linked to active restoration actions, while important for many species, are often difficult to achieve. As a result, the main focus should still be placed on preserving original habitat. See Jones et al (2018).

106 See, among others: Nogués-Bravo et al (2016).

107 Navarro and Pereira (2012).

Case C-226/08*Stadt Papenburg* [2010] ECLI:EU:C:2010:10.
Case C-281/16*Vereniging Hoekschewaards Landschap* [2017] ECLI:EU:C:2017:774
Case C-301/12*Cascina Tre Pini ss* [2014] ECLI:EU:C:2014:214.
Case C-342/05*Commission v Finland* [2007] ECLI:EU:C:2007:341.
Case C-371/98*First Corporate Shipping* [2000] ECLI:EU:C:2000:600.
Case C-383/09*Commission v France* [2011] ECLI:EU:C:2011:369
Case C-387/15*Orleans* [2016] ECLI:EU:C:2016:588.
Case C-399/14*Grüne Liga Sachsen* [2016] ECLI:EU:C:2016:10.
Case C-418/04*Commission v Ireland* [2007] ECLI:EU:C:2007:780.
Case C-461/13*Bund für Umwelt und Naturschutz Deutschland eV* [2015] ECLI:EU:C:2015:433.
Case C-529/15*Folk* [2017] ECLI:EU:C:2017:419.
Case C-683/16*Deutscher Naturschutzring* [2018] ECLI:EU:C:2018:433.
Convention on Biological Diversity 31 ILM 818 (1992).
Court of First Instance of The Hague, ECLI:NL:RBSGR:2011:BR6099(2011).
Court of First Instance of The Hague, ECLI:NL:RBDHA:2013:CA0593(2013).
Directive 92/43/EEC on the conservation of natural habitats and of wild fauna and flora [1992] OJ L 206/7.
Directive 2000/60/EC of the European Parliament and of the Council of 23 October 2000 establishing a framework for Community action in the field of water policy [2000] OJ L 371/1.
Directive 2004/35/CE of the European Parliament and of the Council of 21 April 2004 on environmental liability with regard to the prevention and remedying of environmental liability [2004] OJ L 143/56.
Directive 2008/56/EC of the European Parliament and of the Council of 17 June 2008 establishing a framework of community action in the field of marine environmental policy [2008] OJ L 164/136.
Directive 2011/92/EU of the European Parliament and of the Council of 13 December 2011 on the assessment of the effects of certain public and private projects on the environment [2011] OJ L 26/1.
Opinion Advocate General Kokott in Case C-383/09*Commission v France* [2014] ECLI:EU:C:2011:23.
Treaty on the Functioning of the European Union, O.J. C 326 , 26/10/2012 P. 1–390.

## *Secondary sources*

Stuart K. Allison (2004) 'What "Do" We Mean When We Talk About Ecological Restoration?' 22(4) *Ecological Restoration* 281.
James Aronson and Sacha Alexander (2013) 'Ecosystem Restoration is Now a Global Priority: Time to Roll up our Sleeves' 21 *Restoration Ecology* 293.
James Aronson et al (2011) 'What Role Should Government Regulation Play in Ecological Restoration? Ongoing Debate in São Paulo State, Brazil' 16 *Restoration Ecology* 690.
Anthony D. Barnosky et al (2011) 'Has the Earth's Sixth Mass Extinction Event Already Arrived?' 471 *Nature* 51.
Suvi Borgström and Frederik H. Kistenkas (2014) 'The Compatibility of the Habitats Directive with the EU Green Infrastructure Policy' 23 *European Energy and Environmental Law Review* 36.

Pedro Cardoso (2012) 'Habitats Directive Species List: Urgent Need of Revision' 5 *Insect Conservation and Diversity* 169.

An Cliquet et al (2009) 'Adaptation to Climate Change – Legal Challenges to Protected Areas' 5 *Utrecht Law Review* 171.

An Cliquet and Kris Decleer (2017) 'Halting and Restoring Species Loss: Incorporating the Concepts of Extinction Debt, Ecological Trap and Dark Diversity into Conservation and Restoration Law' 26(2) *Griffith Law Review* 178.

Jordi Cortina-Segarra et al (2016) 'Biodiversity: Speed Restoration of EU Ecosystems' 535 (7611) *Nature* 231.

Michael Curran et al (2014) 'Is There Any Empirical Support for Biodiversity Offset Policy?' 24 *Ecological Applications* 617.

Wybe T. Douma (1998) 'Evolution and Impact of Sustainable Development in the European Union' in Friedl Weiss et al (eds) *International Economic Law with a Human Face*, Kluwer Law International, 271.

Isabelle Doussan and Hendrik Schoukens (2015) 'Biodiversity and Agriculture: Greening the CAP Beyond the Status Quo?' in C.H. Born et al (eds) *The Habitats Directive in its EU Environmental Law Context. European Nature's Best Hope?*, Routledge, 437.

Simon Dufour and Hervé Piegay (2009) 'From the Myth of a Lost Paradise to Targeted River Restoration: Forget Natural References and Focus on Human Benefits' 25 *River Research and Applications* 568.

Lukas Egli et al (2018) 'Winners and Losers of National and Global Efforts to Reconcile Agricultural Intensification and Biodiversity Conservation' 24(5) *Global Change Biology* 2212.

Benis N. Egoh et al (2014) 'Exploring Restoration Options for Habitats, Species and Ecosystem Services in the European Union' 51 *Journal of Applied Ecology* 899.

Yaffa Epstein (2016) 'Favourable Conservation Status for Species: Examining the Habitats Directive's Key Concept through a Case Study of the Swedish Wolf' 28 *Journal of Environmental Law* 221.

Yaffa Epstein et al (2015) 'A Legal-Ecological Understanding of the Favorable Conservation Status for Species in Europe' 9 *Conservation Letters* 82.

European Commission (2000) *Managing Natura 2000 Sites. The Provisions of Article 6 of the 'Habitats' Directive.*

European Commission (2003) *Common Implementation Strategy for the Water Framework Directive (2000/60/EC): Guidance Document No. 10 Rivers and Lakes – Typology, Reference Conditions and Classification Systems.*

European Commission (2007) *Guidance Document on the Strict Protection of Animal Species of Community Interest under the Habitats Directive 92/43/EEC.*

European Commission (2009) *Common implementation strategy for the Water Framework Directive (2000/60/EC): Guidance Document No. 20 on Exemptions to the Environmental Objectives.*

European Commission (2011a) *Communication from the Commission to the European Parliament, the Council, the Economic and Social Committee and the Committee of the Regions, Our Life Insurance, our Natural Capital: An EU Biodiversity Strategy to 2020*, COM(2011) 244 final.

European Commission (2011b) *Commission Staff Working Paper. Impact Assessment. Accompanying the Document EU Biodiversity to 2020.*

European Commission (2013) *Communication from the Commission to the European Parliament, the Council, the European Economic and Social Committee and the Committee of*

*the Regions, Green Infrastructure (GI) – Enhancing Europe's Natural Capital*, COM (2013) 249 final.
European Commission (2014) *Note to the Nature Directors*, Env B02 PM/oe ARES (2014).
European Commission (2016) *Report from the Commission to the Council and the European Parliament under Article 18(2) of Directive 2004/35/EC on Environmental Liability with regard to the Prevention and Remedying of Environmental Damage*, COM(2016)204 final.
European Commission (2017) *Communication from the Commission to the European Parliament, the Council , the European Economic and Social Committee of the Regions – An Action Plan for Nature, People and the Economy*, COM(2017) 198 final.
European Environment Agency (2015) *State of Nature in the EU. Reporting under the EU Habitats and Birds Directives 2007–2012.*
European Environment Agency (2018) *European Waters – Assessment of Status and Pressures.*
Valerie Fogleman (2015) 'The Threshold for Liability for Ecological Damage in the EU: Mixing Environmental and Conservation Law' in C.-H. Born et al (eds) *The Habitats Directive in its EU Environmental Law Context. European Nature's Best Hope?*, Routledge.
Anton Garcia-Ureta and Iñigo Lazkano (2015) 'Instruments for Active Site Management under Natura 2000: Balancing between Stakeholders and Nat Conserv' in C.-H. Born et al (eds) *The Habitats Directive in its EU Environmental Law Context. European Nature's Best Hope?*, Routledge, 71.
Kaspar A. Hallman (2017) 'More than 75 Percent Decline over 27 Years in Total Flying Insect Biomass in Protected Areas' 12(10) *PLOS One*, doi:10.1371/journal.pone.0185809.
Daniel Hering et al (2010) 'The European Water Framework Directive at the Age of 10: A Critical Review of the Achievements with Recommendations for the Future' 408 *Science of the Total Environment* 4007.
Holly Jones et al (2018) 'Restoration and Repair of Earth's Damaged Ecosystems' 285 (1873) *Proceedings of the Royal Society B*, doi:10.1098/rspb.2017.2577.
Dolly Jørgensen (2013) 'Ecological Restoration in the Convention on Biological Diversity Targets' 22 *Biodiversity and Conservation* 2977.
Eric Katz (1992) 'The Big Lie: Human Restoration of Nature' 12 *Research in Philosophy and Technology* 392.
Janne Kotiaho and Atte Moilanen (2015) 'Conceptual and Operational Perspectives on Ecosystem Restoration Options in the European Union and Elsewhere' 52 *Journal of Applied Ecology* 816.
Janne Kotiaho et al (2016a) *Framework for Assessing and Reversing Ecosystem Degradation*, Report of the Finnish Restoration Prioritization Working Group on the Options and Costs of Meeting the Aichi Biodiversity Target of Restoring at Least 15 Percent of Degraded Ecosystems in Finland.
Janne Kotiaho et al (2016b) 'Land Use: A Global Baseline for Ecosystem Recovery' 532(37) *Nature* 37.
Johan Lammerant et al (2013) *Implementation of 2020 EU Biodiversity Strategy: Priorities for Restoration of Ecosystems and Their Services in the EU. Report to the European Commission*, ARCADIS in cooperation with ECNC and Eftec.
Wouter Langhout (2014) 'Why the EU will Fail to Deliver on Ecosystem Restoration', *Birdlife International*, 30 October, www.birdlife.org/europe-and-central-asia/news/why-eu-will-fail-deliver-ecosystem-restoration.

Renaud Lapeyre et al (2015) 'Biodiversity Offsets as Market-Based Instruments for Ecosystem Services? From Discourse to Practices' 15 *Ecosystem Services* 125.

David Moreno-Mateos (2015) 'The True Loss Caused by Biodiversity Offsets' 192 *Biological Conservation* 552.

Brian Moss (2008) 'The Water Framework Directive: Total Environment or Political Compromise' 400 *Science of Total Environment* 33.

Laetitia Navarro and Henrique Pereira (2012) 'Rewilding Abandoned Landscapes in Europe' 15 *Ecosystems* 900.

David Nogués-Bravo et al (2016) 'Rewilding is the New Pandora's Box in Conservation' 26 *Current Biology* R87.

Susan Owens (2003) 'Is there a Meaningful Definition of Sustainability?' 1 *Plant Genetic Resources* 5.

G. Pe'er et al (2017) 'Adding Some Green to the Greening: Improving the EU's Ecological Focus Areas for Biodiversity and Farmers' 10 *Conservation Letters* 517.

Michael M. Redclifft (2006) 'Sustainable Development: An Oxymoron Comes of Age' 12 *Horizonte Antropologicos* 65.

Benjamin J. Richardson (2016) 'The Emerging Age of Ecological Restoration Law' 25 *RECIEL* 277.

Hendrik Schoukens (2014) 'Going Beyond the Status Quo: Towards a Duty for Species Restoration under EU Law?' in V. Sancin and M. Kovič Dine, *International Environmental Law: Contemporary Concern and Challenges in 2014*, GV Založba, 344.

Hendrik Schoukens (2017a) 'Saving the Common Hamster from Extinction through the EU Habitats Directive: A Mandatory Recovery Effort, a Remediation of Past Non-compliance or an Exercise in Futility?' 1 *Nordic Environmental Law Review* 59.

Hendrik Schoukens (2017b) 'Proactive Habitat Restoration and the Avoidance of Adverse Effects on Protected Areas: Development Project Review in Europe after Orleans' 20(2) *Journal of International Wildlife Law and Policy* 125.

Hendrik Schoukens (2017c) 'Non-Regression Clauses in Times of Ecological Restoration Law: Article 6(2) of the EU Habitats Directive as Unusual Ally to Restore Natura 2000?' 13(1) *Utrecht Law Review* 124.

Hendrik Schoukens and Harm Dotinga (2015) 'Natura 2000 and Fisheries: A Question of Competence or Willingness?' in C.-H. Born et al (eds) *The Habitats Directive in its EU Environmental Law Context. European Nature's Best Hope*, Routledge, 375.

SER (2004) *The SER International Primer on Ecological Restoration.* Society for Ecological Restoration.

Lorenzo Squitani (2012) 'The Development of Ecological Corridors: Member States' Obligations under the Habitats and Birds Directive' 9 *Journal of European Environmental & Planning Law* 180.

Anastasia Telesetsky, An Cliquet and Afshin Akhtar-Khavari (2017) *Ecological Restoration in International Environmental Law*, Routledge.

# 9 Public participation and socio-economic justice in eco-restoration law and governance

## The UN Environment–Ogoniland case study

*Uzuazo Etemire and Menes Abinami Muzan*

### 9.1 Introduction

Nigeria's Niger Delta region, of which Ogoniland is part, is widely considered as one of the world's richest biodiversity hotspots.[1] Similar to other Niger Delta communities, Ogoniland has had a tragic history of massive ecological pollution, majorly caused by recurring oil spills from the activities of multinational oil companies such as Shell, with the complicity of the Nigerian government which holds a major stake in the industry.[2] This pollution has damaged the unique ecosystem upon which the Ogoni people depend for social, cultural and economic sustenance.[3] But then, the Nigerian economy depends heavily on its oil industry, which accounts for the bulk of the nation's export earnings and revenue.[4] The tensions between these economic and environmental agendas led the Nigerian government to invite United Nations (UN) Environment (formerly the UN Environment Programme) to conduct a thorough environmental assessment of Ogoniland and propose a strategy for the restoration of its environment. This study, published in 2011, provides the first systematic scientific study of the extent of oil contamination and ecological damage in Ogoniland.[5]

In its report, UN Environment designed and proposed a 'multi-stakeholder' participatory approach for eco-restoration in order to avoid the legal pitfalls and institutional weaknesses of the traditional state-driven approach to environmental governance in Nigeria. Furthermore, UN Environment concluded that 'the environmental restoration of Ogoniland could prove to be the *world's most* wide-ranging and long term oil clean-up exercise ever undertaken'[6] which is expected to take 'around 25–30 years to complete'.[7] Indeed, the implementation of the project over such an extensive timeframe is likely to produce new knowledge and reveal best practices that will ultimately be beneficial to communities around the

1 Afinotan and Ojakorotu (2009), 194; see generally Eyinla and Ukpo (2006).
2 Konne (2014), 183
3 Idowu (1999), 166–167
4 See CIA (2013).
5 UNEP (2011), 8.
6 UNEP (2011), emphasis added.
7 Ibid, 226.

world with similar challenges, and to experts who work in this field. While the implementation was stalled for several years,[8] sustained pressure on the Nigerian government from civil society organisations[9] finally led to the beginning of restoration in 2016.[10] Beyond its potential to engender reforms in environmental remediation laws in Nigeria, this case study has international significance regarding 'plausible pathways' for achieving globally agreed targets such as the UN 2050 vision of maintaining ecosystem services.[11]

Generally speaking, eco-restoration seeks to systematically renew, improve and restore damaged or degraded ecosystems across extensive landscapes and water bodies, in a largely holistic manner, through deliberate human intervention.[12] It is distinguishable from *environmental* restoration, which is more limited in scope, such as the clean-up of a discrete oil-spill site or former mine in order to improve local aesthetics and functionality.[13] Although the concept of eco-restoration has long been obscured in the field of environmental law by the dominant focus on the philosophy of sustainable development, restoration is gradually receiving more recognition in view of the need to adequately address past environmental losses so as to ensure that future environmental conservation and sustainable development will be feasible.[14]

The aim of this chapter is to critically assess UN Environment's stakeholder-driven governance model as an ingredient for successful eco-restoration. This model is evaluated in contrast to the traditional, state-driven approach dominated by government agencies and command-and-control regulation. The chapter also considers the importance of community acceptability and social justice in ensuring a holistic and successful eco-restoration project. Section 9.2 discusses the theoretical basis for the socially inclusive, stakeholder-driven model for eco-restoration governance. On this basis, section 9.3 analyses the status and challenges of the state-based regulatory framework for restoration governance in Nigeria. Section 9.4 examines UN Environment's stakeholder-driven Ogoniland eco-restoration project, with a particular focus on its implementation and its progress in meeting not just ecological goals, but also ensuring social and economic justice for Ogoni communities. Overall, the thesis of this chapter is that, given the engagement with a broader group of stakeholders – beyond the traditional state actors and developers – the implementation of the UN Environment strategy will enable eco-restoration goals to be achieved in the context of the broader social and economic ambitions of the people of Ogoniland.

8 Alabi (2012), 167–169.
9 Amnesty International (2014).
10 UNEP (2016).
11 Secretariat of the Convention on Biological Diversity (2014), 134–135.
12 See Hayes (2002), 115; Suding et al (2015), 638; see also Telesetsky (2013), 494, 503; Telesetsky, Cliquet and Akhtar-Khavari (2017).
13 SER (2004). See also Clewell and Aronson (2007), 216.
14 See Richardson (2016), 1.

## 9.2 Participatory approach to ecological restoration: theoretical basis

The state-based approach to restoration governance largely has its roots in the political dogma of rational elitism. It postulates that public decision-making should be left almost completely to experts or 'elites' with specialised training and experience, such as government administrators and heads of economic enterprises.[15] In such a system, environmental governance is considered as 'complex and technical, requiring primarily technical and administrative expertise',[16] without the need for any substantive input from the general public. While this oligarchical political theory generally does not prioritise wide public participation as a solution to societal problems, it may allow it in a limited, *albeit* inadequate, manner.[17] As further argued below, this approach is substantially inadequate for ensuring effective eco-restoration governance and for dealing with the social and ecological acceptability and justice issues that accompany eco-restoration projects.

Hence, there is a compelling need to seek the alternative path of democratising restoration governance beyond the domination of 'expert' regulators and scientists. As Richardson rightly argues, public participation 'is an indelible element of successful restoration governance'.[18] This accords with Principles 7 and 10 of the widely adopted 1992 Rio Declaration on Environment and Development (Rio Declaration),[19] as they reflect the importance of wide stakeholder involvement in processes aimed at restoring damaged ecosystems.

Indeed, this is what 'environmental democracy'[20] (a sub-theme of the classical theory of participatory democracy)[21] is all about: it is a well-established political aspiration which, as against rational elitism, recognises the public – including community members and nongovernmental organisations (NGOs) – as having a critical role in ensuring effective eco-restoration governance, and thus must be allowed to participate and partner with the traditional actors – government agencies and economic developers – to that end.[22]

To be sure, wide public and stakeholder partnership at the relevant level is critical in engendering successful eco-restoration governance from a number of perspectives. For one, considering the expensive nature of landscape-scale eco-restoration, it is potentially costly and labour intensive.[23] While relevant public authorities and those responsible for the source of pollution may bear most of the restoration costs, members of affected communities and interested NGOs can also make valuable contributions to the process by making available their expertise and local knowledge, and participating in implementation efforts that include tree

15 See Bealey (1996), 323.
16 Richardson and Razzaque (2006), 170.
17 Barton (2002), 85–86; see also Checkoway and Til (1978), 27.
18 Richardson (2016), 12.
19 Adopted by the UN Conference on Environment and Development (UNCED), 3–14 June 1992; (1992) 31 ILM 874.
20 See generally Mason (1999).
21 See Kweit and Kweit (1981), 7, 35; Rosenbaum (1978), 43–48.
22 Bándi (2014), 3–7.
23 Richardson (2016), 8, 11

planting, removal of invasive species and reintroduction of ecologically compatible species. Given the very long timeframes of some eco-restoration projects, over many decades, sustained local participation is particularly important to their achievement.

In support of the above point, it has for instance been noted that many NGOs 'can often be in a stronger position than governments',[24] especially in terms of their diverse 'experience, expertise and capacity'.[25] They can easily bring this to bear beneficially on the restoration process and its governance. There are also examples of indigenous people making useful contributions to restoration projects in their communities through their skills and traditional knowledge about ecological processes.[26] While a part of this civil and community contribution to the project can be delivered by volunteers, some other aspects may deserve to be financially remunerated. In other words, public participation in eco-restoration projects can ensure adequate manpower for a project and deliver economic benefits to community members who may have lost their means of livelihood to ecological damage. Indeed, the latter prospect of economic justice is likely to improve the social acceptability of the project.

Importantly, and in line with the dogma of rational elitism, strong scientific expertise is central to effective eco-restoration.[27] Scientific experts can help determine such issues as the impact on human health of the degraded ecosystem and options for treatment; the extent of ecological damage; safe and effective restoration measures; and environmental risks as well as uncertainties that may derail a restoration project. Yet social considerations, ascertained primarily through the involvement of the concerned public, are equally vital for the success of an eco-restoration project.[28] For instance, the issue of what historical environmental baseline an eco-restoration project should target touches on 'human values and preferences for one time period and natural conditions';[29] while this is essentially a social choice for the affected community, major scientific input is required in assessing the feasibility and implications of a particular baseline target, or other key dimensions of a restoration project, and bringing them to fruition. Hence, a framework for close and cordial collaboration between scientific experts and concerned communities and NGOs is important to ensuring a successful and effective eco-restoration programme.

No doubt, some communities may lack knowledge about environmental degradation and the options for, and benefits of, restoring an ecosystem. Thus, community participation must also include measures for community education about eco-restoration. This would provide a platform to correct misconceptions and resolve doubts in the minds of the concerned public, and attract their support for a project. Active participation in eco-restoration governance would also help

24 Willetts (1982), 185; Princen and Finger (1994), 34.
25 United Nations (1992), Ch. 27.
26 See Burrows, Burbidge and Fuller (2004); Anderson (2005).
27 Richardson (2016), 10.
28 See Barton (2002), 101; Lee and Abbot (2003), 80, 84.
29 Richardson (2016), 10.

improve the environmental knowledge of participants, in line with the thoughts of Rousseau as expressed by Pateman.[30] By engaging with and ensuring their participation in eco-restoration projects, communities could 'learn about new problems and solutions'[31] and 'discover their own real interests' and needs,[32] as the participatory process and educative engagements force them to reflect more deeply about their 'preferences and priorities' and their 'value and beliefs'.[33]

Additionally, eco-restoration projects, given their scale, are potentially disruptive of major social activities, and thus could deprive locals of their economic means of livelihood and generally inconvenience the relevant communities for a protracted period of time. '[R]estoration projects may displace people [...] some land [and river] uses may need to cease or be reduced in intensity', explains Richardson from his case studies.[34] These changes, on a social level, may also affect custodians of traditional practices and cultural ceremonies if there is reduced community access to recovering ecosystems, or if their aesthetic qualities change. Business-wise, this will impact farmers, fishers, hunters, traders and landowners. These effects could lead to considerable opposition to an eco-restoration project by affected local groups, and generate conflict between the latter and proponents of such projects, as has been witnessed in both the United States[35] and the United Kingdom.[36] Yet restoration projects that improve soil fertility and water quality, and deliver other functional benefits to local communities, may win their support through the prospect of more economically productive landscapes.

To avoid or minimise such potential opposition and conflict, and secure peace and the cooperation of the affected communities, community participation in a restoration project is vital as it helps build understanding and broader agreement among stakeholders.[37] In support, Wengert argues that it will enable better 'understanding and tolerance', encourage 'modifications of values and opinions', and increase 'confidence and trust'.[38] Furthermore, by ensuring public participation in eco-restoration, some of the causes of conflict, such as the imposition of 'disproportionate costs on one stakeholder or community' or 'inequities in the distribution of the [...] benefits', might be resolved expeditiously at an early stage in project planning. With these safeguards in place, and the consequential increase in the acceptability of and community sense of 'ownership' over a project, the potential for conflict and undue opposition to eco-restoration should be significantly weakened.

As supported by the views of several scholars, a related justification for involving the public in eco-restoration efforts is that they promote the legitimacy and

30 See Pateman (1970), 42–43, 31–33.
31 Richardson (1983), 55.
32 Ibid, 56; see also Bachrach (1975), 40.
33 Burton (2009), 263, 266.
34 Richardson (2016), 10–11.
35 See Hamann (1997).
36 Vaughan (2014).
37 Fiorino (1990), 226, 234.
38 Wengert (1976), 23, 26–27

credibility of restoration governance entities in the opinions of the affected communities.[39] This is particularly important for developing countries, such as Nigeria, where the dearth of environmental democratic norms in public governance has caused polluted communities to lose faith in their governments and constantly doubt the sincerity and legitimacy of the actions of restoration authorities (even if aimed at fostering the communities' interests). Related to this, wide stakeholder participation would help improve transparency and accountability in eco-restoration governance. This would potentially prevent financial corruption and 'agency capture' by vested interests (e.g. oil companies) – common vices in Nigeria – that could easily undermine the successful execution of a restoration project.[40] These pernicious practices thrive in a milieu of secrecy and public exclusion, which could be counteracted by allowing more involvement from the public in restoration projects.

Community participation in eco-restoration governance also brings unwelcome risks, such as inflaming rather than abating community grievances over natural resource use and responsibilities for repairing degradation. In such situations, the solution is neither 'little or no participation' as the rational elitist may say; nor is it to blindly follow Alfred E. Smith's advice that '[t]he only cure for the evils of democracy is more democracy'.[41] The latter view 'could well be adding fuel to the fire',[42] whilst the former would erode the benefits of participation. Rather, the solution lies in the slight moderation of participation democracy – that is, ensuring respect for and systematic introduction of certain beneficial tenets of other political traditions into the body of participatory democracy, to enhance its output for the greater good. Huntington's thoughts on the need to curtail the excesses of participatory democracy through moderation are instructive here:

> Greek philosophers argue that the best practical state – the 'mixed regime' – would combine several different principles of government in its constitution [...] A value [i.e. democracy] which is normally good in itself is not necessarily optimized when it is maximized [...] There are [...] potential desirable limits to the extension of political democracy. Democracy could have a longer life if it has a more balanced existence.[43]

In the context of eco-restoration, the above concern is exemplified by a situation where democratic majorities may easily and legally decide not to prioritise particular rights and interests, especially those of unpopular minorities.[44] This is of concern in developing countries like Nigeria, and even developed nations, where the degraded areas requiring restoration are sometimes located in territories of minority or vulnerable ethnic groups.[45] A frequent complaint of these

39 See Bodansky (2007) 711; Ebbesson (1997), 51, 75–81; Gellhorn (1974), 424.
40 See Etemire (2015), 85, and Barton (2002), 104.
41 As quoted in Huntington (1975) 36.
42 Ibid.
43 Ibid, 36–37.
44 Dryzek (2000), 10–11.
45 Adeola (2001), 42. See generally UNDESA (2009), 21–22.

groups is that their voices and interests are drowned out by those of the majority ethnic groups whose members occupy most of the key decision-making positions.[46]

The point is that democratic frictions of the above nature can be avoided by partly inculcating in a nation's legal and political framework the norms of a political thought like *liberalism* for instance, particularly its core tenets of 'freedom' and 'equality' for *each* individual irrespective of popular democratic opinion.[47] This liberal idea is based on the notion that certain basic rights and interests – including those on the environment and self-preservation – accrue to every individual and must therefore be respected, promoted and protected, and not relegated.[48] To be sure, liberalism – which is wrapped in the general philosophy of 'individualism' that was central to the ideologies of liberal thinkers like John Locke and Immanuel Kant[49] – holds that 'the state must be restricted in scope and restrained in practice to ensure the maximum possible freedom of every citizen',[50] for a proper political order is 'one in which people are able to develop their natural interests free from arbitrary use of political authority and coercive power'.[51] In other words, eco-restoration projects should function within a larger constitutional context in which the basic rights and interests of all members of society enjoy minimum guarantees. Indeed, this is a perspective that will give local minorities better standing and an amplified voice for them to protect their social and ecological interests in the course of eco-restoration, especially where the democratic majority may be averse to critical aspects of their course.

It is against the backdrop of the foregoing survey of the role of community and stakeholder participation that we now analyse the state-based approach to restoration governance in Nigeria.

## 9.3 State-based structures for restoration governance in Nigeria

Aspects of eco-restoration work often require the leadership of government regulators. Across the world, the most common model for restoration governance is largely state-based and state-driven, especially in developing countries like Nigeria. On the one hand, there are state laws and policies enacted to ensure (a limited form of) restoration; and on the other hand, there are government institutions established to implement those legal regimes and enforce them against those responsible (where ascertainable) for environmental degradation. Within this traditional restoration governance structure, there is usually little or no space for other stakeholders – beyond government officials and the polluters – to participate meaningfully in the process.

46 See Etemire (2014), 3–4.
47 Dryzek (2000).
48 Ibid.
49 Larmore (1999), 603.
50 Held (2006), 262.
51 Ibid, 263–264.

Regrettably, that is the way the major Nigerian legal instruments mandating oil companies to restore oil-polluted environments[52] are often designed and implemented. For example, the Oil Spill Recovery, Clean-up, Remediation and Damage Assessment Regulations of 2011[53] provide for an oil-spill Joint Investigation Visit (JIV).[54] The JIV team – comprising representatives of the polluter, members of the affected community and the government[55] – is charged with investigating the occurrence of oil spills and ascertaining their cause and the area affected.[56] It is largely on the basis of the JIV's findings that issues of liability and restoration of the environment are decided. Although this procedure offers an opportunity – albeit limited – for public participation, it is nevertheless usually honoured in the breach and in a manner that excludes meaningful community participation.[57] It is thus no surprise that this strategy, which is firmly in the grip of government regulators, has so far failed to achieve adequate restoration of oil-degraded environments in the Niger Delta.[58]

To be sure, there are, notionally, laws and policies in Nigeria that may contribute to ensuring the restoration of degraded environments. However, these are mostly focused on the more limited idea of *environmental* restoration than on the broader notion of *ecological* restoration. In cases of oil pollution, the latter approach is exemplified by relevant provisions of the Petroleum (Drilling and Production) Regulations of 1969.[59] Similarly, the NOSPRA Act, which sets the parameters for the detection of and response to oil spillages in Nigeria, is mainly concerned with the 'clean-up'[60] of impacted sites and, like the aforementioned petroleum regulations, does not address far-reaching eco-restoration issues such as the restoration of flora and fauna populations depleted by the long-term pollution.

Even at that, there are still fundamental questions about the liability of oil companies for remediating oil-damaged areas in the Niger Delta (to the extent provided in the laws) and the general problems that members of affected communities encounter in their search for justice against the polluting companies. Particularly, it

52 For instance, see the National Oil Spill Detection and Response Agency (Establishment) Act of 2006, Federal Republic of Nigeria Official Gazette No. 72 Vol. 93, 29 December 2006 (hereinafter 'NOSDRA Act').

53 Made by the National Oil Spill Detection and Response Agency (NOSDRA) under powers conferred in it by section 26 of the NOSDRA Act.

54 Rim-Rukeh (2015), 259–271.

55 Section 5 of the Oil Spill Recovery, Clean-up, Remediation and Damage Assessment Regulations of 2011.

56 See Amnesty International (2013), 14.

57 Ibid, 40–41.

58 Juras (2012).

59 L.N. 69 of 1969; see section 25 which only obliges the polluter to 'control and, if possible, end it [i.e. the pollution]', and section 46 (c) where a polluter with a terminated license is only mandated to 'take reasonable steps to restore as far as possible to their original condition *the surface of the relevant area* [...] damaged in the course of his operation'.

60 See section 5 (b) and (c); section 6 (3).

is the condition that private litigants comply with certain unduly stringent legal requirements of the civil liability regime that, in many cases, actually stands in the way of access to justice, in terms of ensuring a level of restoration and (adequate) victim compensation. One such is the requirement that litigants must bring their cases within the restrictive timeframe of the relevant Nigerian statute of limitation applicable to the case, which generally runs from the date the act or omission complained of first occurred.[61] For example, as prescribed by its enabling statute, an aggrieved party can only bring legal action against the Nigerian National Petroleum Corporation (NNPC) within a year from the date when the act or omission complained of first occurred.[62] This differs from the more reasonable alternative adopted by many jurisdictions, where the limitation runs from the date the victim first knew of the damage and or the entity responsible.[63] Without the latter standard, victims may only become aware of their rights after expiration of the limitation period.

Another critical issue is that, before filing an action, the aggrieved party is required to serve a *pre-action notice* on the prospective defendant, the rationale being to give an opportunity for out-of-court settlement of disputes.[64] In Nigeria, pre-action notice of one month or more is a prevalent practice of government agencies, including those with environmental responsibilities.[65] If the litigant fails to give the notice, the trial court will be deprived of its competence to hear the matter at all, except in the unlikely event that the defendant fails to raise it, in which case it would be considered a mere procedural irregularity.[66] Yet environmental risk is of such a nature that a timely injunction, even given *ex parte*, is what may be required to avert possibly irreversible environmental damage, a goal which can easily be frustrated by a law requiring pre-action notice in a blanket and mandatory manner.[67] However, as noted by Fagbohun,[68] there are better approaches adopted in some other jurisdictions, which include: 1) the court staying proceedings until the notice is given, instead of dismissing or striking out the suit;[69] 2) approving that some public interest suits can be brought without the notice;[70] or 3) not requiring the notice where the environment is under threat of severe or irreversible damage.

61 For an example of the restrictive stance of the Nigerian courts on the issue of limitation of action, see *Gulf Oil Company (Nig) Lts v Oluba* (2003) FWLR (pt 145) 712.

62 Section 12(1) of the NNPC Act, Cap N123, Laws of the Federation of Nigeria, 2004.

63 For instance, see section 11(4) and 14 of the Limitation Act, 1980 (UK).

64 Tobi (1998), 191.

65 For example, section 32 (1) of the NESREA Act and section 12 (2) of the NNPC Act both provide for a one-month pre-action notice before any action can be commenced against the respective bodies.

66 See *Aro v Lagos Island Local Government Council* (2000) FWLR (pt. 13) 2132, and the more recent Nigerian Supreme Court decision in *Mobile Producing (Nig) Unlimited v LASEPA, FEPA & Ors* ((2002) 18 NWLR (pt. 789) 1.

67 Fagbohun (2012), 66–67.

68 Ibid, 68.

69 *Pymatuning Water Shed Citizens for a Hygienic Environment v Eaton,* 644 F. 2d 995 (3rd Cir. 1981) (USA).

70 *Natural Resources Defence Council Inc. v Train,* 510 F. 2d 692 (D.C. Cir. 1974) (USA)

In addition, litigants must show that they have *legal standing* to bring the action. Yet the standing rule in Nigeria, especially on environmental matters, is unduly restrictive as it warrants that standing be accorded only to plaintiffs whose direct *private* legal right has been or is likely in danger of being adversely affected by the act or omission complained of.[71] In other words, an eligible environmental plaintiff must have a direct personal and proprietary relationship with the subject matter of the litigation, or s/he must have suffered special damages peculiar to him/herself from the interference with a public right.[72] Indeed, this restrictive rule has rendered unsuccessful many lawsuits against operators in the Nigerian oil industry such as Shell and the NNPC.[73] According to Ogowewo, this standing rule – which seasoned commentators agree is in dire need of liberalisation and 'modernisation'[74]– has a 'court-closing' effect, as it 'immunizes from judicial review a substantial aspect of the [non-] exercise of governmental power',[75] and reduces the likelihood of a successful public interest suit, or oil companies being held accountable through private action for causing damage to a public environmental media and therefore liable to pay compensation to the affected community and execute the necessary restoration work.

Another significant hurdle in pursuing civil claims in Nigeria is the requirement to establish clear *causation*. The plaintiff must clearly show that the pollution caused the personal injury suffered, which is directly traceable to the unlawful acts or omission of the defendant. This legal threshold requires clear scientific evidence which is not always certain or precise (as to whether, or the extent to which, the pollution and the injury resulted from the defendant's activities or other sources and factors), and is as financially expensive as it is time consuming to collect. This has been a significant barrier to access to environmental and economic justice for victims of oil pollution in Nigeria,[76] especially as affected local people are usually too poor to secure the required scientific evidence, and even where the finance is available, the environmental damage and human injury may have occurred long after the pollution, thereby making the necessary causal links difficult to prove. New approaches to lessening the onerous burden of strict proof of causation on plaintiffs, within the bounds of fairness to all parties, are being adopted[77] and suggested,[78] but they are yet to be adopted in Nigeria.

71 *Adesanya v President of the Federal Republic of Nigeria* (1981) 1 All NLR 1. See also Frynas (2000), 207.

72 Fagbohun (2012), 70.

73 *Oronto Douglas v Shell Petroleum Development Company Nigeria Limited and Ors* (Unreported) Suit No: FHC/L/CS/573/96, 17 February 1997. The Court of Appeal's decision is reported in (1999) 2 NWLR 466.

74 As noted by Tobi JCA in *Busari v Oseni* (1992) 4 NWLR (pt. 237) 557.

75 Ogowewo (2000), 542.

76 See, for instance, *Shell v Graham Otoko* (1990) 6 NWLR (pt. 159) 693 at 724–725; *Atunbi v Shell BP* (Unreported) Suit No. UHC/48/78, 25 November 1974.

77 For instance, see the case of *Fairchild v Glenhaven Funeral Service Ltd* [2002] UKHL 22, where the less burdensome 'but for' test was applied to the injury case. See generally Bates et al (2004).

78 For instance, for the suggestion of plausible presumption as sufficient evidence in some cases, see Bocken (1997), 144.

Indeed, the existence of the above legal limitations to accessing socio-economic and ecological justice in Nigeria is not unconnected with the lack of adequate public input in environmental law-making and reform processes in the country,[79] as alluded to by Nigeria's Deputy Senate President.[80] Many of Nigeria's environmental laws in the oil sector were made under military rule, at a time when public participation was an even lower consideration as compared with Nigeria's current and more democratic regime. Such inadequate public input in the design of environmental laws and its consequences is part of what the 2010 UN Environment Guidelines for the Development of National Legislation on Information, Public Participation and Access to Justice in Environmental Matters (Bali Guidelines)[81] aims to address.[82] The Bali Guidelines were developed to guide states ('primarily developing counties') with input from civil society from around the world.

Furthermore, even the governance of *environmental* restoration (which may be considered a step towards eco-restoration) in Nigeria is structured in such a way that the process lacks rigorous, independent (governmental) oversight. Thus, restoration obligations imposed on oil companies are not adequately applied or enforced due to lack of independent oversight by the relevant government regulators. The fact is that oil production is undertaken via joint venture arrangements between the Nigerian government through the NNPC (usually the major stakeholder) and multinational oil companies. Considering that both entities are essentially business partners and are (directly and indirectly) responsible for environmental degradation in Ogoniland, the reality is that the relevant government agencies have neither been able to effectively play their role as regulators in relation to the activities of oil companies,[83] nor able to ensure that their consortium with these companies acts in an environmentally responsible manner. It was on this basis that the African Commission on Human and Peoples' Rights in 2001 held the Nigerian government liable for the oil pollution of Ogoniland by Shell and NNPC, and required it to remediate the damaged environment, in the case *Social and Economic Rights Action Centre (SERAC) and another v Nigeria* (*Ogoniland case*).[84]

Effectively holding oil companies to account for the environmental effects of their operations could easily affect the huge economic gains which the government derives from the oil industry. Thus, given the relevance of the oil companies to the national economy, and the economic power they invariably wield in Nigeria, government regulatory agencies have become susceptible to the phenomenon of 'agency capture'[85] – a form of failure in governance where a regulatory agency is influenced and

79 Etemire (2015), 222–223.
80 Ekweremadu (2012).
81 Adopted by the Governing Council of UNEP, UNEP/GCSS.XI/11, Decision SS.XI/5, Part A, 26 February 2010.
82 For instance, see Guideline 13 of the Bail Guidelines.
83 Konne (2014), 195.
84 (2001) AHRLR 60. The Nigerian government however took no immediate steps to comply with this 'landmark' decision (until its recent efforts to implement in the 2011 UN Environment report on Ogoniland).
85 Amnesty International (2009), 48.

controlled by the company it was setup to regulate, against its original public interest agenda.[86] It is in this light that the argument has been rightly made that wide stakeholder participation in restoration governance is essential to ensuring transparency, countering 'agency capture' and insulating the process of governance against unwholesome influence from any particular stakeholder.[87]

What is more, the governance structure in Nigeria – and many other developing countries – for ensuring restoration of degraded environments is characterised by inadequate resources and capacity on the part of its major driver and enforcer: the relevant state agencies. For instance, the Department of Petroleum Resources (DPR), among others, generally possesses federal responsibility for enforcing compliance with environmental regulations and guidelines in the oil industry, including ensuring restoration upon environmental damage.[88] Yet the UN Development Programme (UNDP) in its assessment has reported that 'its [i.e. the DPR's] effectiveness has been seriously hampered by inadequate personnel, funding, equipment and logistics support [...] The sordid environmental condition [in the Niger Delta] obviously reflects institutional failures'.[89] A World Bank report equally points to the fact that '[l]acking monitoring and basic office equipment, the Department [DPR] is currently not able to perform its duties and is limited to obtaining Oil Company spill reports'.[90]

This situation has forced regulators like NOSDRA and other agencies to be inappropriately reliant on the oil companies directly responsible for pollution in order to perform their responsibilities, as revealed by research from Amnesty International[91] and UN Environment.[92] This sets the stage for the regulator to be 'captured', controlled and compromised by the polluter for its own interest and in a manner inimical to ensuring successful restoration. This situation heightens the need for other stakeholders to be involved in the governance and execution of restoration projects. Their proper involvement may well help to boost transparency in the process and contribute some of the much-needed expertise that would reduce the pernicious influence of oil polluters over the restoration process.

Considering this state of affairs, it is obvious that a new approach to restoration governance that will more effectively ensure successful eco-restoration is required. Section 9.4 analyses UN Environment's proposed eco-restoration project in Ogoniland and its stakeholder-driven approach as arguably a better alternative to Nigeria's maligned state-based regime.

86 Levine and Forrence (1990), 167–197; see also Spyke (1999), 275.
87 Ayres and Braithwaite (1992).
88 See www.dpr.gov.ng.
89 UNDP (2006), 310.
90 World Bank (1995), 55.
91 Amnesty International (2013), 44 (footnotes omitted); UNEP (2011), 12.
92 UNEP (2011), 12.

## 9.4 The UN Environment–Ogoniland eco-restoration project

A fundamental and unique feature of the Ogoniland eco-restoration project and its proposed governance structure is its largely stakeholder-driven approach, which is reasonably responsive to the issues of social acceptability of the project and justice for the Ogoni people. Given the conclusions of section 9.2 above, this stakeholder-driven structure is fundamental to the successful execution of the Ogoniland eco-restoration project. In particular, to effectively govern and execute the project, the UN Environment report involves, and lays responsibility on, three key stakeholders: the Nigerian government, the oil companies operating in Ogoniland, and the Ogoni local community.[93] UN Environment itself is also identified as a stakeholder that provided the assessment, designed the governance structure and is envisioned to continue to play an oversight role in the execution of the project.

In analysing how the restoration project and its governance structure constitutes a paradigm shift from the norm in Nigeria – and could benefit other countries – especially from the standpoint of public participation, it is vital to look at the inception of the project. In this connection, one must appreciate the positive implications of the Nigerian government's involvement of UN Environment in the Ogoniland restoration efforts as a neutral and credible party that helped to ameliorate the tensions that existed between the regulators, the oil companies and the Ogoni community, which had prevented meaningful dialogue between them.[94] In addition, UN Environment designed the project with the involvement of a wide range of stakeholders, including international and local (scientific) experts as well as community members from Ogoniland.[95] This wide stakeholder involvement has not only helped to expand the focus of restoration from the limited traditional state-based *environmental clean-up* to the broader *ecological* perspective,[96] it has improved the social acceptability of the project which is necessary for its overall success.[97]

Furthermore, considering the shortcomings in existing Nigerian institutions with environmental restoration mandates, UN Environment recommended the establishment of an 'Ogoniland Environmental Restoration Authority' to oversee implementation of the Ogoniland restoration project.[98] It was also suggested by UN Environment that the Authority be managed by a mix of government officials and other relevant, non-state stakeholders.[99] To give effect to this, the Nigerian government has set up the Hydrocarbon Pollution Remediation Project (HYPREP) to oversee the restoration exercise and, in August 2016, inaugurated its 13-member Governing Council and 12-member

93 For example, UNEP (2011), 224, 230.
94 UNEP (2011), 6–8.
95 Ibid, 8.
96 Ibid, 14, 211, 214, 224.
97 Richardson (2016), 10–12
98 UNEP (2011), 211.
99 Ibid, 224.

Board of Trustees.[100] Commendably, both governing bodies are made up of high-level government officials, UN Environment (holding observer status), and representatives of relevant oil industry operators, NGOs, the Ogoni community and other Niger Delta communities.[101] This wide stakeholder involvement has already improved the transparency and accountability in the system, as well as reducing the chances of the project being captured and controlled by a single interest.

This inclusive form of eco-restoration governance has boosted the social acceptability of the project, and enabled HYPREP, in accordance with the recommendations of the UN Environment report, to prioritise the restoration of the livelihoods of the Ogoni people – with particular regard to their socio-economic status – along with restoration of the actual ecological damage. This is a core issue that does, or should, differentiate eco-restoration agendas in countries of the Global South such as Nigeria from those of the Global North. Indeed, Richardson rightly notes that eco-restoration should not only focus on improving nature but 'should also [be devoted to] improving human society'.[102] This is because restoring damaged ecosystems is not viable in the long term if human beings, who caused such damage and retain this ability, are left in their 'broken' state; for in a bid to improve their wellbeing, they will likely engage in destructive, short-sighted activities that will undo the gains of the restoration work.[103] To be sure, in some cases, the eco-restoration governance agenda in developing countries must prioritise and place a stronger emphasis on this social (as against environmental) restoration aspect of eco-restoration work, compared to such projects undertaken in the Global North where poverty and deprivation are less prevalent social problems.

This is so because ecological degradation in the Global South has brought significant damage to communities' predominant means of livelihood – such as farming, fishing and hunting – resulting in severe levels of unemployment and poverty beyond the experience of people in the Global North affected by ecological damage.[104] Furthermore, to ensure adequate support and commitment by affected communities in the Global South, the eco-restoration agenda must place more emphasis on community environmental education about the relevance, direction, opportunities and changes that restoration projects will engender. The obvious reason for this is the fact that ecological illiteracy within affected communities in many developing countries is lower than that of citizens of countries of the Global North with access to greater educational opportunities.[105] Moreover, as with most indigenous people, many affected communities in the Global South tend to have a more subsistence livelihood, depending more directly on the wild environment for not just food and water, but also for herbal medication,

100 Financial Watch (2016).
101 Ibid.
102 Richardson (2017), 172.
103 Ibid.
104 See Amnesty International (2009), 9, 34.
105 Etemire (2015), 82; Amnesty International (2009), 10, 21.

fuelwood, building materials and cultural/spiritual inspiration,[106] than their counterparts in the more urbanised North.

It is therefore commendable to find HYPREP placing the socio-economic concerns of the affected communities of Ogoniland at the forefront of its agenda for eco-restoration in implementing the UN Environment report.[107] Currently, HYPREP is sensitising and raising awareness across all the four local government areas of Ogoniland; many affected Ogoni communities have been visited and thousands of 'Ogoni leaders, women and youths' have been educated about the eco-restoration project and the impact of oil contamination through community meetings and focus group discussions,[108] in accordance with UN Environment's recommendation to enlighten the affected public in this direction.[109] The HYPREP Project Coordinator, Dr Marvin Dekil, has also convened regular roundtable meetings for 'disseminating information about HYPREP activities to the public as well as receiving valuable feedback from stakeholders to enrich the project delivery'.[110] As part of this enlightenment campaign, illegal artisanal refiners are being engaged and educated on 'the need to stop their illegal activities which is harmful to their health and the environment'.[111] Ogoni communities are beginning to take a more proactive stand against those engaged in such illegal oil activities.[112] This accords with UN Environment's recommendation for community involvement in stopping illegal sources of continuous pollution to ensure the success of the restoration work on the environment when it fully commences.[113]

HYPREP has also commenced the implementation of UN Environment's recommendation to undertake a Health Impact Assessment Study to ascertain and generate data on the impact of oil pollution on the health of the people in Ogoniland, and to act upon it.[114] This is being executed within the ongoing HYPREP-coordinated Health Outreach programme, under which 2,848 people in Ogoniland have received medical treatment, another 2,963 have received qualitative health education, and another 65 have had surgery performed for free.[115] Also, many ailments suffered by people in Ogoniland have been traced to contaminated water sources.[116] As recommended by UN Environment,[117] HYPREP is working to provide Ogoni communities with adequate access to potable water in line with World Health Organization (WHO) standards, after an initial assessment revealed obsolete/abandoned water infrastructure; water quality

106 UNEP (2011), 32.
107 HYPREP (2018).
108 Ibid, 39.
109 UNEP (2011), 16.
110 HYPREP (2018), 36
111 Ibid, 30.
112 Goodwin (2017).
113 UNEP (2011), 13, 17.
114 Ibid, 14.
115 HYPREP (2018), 19.
116 UNEP (2011), 215.
117 Ibid, 215.

below WHO standards; and difficult access to potable water 'as distance to reticulation points in some cases is more than a mile'.[118]

Furthermore, HYPREP and the Nigerian government have focused keenly on the *economic* restoration of Ogoni people. For example, among the bodies engaged by HYPREP to carry out the Health Outreach programme is the Ogoni Doctors Forum.[119] Technical experts from several Ogoni communities are also being trained for the clean-up and restoration exercise. Furthermore, there are concrete plans by HYPREP to develop a robust training and 'enterprise development programme [...] with the core participation of the community'; already, a Memorandum of Understanding has been signed by the Nigerian Minister for Environment with the UN Institute for Training and Research to develop a programme for training many women and youths, amongst others, 'in various skill sets to boost the local economy of Ogoniland'.[120] Additionally, the Office of the Nigerian Vice President is developing a programme for the establishment of Modular Refineries across the Niger delta region, which will also benefit the Ogoni economy.[121] These steps are helping to actualise UN Environment's recommendation to ensure that Ogoni people are involved in, and benefit socio-economically from, the eco-restoration activities.[122]

All these elements of community restoration ultimately help to advance public participation in the restoration project itself, secure the necessary community acceptance and support for the project, and substantially deliver to them the socio-economic justice that they have for so long been largely denied by the 'broken' state-based restoration governance and pollution liability system of Nigeria.

## 9.5 Conclusion

This chapter has critically appraised the value of the stakeholder-driven governance model as an ingredient for successful eco-restoration projects, especially against the backdrop of, and as a better alternative to, the traditional state agency-driven approach to eco-restoration. The discussion broadly revealed the inadequacies and challenges of the state-based approaches to eco-restoration in Nigeria, which has for long prevented those communities most affected by oil pollution from accessing adequate ecological, social and economic justice and remedies. It is in this light that the idea of the stakeholder-based approach was also considered, against its theoretical rationale in the scholarly literature, as having the strength to overcome the limitations of the state-based system that has catered mainly to the very industry that generated the environmental degradation in the first place.

Beyond shattering the orthodoxy of the state-based approach, and pointing to the virtues of a stakeholder-driven system as the future of eco-restoration for any

118 HYPREP (2018), 18.
119 Ibid, 20.
120 Ibid, 23.
121 Ibid, 30.
122 UNEP (2011), 229.

effective regulatory system, this case study is of value internationally in highlighting that eco-restoration is not just about recovery of ecological systems and their biodiversity, but also about restoring the affected human society and socio-economic livelihoods. While the core focus of the UN Environment restoration strategy was on the clean-up and remediation of the damaged ecosystem of Ogoniland, its recommendations, which are currently being implemented by HYPREP, demonstrate that holistic eco-restoration includes – and requires for its success – the restoration of the people of the affected community, given how severely local people have been impacted by the scourge of oil pollution for many decades in the Niger Delta.

Thus, the need to prioritise socio-economic justice and benefits for the affected Ogoni communities – largely denied them under the state agency-administered system – was made an integral part of the eco-restoration project through stakeholder involvement in the design of the project. This strategy has substantially helped to boost the social acceptability and support of the project by the affected communities. To be sure, the establishment of this relationship between ecological restoration and community restoration is vital from two perspectives: 1) the community members constitute such an integral part of the ecosystem that the restoration of the latter can hardly be deemed complete without the restoration of the former; and 2) community members are sometimes pushed by socio-economic hardship into ecologically harmful practices as a means of survival, such that if this hardship is not being progressively addressed, any success achieved in restoring the ecosystem may be considered temporary at best.[123]

Indeed, the findings in this chapter can serve well as the basis for reforms to the current legal regimes and institutional practices relevant to the restoration of damaged ecosystems across Nigeria and other societies with similar ecological challenges.

## References

### *Primary sources (law materials)*

*Adediran v Interland Transport* (1991) 9 NWLR (pt. 214) 155.
*Adesanya v President of the Federal Republic of Nigeria* (1981) 1 All NLR 1.
*Aro v Lagos Island Local Government Council* (2000) FWLR (pt. 13) 2132
*Atunbi v Shell BP* (Unreported) Suit No. UHC/48/78, 25 November, 1974.
*Busari v Oseni* (1992) 4 NWLR (pt. 237) 557
*Fairchild v Glenhaven Funeral Service Ltd* [2002] UKHL 22.
*Gulf Oil Company (Nig) Ltd v Oluba* (2003) FWLR (pt. 145) 712.
*Mobile Producing (Nig) Unlimited v LASEPA, FEPA & Others* (2002) 18 NWLR (pt. 789) 1.
National Oil Spill Detection and Response Agency (Establishment) Act of 2006, Federal Republic of Nigeria Official Gazette No. 72 Vol. 93, 29 December 2006.
*Natural Resources Defence Council Inc. v Train*, 510 F. 2d 692 (D.C. Cir. 1974)

123 Etemire and Muzan (2017), 294–295

Nigerian National Petroleum Corporation Act, CAP N123, Laws of the Federation of Nigeria, 2004.

*Oronto Douglas v Shell Petroleum Development Company Nigeria Limited and Others*, Unreported Suit No: FHC/L/CS/573/96, 17 February 1997.

Petroleum (Drilling and Production) Regulations 1969 L.N. 69 of 1969.

Petroleum Act Cap P10, Vol. 12, Laws of the Federation of Nigeria, 2010 .

*Pymatuning Water Shed Citizens for a Hygienic Environment v Eaton*, 644 F. 2d 995 (3rd Cir. 1981).

Rio Declaration on Environment and Development, Adopted by the UN Conference on Environment and Development (UNCED), 3–14 June 1992; (1992) 31 ILM 874

*Shell v Graham Otoko* (1990) 6 NWLR (pt. 159) 693.

*Social and Economic Rights Action Centre (SERAC) and another v Nigeria* (2001) AHRLR 60.

## Secondary sources

Francis Adeola (2001) 'Environmental Injustice and Human Rights Abuse: The States, MNCs, and Repression of Minority Groups in the World System' 8(1) *Human Ecology Review* 39.

Ali Adoyi (2016) 'MOSOP Calls for End of Hostilities in Ogoni Communities', *Daily Post*, 21 December, http://dailypost.ng/2016/12/21/mosop-calls-end-hostilities-ogoni-communities.

Lawrence A. Afinotan and Victor Ojakorotu (2009) 'The Niger Delta Crisis: Issues, Challenges and Prospects' 3(5) *African Journal of Political Science and International Relations* 194.

Saheed Alabi (2012) 'Country Report: Nigeria Recent Developments in the Niger Delta of Nigeria' 1 *IUCN Academy of Environmental Law e-Journal* 167.

Amnesty International (2009) *Nigeria: Petroleum, Pollution and Poverty in the Niger Delta*, Amnesty International Publications.

Amnesty International (2013) *Bad Information: Oil Spill Investigations in the Niger Delta*, Amnesty International Publications.

Amnesty International (2014) *No Progress: An Evaluation of the Implementation of UNEP's Environmental Assessment of Ogoniland, Three Years On*, Amnesty International Publications.

M. Kat Anderson (2005) *Tendering the Wild: Native American Knowledge and the Management of California's Natural Resources*, University of California Press.

Ian Ayres and John Braithwaite (1992) *Responsive Regulation*, Oxford University Press.

Peter Bachrach (1975) 'Interests, Participation, and Democratic Theory', in Roland J. Pennock and John W. Chapman (eds) *Participation in Politics*, Lieber-Atherton.

Gyula Bándi (2014) 'Introduction into the Concept of Environmental Democracy' in G. Bándi (ed.) *Environmental Democracy and Law: Public Participation in Europe*, Europa Law Publishing.

Barry Barton (2002) 'Underlying Concepts and Theoretical Issues in Public Participation in Resource Development' in Donald N. Zillman, Alastair R. Lucas and George (Rock) Pring (eds) *Human Rights in National Resource Development: Public Participation in the Sustainable Development of Mining and Energy Resource*, Oxford University Press.

John Bates, William Birtles and Charles Pugh (2004) *Liability for Environmental Harm*, Lexis Nexis.

Frank Bealey (1996) 'Democratic Elitism and the Autonomy of Elites' 17(3) *International Political Science Review* 319.

Hubert Bocken (1997) 'The Compensation of Ecological Damages' in Peter Wetterstein (ed.) *Harm to the Environment: The Right to Compensation and the Assessment of Damages*, Clarendon Press.

Dan Bodansky (2007) 'Legitimacy' in Dan Bodansky, Jutter Brunnee and Ellen Hey (eds) *International Environmental Law*, Oxford University Press.

Neil Burrows, Andrew Burbidge and Phil Fuller (2004) 'Integrating Indigenous Knowledge of Wildland Fire and Western Technology to Conserve Biodiversity in an Australian Desert', www.millenniumassessment.org/documents/bridging/papers/burrows.neil.pdf.

Paul Burton (2009) 'Conceptual, Theoretical and Practical Issues in Measuring the Benefits of Public Participation' 15(3) *Evaluation* 263.

CIA (2013) *The World Fact Book: Nigeria*, Central Intelligence Agency, www.cia.gov/library/publications/the-world-factbook/geos/ni.html.

Barry Checkoway and Jon V. Til (1978) 'What Do We Know about Citizen Participation? A Selective Review of Research' in Stuart Langton (ed.) *Citizen Participation in America*, Lexington Books.

Andre Clewell and James Aronson (2007) *Ecological Restoration: Principles, Values and Structure of an Emerging Profession*, Society for Ecological Restoration International, Island Press.

John S. Dryzek (2000) *Deliberative Democracy and Beyond: Liberals, Critics, Contestations*, Oxford University Press.

Jonas Ebbesson (1997) 'The Notion of Public Participation in International Environmental Law' 8 *Yearbook of International Environmental Law* 51.

Kaniye Ebeku (2003) 'The Right to a Satisfactory Environment and the African Commission' 3 *African Human Rights Law Journal* 149.

Augustine Ehikioya (2016) 'We will Restore Ogoni Ecosystem, Says Buhari', *The Nation*, 5 August, http://thenationonlineng.net/will-restore-ogoni-ecosystem-says-buhari/.

Ike Ekweremadu (2012) 'Opening Address', International Conference on Law Reform and Law-making Process in Nigeria, Abuja, Nigeria, 16 July.

Uzuazo Etemire (2014) 'The Status of Indigenous Peoples under the Convention on Biological Diversity Regime: The Right to Biological Resources and the Protection of Traditional Knowledge' 3 *Environmental Law and Practice Review* 1.

Uzuazo Etemire (2015) *Law and Practice on Public Participation in Environmental Matters: The Nigerian Example in Transnational Comparative Perspective*, Routledge.

Uzuazo Etemire and Menes A. Muzan (2017) 'Governance and Regulatory Strategies beyond the State: Stakeholder Participation and the Ecological Restoration of Ogoniland' 26(2) *Griffith Law Review* 275.

Patrick Eyinla and Joseph Ukpo (2006) *Nigeria: The Travesty of Oil and Gas Wealth*, Catholic Secretariat of Nigeria.

Olanrewaju Fagbohun (2012) *Mournful Remedies, Endless Conflicts and Inconsistencies in Nigeria's Quest for Environmental Governance: Rethinking the Legal Possibilities for Sustainability*, Nigerian Institute of Advanced Legal Studies.

Financial Watch (2016) 'Buhari Inaugurates Board of Trustees for Implementation of the UNEP Report on Ogoniland', 4 August, www.financialwatchngr.com/2016/08/04.

Daniel J. Fiorino (1990) 'Citizen Participation and Environmental Risk: A Survey of Institutional Mechanisms' 15(2) *Science, Technology and Human Values* 226.

Jedrzej Frynas (2000) *Oil in Nigeria: Conflict and Litigation between Oil Companies and Village Communities*, Lit Verlag.
Ernest Gellhorn (1974) 'Standing Participation and Who Will Pay?' 26 *Administrative Law Review* 423.
Ann Goodwin (2017) 'At Ogoni Day, MOSOP Cautions Indigenes Against Bunkering', *The Guardian*, 5 January.
Jack Hamann (1997) 'Wolves' Return to Yellowstone Sparks Controversy', *CNN News*, 12 November, http://edition.cnn.com/EARTH/9711/12/yellowstone.wolves/.
David Hayes (2002) 'Land Conservation and Restoration: Moving to the Landscape Level' 21(2) *Virginia Environmental Law Journal* 115
David Held (2006) *Models of Democracy* 3rd edn, Polity Press.
Samuel P. Huntington (1975) 'The Democratic Distemper' 41 *Public Interest* 9.
HYPREP (2018) *Hydrocarbon Pollution Remediation Project Activities Yearbook 2017*, Hydrocarbon Pollution Remediation Project.
Amos Idowu (1999) 'Human Rights, Environmental Degradation and Oil Multi-National Companies in Nigeria: The Ogoniland Episode' 17(2) *Netherlands Quarterly of Human Rights* 161.
Tunde Imoobe and Iroro Tanshi (2009) 'Ecological Restoration of Oil Spill Sites in the Niger Delta, Nigeria' 11(2) *Journal of Sustainable Development in Africa* 54.
Alexander Juras (2012) 'The Bali Guidelines for Development of National Legislation on Access to Information, Public Participation and Access to Justice in Environmental matters', First Meeting of Focal Points appointed by Governments of the Signatory Countries of the Declaration on the Implementation of Principle 10, Santiago de Chile, 6–7 November.
Barisere Konne (2014) 'Inadequate Monitoring and Enforcement in the Nigerian Oil Industry: The Case of Shell and Ogoniland' 47 *Cornell International Law Journal* 181.
Mary Kweit and Robert Kweit (1981) *Implementing Citizen Participation in a Bureaucratic Society: A Contingency Approach*, Praeger.
Charles Larmore (1999) 'The Moral Basis of Political Liberalism' 96(12) *Journal of Philosophy* 599.
Maria Lee and Carolyn Abbot (2003) 'The Usual Suspects? Public Participation under the Aarhus Convention' 66 *Modern Law Review* 80.
Michael Levine and Jennifer Forrence (1990) 'Regulatory Capture, Public Interest, and the Public Agenda: Towards a Synthesis' 6 *Journal of Law, Economics and Organisation* 167.
Michael Mason (1999) *Environmental Democracy*, Palgrave Macmillan.
Justice C. Nwobike (2005) 'The African Commission on Human and Peoples' Rights and the Demystification of Second and Third Generation Rights under the African Charter: *Social and Economic Rights Action Centre (SERAC) and the Centre for Economic and Social Rights (CESR)* v. *Nigeria*' 1(2) *African Journal of Legal Studies* 129.
Tunde Ogowewo (2000) 'Wrecking the Law: How Article III of the Constitution of the United States Led to the Discovery of a Law of Standing to Sue in Nigeria' 26 *Brooklyn Journal of International Law* 527.
Eghosa Osaghae (1998) *Crippled Giant: Nigeria Since Independence*, Hurst and Co.
Carole Pateman (1970) *Participation and Democratic Theory*, Cambridge University Press.
Thomas Princen and Mattias Finger (1994) *Environmental NGOS in World Politics*, Routledge.
Ann Richardson (1983) *Participation*, Routledge & Kegan Paul.

Benjamin J. Richardson (2016) 'The Emerging Age of Ecological Restoration Law' 25(3) *Review of European Community and International Environmental Law* 1.

Benjamin J. Richardson (2017) 'Restoring Layered Geographies: Ecology, Society and Time' 26(2) *Griffith Law Review* 172.

Benjamin J. Richardson and Ted Lefroy (2016) 'Resourcing Ecological Restoration: The Legal Context for Commercial Activities' 24(5) *Restoration Ecology* 686.

Benjamin J. Richardson and Jona Razzaque (2006) 'Public Participation in Environmental Decision-Making' in Benjamin J. Richardson and Stepan Wood (eds) *Environmental Law for Sustainability*, Hart Publishing, 165.

Akpofure Rim-Rukeh (2015) 'Oil Spill Management in Nigeria: SWOT Analysis of the Joint Investigation Visit (JIV) Process' 6 *Journal of Environmental Protection* 259.

N.M. Rosenbaum (1978) 'Citizen Participation and Democratic Theory' in Stuart Langton (ed.) *Citizen Participation in America*, Lexington Books.

Secretariat of the Convention on Biological Diversity (2014) *'Global Biodiversity Outlook 4: A Mid-term Assessment of Progress Towards the Implementation of the Strategic Plan for Biodiversity 2011–2020'*.

SER (2004) *The SER International Primer on Ecological Restoration*, Society for Ecological Restoration (SER) International Science and Policy Working Group.

Shell Nigeria (n.d.) 'Frequently Asked Questions on the UNEP Environmental Assessment of Ogoniland', www.shell.com.ng/sustainability/environment/unep-environmental-assessmen-of-ogoniland/unep-faq.html.

Nancy Spyke (1999) 'Public Participation in Environmental Decision-Making at the New Millennium: Structuring New Spheres of Public Influence' 26 *Boston College Environmental Affairs Law Review* 263.

Katharine Suding et al (2015) 'Conservation: Committing to Ecological Restoration' 348 (6235) *Science* 638.

Anastasia Telesetsky (2013) 'Ecoscapes: The Future of Place-based Ecological Restoration Laws' 14 *Vermont Journal of Environmental Law* 494.

Anastasia Telesetsky, An Cliquet and Afshin Akhtar-Khavari (2017) *Ecological Restoration and International Environmental Law*, Routledge.

Niki Tobi (1998) 'Environmental Litigation' in S. Simpson and O. Fagbohun (eds) *Environmental Law and Policy*, Faculty of Law, Lagos State University.

UNDESA (2009) *State of the World's Indigenous People*, United Nations Department of Economic and Social Affairs.

UNDP (2006) *Niger Delta Human Development Report*, United Nations Development Programme.

UNEP (2010) *UNEP/GCSS.XI/11, Decision SS. XI/5, Part A*, 26 February, United Nations Environment Programme Governing Council.

UNEP (2011) *Environmental Assessment of Ogoniland*, United Nations Environment Programme.

UNEP (2016) 'Nigeria Launches $1 Billion Ogoniland Clean-up and Restoration Programme', 2 June, United Nations Environment Programme, www.unep.org/newscentre/default.aspx?DocumentID=27076&ArticleID=36199.

United Nations (1992) *Agenda 21*, 31 ILM 874.

United Nations (2010) *Guidelines for the Development of National Legislation on Information, Public Participation and Access to Justice in Environmental Matters, Adopted by the Governing Council of UNEP*, UNEP/GCSS.XI/11, Decision SS. XI/5, Part A, 26 February.

Adam Vaughan (2014) 'Rewilding Britain: Bringing Wolves, Bears and Beavers Back to the Land', *The Guardian*, 19 September.

Norman Wengert (1976) 'Citizen Participation: Practice in Search of a Theory' 16 *Natural Resources Journal* 23.

Peter Willetts (1982) 'The Impact of Promotional Pressure Groups on Global Politics', in Peter Willets (ed.) *Pressure Groups in the Global System*, Pinter.

World Bank (1995) *Defining an Environmental Development Strategy for the Niger Delta*, Vol. II, Industry and Energy Operations Division, West Central Africa Department, World Bank.

# 10 Motivating ecological restoration by private landowners through special purpose districts

*Anastasia Telesetsky*

## 10.1 Introduction

A key concept for the success of ecological restoration (hereafter 'eco-restoration') is adoption of a shared restoration ethos by a broad cross-section of society that leads to active intervention to achieve restoration outcomes.[1] Without the engagement of private actors and particularly private landowners in many jurisdictions, large-scale ecological restoration is not viable. While some national governments have made restoration commitments at the scale of large landscapes, such as Rwanda in its Bonn Challenge commitment,[2] many states are limited in their ability to assert jurisdiction over large landscape areas because of the presence of strong private property regimes. When governments are politically reluctant to mandate corrective action by private landowners on lands that are ecologically degraded, the potential for eco-restoration depends entirely on the motivation of private landowners who unilaterally control large areas of a state.[3]

This chapter has two objectives. First, it will propose a typology of private actors who might be engaged in active restoration on private lands, and explore some of the motivations for engaging in restoration that is not legally mandated. As this chapter illustrates through a variety of examples, certain private landowners, including Indigenous peoples, non-governmental organizations (NGOs), and

1 Ecological restoration is understood in this chapter as active restoration interventions designed to assist recovery of ecological functions, structures, and systems that have been historically associated with a specific ecosystem. Eco-restoration does not require a system to be restored to an 'original' ecosystem, which may not be possible where an ecosystem is so altered that it retains few or none of its original characteristics.

2 Bonn Challenge, Rwanda (2011) www.bonnchallenge.org/content/rwanda (committing to 2 million hectares of restoration through increased focus on a combination of agroforestry and natural forest regeneration); see also Kigali Declaration of Forest Landscape Restoration (1 September 2016), www.bonnchallenge.org/blog/kigali-declaration-forest-landscape-restoration-africa (including public forest restoration commitments from government officials in 13 African states).

3 In the United States, the federal and state governments own about 807 million acres of the 2.27 billion acres of land. Native American tribal groups own about 45 million acres. This is 38 percent of the U.S. land mass, leaving approximately 62 percent in private ownership. Public Land Ownership by State (1995) www.nrcm.org/documents/publiclandownership.pdf (municipal lands are not included in these calculations).

corporations, may undertake eco-restoration activities on their land for a variety of reasons ranging from moral responsibility to the land to reputational concerns. There is, however, a much larger group of private landowners – the 'residual landowners' – who are not presently engaged in eco-restoration on their lands for a spectrum of reasons. For some individuals, there is a lack of awareness of the types of problems that eco-restoration is attempting to address, such as the fragmentation of ecosystems. For other individuals with knowledge about eco-restoration, there may be a lack of financial resources or skills to undertake a potentially expensive and technically complex project that will require ongoing monitoring. For yet another group of individuals with both awareness and resources, there may be the pull of the status quo bias, resulting in inertia.[4] The combination of these reasons for inaction explain why the existence of private lands has been repeatedly identified as a 'major restoration challenge'.[5]

Given the difficulty in achieving broad-scale restoration at a landscape level when many private owners are presently unwilling or unable to make long-term investments in eco-restoration, what role could law play in motivating 'residual landowners' to engage in eco-restoration? Are there governance models that might increase landowner involvement, particularly among 'residual landowners'? This chapter proposes a closer look at the formation of 'special purpose districts', a form of independent government that is geographically based but separate from existing local government, as one possible structure for achieving landscape-level eco-restoration goals. Recognizing that there are many forms of specialized local governance frameworks that could be explored,[6] this chapter focuses on examples from the United States, with other precedents from Australia, to illustrate its argument,

## 10.2 Typology of private actors and ecological restoration

### *10.2.1 Overview*

There is a tremendous diversity of private actors involved in projects with an eco-restoration component, from small, informal community groups to major global corporations. In order to provide some framework to think about where there may be opportunities for improved governance through creating or extending special purpose districts, this section provides a typology of six groups of private actors, drawing on a variety of examples of privately supported restoration projects. Table 10.1, at the end of the section, summarizes and compares the motivations and likely efficacies of these six actors.

A couple of observations will help to frame the following sections in order to connect the typology of private actors to the material that follows proposing

4 Telesetsky (2017).
5 Sayles (2018), 67.
6 One model from France is the approximately 15,000 *syndicats intercommunaux á vocation unique* that operate in a similar manner to special purpose districts.

special purpose districts as an institutional tool for improving landscape-level eco-restoration outcomes. First, the description of most of the actors below highlights efforts of actors and groups who are actively attempting to undertake eco-restoration. For many if not most of these groups of actors, there will remain gaps between their long-term visions of eco-restoration and the means to achieve this restoration. Well-intentioned actors may not have the resources or skills to achieve successful outcomes in restoration. This chapter anticipates that groups or individuals that already have motivation to restore would be the most likely candidates to lead efforts to establish special purpose districts.

The second observation is that, even though the typology below seems to suggest that the group that requires the most policy attention is the 'residual landowners' because they represent the most potential for future investments in eco-restoration, involvement from these landowners is highly dependent on the other five groups fostering a normative culture of restoration within a specific landscape. Without this creation of a 'culture' or 'community of practice', disengaged landowners who have awareness of the need for eco-restoration and the resources to undertake such restoration will remain disengaged. Even with the creation of such a 'culture', they may still opt out of participating in a wider effort, or might even actively resist the creation of a community. Where the preconditions for special districts to further environmental stewardship and restoration are lacking, it is still possible that some positive initiatives may ensue from less elaborate regimes, such as voluntary land-care groups.

### *10.2.2 Group one: 'big personas' – committed individuals*

The first discernible group of actors is the collection of influential individuals who engage in active restoration projects, typically on their own property, or are financially very generous in supporting the restoration efforts of others. These philanthropic individuals have been categorized as 'eco-barons'.[7] While it is not possible to ascribe a single motivation to explain why they participate in or underwrite restoration, public comments by such individuals in newspapers, essays, and websites point to an ethic of responsibility. Because land tenure is secure and financial or labour resources are less limited than for other groups, many of these individuals are willing and able to experiment with a variety of restoration approaches.

To illustrate the types of individuals and groups that might occupy this 'active restoration' space, the narratives of Aldo Leopold, M.C. Davis, Ted Turner, and Wenliang Wang are informative. While each of these big personas works in a different region and focuses on different objectives in their eco-restoration projects, these individuals all share the value of investing in the long-term ecological health of a region through active restoration. Even though each of the personas described below came to become involved in restoration for distinct personal reasons,

7 Edward Hume (2009).

most of them are directly involved in the planning for a restoration process or outcome.

Starting with the narrative of Aldo Leopold, who was professor of wildlife management at the University of Wisconsin, demonstrates that 'big' impact does not depend on having access to financial resources. Leopold is characterized as a big persona because of his lifelong commitment to understanding the ecological legacy that a small group of like-minded individuals might be able to restore over a lifetime. In Baraboo, Wisconsin, Leopold bought a degraded farm along the Wisconsin River in 1935 for $8 an acre. With the tireless assistance of his family and friends, Leopold replanted the land with 40,000 conifers and hardwoods as well as re-establishing native prairie.[8] For Leopold, the process of ecological theory and experimentation on the Baraboo property offered a subject of great personal and professional interest while bringing his family together on a shared project.

M.C. Davis and Ted Turner represent different big personas as major individual landholders with substantial financial resources to back their visions of restoration. Based in Freeport, Mississippi, multi-millionaire M.C. Davis uses his revenue from trading in oil and timber to restore the Piney Woods, a longleaf pine forest ecosystem that has 97 percent disappeared from the American Southeast.[9] He has been investing about US$1 million a year in replanting 51,000 acres that he purchased specifically for habitat restoration in hopes that black bears would eventually re-inhabit the area. Inspired by a community presentation that he heard on the disappearance of bears in the Southeast, Davis' large-scale habitat restoration project is motivated in part by a strong personal desire connected to a single species.

Ted Turner, the second largest private landowner in the United States with 2 million acres of land including 18 operating ranches, has committed himself to reviving a number of endangered native species across his landholdings. While the ranches are intended to be economically self-sustaining, Turner also expects them to operate in an 'ecologically sensitive manner while promoting the conservation of native species'.[10] To promote conservation of native species, Turner has invested in restoration efforts including reintroduction of species and restoration of key habitat through the Turner Endangered Species Fund and the Biodiversity Division of Turner Enterprises. Turner has been willing to experiment with restoration projects such as restoring a viable population of Aplomado Falcon to the Chihuahuan grasslands on one of his New Mexican ranches that would count toward the federal recovery of the species under the Endangered Species Act.[11] While the Aplomado Falcon did not recover as hoped for, the five-year effort reflected a personal commitment from Turner and his Endangered Species Fund team.

8 Aldo Leopold Foundation Staff, interview (June 2014).
9 Hiss (2014).
10 Turner Ranches, Turner Enterprises, www.tedturner.com/turner-ranches.
11 Turner Endangered Species Fund, 'Aplomado Falco', http://tesf.org/project/aplomado-falcon.; Endangered Species Act 16 U.S.C. §§ 1541–1544.

Turner and his team have successfully restored 250 bighorn sheep that became the largest desert population of this species in New Mexico and the largest population on private land in the country.[12] Ongoing projects continue for black-footed ferrets, bolson tortoises, Chiricahua leopard frogs, prairie dogs, and numerous other threatened species.

Notably, Turner sees himself as a leader in private restoration efforts. He now uses his personal influence, social capital, and the experts that he employs to influence other big personas to follow his example. Because over 60 percent of the United States is privately owned and 80 percent of the country's endangered and threatened species exist on land that is partially or solely in private hands,[13] Turner is seeking action from other large landowners. To this end, he is a major contributor to an organization called the Western Landowners Alliance. As the Turner Endangered Species Fund website observes, '[i]t is now incumbent on us to consider new collaborations that can increase the number of private landowners motivated by an approach to land management that includes a focus on imperilled species'.[14] In July 2011, landowners representing 8 million acres of land owned or leased in nine western states and Alberta met to discuss landscape-level conservation and restoration.[15] While the Founders' Group of the Western Landowners Alliance was composed primarily of landowners with large landholdings who were motivated to conserve a particular Western 'big land' rural lifestyle,[16] this movement led by big personas has evolved to encourage participation from all landowners in landscape-level conservation and restoration.

The fourth big persona differs from the other individuals because his investment in restoration extends beyond lands that he owns or uses. Wenliang Wang, a businessman whose company manages Dandong Port in China, has financially underwritten a variety of restoration projects. For example, in 2015, Wang was recognized for his ongoing $2 million-a-year investment to restore and maintain 50,000 acres of wetlands owned by his company that form part of the Dandong Yalu River Estuary Wetland National Nature Reserve in China.[17] In 2016 he committed US$5 million to restore dying mangroves in the Rookery Bay National Estuarine Research Reserve in Florida, which he had never seen. Allegedly, he was willing to invest in the project because the organization backing the restoration efforts already had a plan and permits for the work.[18] Also notably, Wang is

12 Turner Endangered Species Fund, 'Desert Bighorn Sheep', http://tesf.org/project/desert-bighorn-sheep.
13 Turner Endangered Species Fund, 'Western Landowners Alliance', http://tesf.org/project/western-landowners-alliance-wla.
14 Ibid.
15 Western Landowners Alliance, 'About', www.westernlandownersalliance.org/about; Western Landowners Alliance, 'Working Lands Health', www.westernlandownersalliance.org/programs/water (typical landscape restoration efforts by members have focused on riparian restoration and wetland restoration).
16 Western Landowners Alliance, 'About', www.westernlandownersalliance.org/about.
17 Cendrowski (2015).
18 Staats (2016).

investing in what some ecologists deem a novel technique of 'ecological mangrove restoration' that changes the hydrology of the area.[19]

What unifies the approach of these private actors is that generally few parties are involved in the decision-making, and the individuals exercise considerable autonomy in their decision-making that influences the type of restoration work. As these individuals are driven by their own personal passion to engage in or support eco-restoration efforts, they would be likely candidates for leading the formation of a special purpose restoration district, discussed in section 10.3. This group of committed individuals is likely to play a major role in generating early momentum in organizing a district by, for example, initially funding a landscape-level plan or blueprint for restoration efforts or funding the drafting of legislation to create a district.

### *10.2.3 Group two: Indigenous peoples*

Across the globe, numerous Indigenous communities are investing in landscape restoration on lands that they own or have a cultural interest in preserving. For some communities, restoration signals an assertion of Indigenous sovereignty over lands traditionally relied upon. In addition, restoration activities might be considered an investment in the social future of a community by bringing together its members, who may have been fragmented by forces of modernity, and reviving their culturally significant resources (e.g. grasses for weaving or herbs for medicine).

There are numerous examples of Indigenous-led eco-restoration, but two stories highlight the underlying motivations behind large-scale restoration projects led by Indigenous peoples. First, in Australia, where 31 percent of the land is under Aboriginal control or use rights,[20] eco-restoration undertaken by Aboriginal communities strengthens their culture because without 'healthy country' such people lose an essential part of their identity.[21] In 1995, the Makmak clan received title back to their land from the government, but the land had become severely degraded due to weeds, leading to disruptions in floodplain flows and natural fire processes. Over the following years, the clan, through an extensive program of fire, herbicides, ground control, and biological control, ecologically restored the floodplain so that cultural traditions could be resumed, including access to subsistence foods. Aboriginal commentators emphasized that '[t]he restoration happening in these threatened areas has the potential to offer guidelines for new forms of restoration that seek to protect the integrity of the relationships between people and place, and to protect sacred places'.[22] This example

19 Ibid.

20 Richardson and Lefroy (2016), 670.

21 United Nations Declaration on the Rights of Indigenous People, UNGA 61/295 (2008) article 25:

> (Indigenous peoples have the right to maintain and strengthen their distinctive spiritual relationship with their traditionally owned or otherwise occupied and used lands, territories, waters and coastal seas and other resources and to uphold their responsibilities to future generations in this regard).

22 Daiyi and Ford (2002).

suggests that at the community level there exists an intrinsic motivation to restore Indigenous lands so as to revive the cultural identity of their traditional owners. This corroborates the 'sense of place' literature that shows how people's emotional and cultural bonds to specific landscapes can be crucial for how such places are environmentally managed.[23]

The connection between people and place is equally critical for other Indigenous peoples even where they do not have recognized land tenure. The native Hawaiian people are volunteering their time and investing resources to restore Kaho'olawe island in the Hawaiian archipelago. The island that was formerly used for a penal colony, sheep and cattle ranches, and munition testing, was conveyed by the U.S. government to the State of Hawaii in 2003. The land has not yet been returned to the Native Hawaiians, but there is an expectation that the land will be managed by a Native Hawaiian entity 'to preserve and practice all rights traditionally exercised for cultural, spiritual, and subsistence purposes'.[24] In the meantime, because of the cultural and ecological significance of the island to them, some Native Hawaiians have volunteered thousands of hours on watershed restoration and revegetation. The act of eco-restoration is a celebration of native Hawaiian culture so that the process of restoration is as significant as its outcomes. As the island's Reserve Commission observes:

> The healing of Kanaloa (Kaho'olawe) is a physical and spiritual renewal that is deeply rooted in the revival of cultural practices, traditions, and rituals. As a result, activities conducted by the Kaho'olawe Island Reserve Commission Restoration Program are guided by the need to achieve a more holistic understanding between man and nature and to place strong emphasis on healing as well as environmental restoration.[25]

Like the big personas of restoration, the Indigenous groups rely on having some autonomy in decision-making. The ability to exercise this autonomy, however, may be complicated by the status of land tenure. In most cases there will be little need for external programs to catalyse restoration projects, because most of these groups who are undertaking restoration have some level of intrinsic motivation based on stewardship and recovering cultural resources that form a group's identity. The financial and resource capacity to undertake programs, however, may be limited and these groups may rely on external support to cover costs that exceed their capacity. Indigenous groups who live on lands that are surrounded by or interspersed with private lands may assume leadership roles within their communities to seek broader restoration efforts, particularly where there is a desire to restore a historical range for a species culturally important to the group. For example, in 2014, a number of Native American tribes located in Montana

23 Masterson et al (2017).
24 Kaho'olawe Island Reserve Commission, 'Core Programs' http://kahoolawe.hawaii.gov/coreprograms.shtml#restoration.
25 Ibid.

entered a treaty with First Nations people of Canada to cooperate on transboundary restoration efforts for wild bison, a sacred animal for all of the treaty parties.[26] The likelihood of Indigenous leadership on the formation of a restoration district will depend largely on the relationship between Indigenous peoples and others living on a given landscape. Where there are pre-existing relationships of trust, Indigenous leaders may play a crucial role in the formation of special purpose districts. Where there are pre-existing tensions between Indigenous people and other community members, however, Indigenous groups may not want to use their political capital to foster broader participation.

### *10.2.4 Group three: corporate entities*

In some countries there is increasing corporate-led and corporate-funded restoration, as discussed by Froukje Platjouw in Chapter 7 of this volume. Some of these corporate efforts are in response to functional needs. For example, landscape rehabilitation that may or may not deliver a return of key ecological values may be legally required for certain corporate sectors such as surface mining industry,[27] and landscape restoration may be mandated by common law or statute for unlawful environmental damage.

Other projects may include more of a corporation's own vision of itself as a social actor. These types of non-mandated projects might be focused on the promotion of corporate social responsibility or the protection of a company's social charter to operate within certain communities. For certain types of companies, landscape-level restoration may also make business sense if restoration efforts can contribute to ecosystem services upon which the company depends, such as in the forestry and farming sectors.[28]

For example, CEMEX, a Mexican cement company operating in the borderlands between the United States and Mexico, is investing in environmental restoration for reasons of corporate social responsibility and social licence. CEMEX has a created an office called 'CEMEX Nature' that serves to promote 'culture of appreciation and respect for nature, not only among our own stakeholders, but also within the global community'.[29] At the 13th Conference of the Parties to the Convention on Biological Diversity in 2016, CEMEX managers pledged to seek concrete actions to solve biodiversity challenges as part of the Cancun Business and Biodiversity Pledge.[30] CEMEX operates a number of quarry rehabilitation projects, plus larger landscape-level restoration projects including the

26 Brown (2014).

27 See e.g. Surface Mining Control and Reclamation Act, 30 U.S.C. 1201–1328 (requiring mine operators that cause adverse effects on fish, wildlife, and environmental values to achieve enhancement of these resources where possible).

28 For example, restoration of watersheds can make sense for a water purveying company who relies upon the natural filtering capacity of the land.

29 Cemex Nature, 'About', www.cemexnature.com.

30 Cancun Business and Biodiversity Pledge (December 2016), www.cbd.int/business/pledges/pledge.pdf.

15-year-old effort in the El Carmen Reserve, a private 140,000 hectare trans-boundary conservation area owned by CEMEX that protects a biological corridor inhabited by desert bighorn sheep.[31]

While eco-restoration is not a core competence of most businesses, a few smaller businesses fund restoration work. For example, the owners of the Kaboega Farm, a property of 6,550 hectares in South Africa, ceased all agricultural activity 20 years ago on their property. Recognizing that they lived in a unique location that included biologically unique temperate rainforest fragments, the former farm owners now pursue a business plan based on restoring the four different ecosystems on their land by attracting visitors to see their restoration work-in-progress.[32] Business plans based exclusively on restoring endangered landscapes are risky. For example, the Earth Sanctuaries company, created in 1988 to recover native Australian ecosystems through the formation of 11 different sanctuaries, ultimately went bankrupt even though the company had received many accolades.[33]

While decisions to invest in restoration seem to reflect both intrinsic and extrinsic motivation on the part of the leadership of companies, some corporate entities appear to be pursuing external recognition for their restoration efforts by seeking certification of their restoration projects. While corporations typically do not explain their motivations here, some reasons might include improved marketing for the company or internal accountability checks to ensure corporate funds are spent on meaningful projects. The non-profit Wildlife Habitat Council offers certification of habitat projects, species reintroduction projects, and education projects, as well as project guidance to help potential clients understand what types of programs are likely to be successfully certified.[34] While corporations can inject considerable financial capital into restoration projects, they typically lack the necessary technical skills and expertise to undertake environmental restoration, and rely on outside consultants. The business sector might thus contribute most effectively through collaborations with other stakeholders such as academic institutions or NGOs with such expertise.

Unlike the two groups discussed above, corporations as private landowners are unlikely to contribute core leadership to the formation of a special restoration district unless there is an external driver related in some fashion to improved business outcomes. For example, a business that is a major landowner in a region proposed as a restoration district might take a leadership role in order to be perceived as fulfilling its social contract within the broader community. Depending on the nature of the business' interest in its landholdings, some businesses will be

31 Cemex Nature, 'Meet El Carmen' (21 October 2015), www.cemexnature.com/el-carmen-p/el-carmen/

32 Reid (2016).

33 Richardson (2017), 11.

34 See e.g. Wildlife Habitat Council, 'Deserts Project Guidance', www.wildlifehc.org/wp-content/uploads/2015/10/WHC-Deserts-Project-Guidance_FINAL.pdf (indicating that an appropriate conservation and educational objective that might be eligible for certification would be 'Restoring ... a diverse mosaic of desert habitat that will provide for both plant and animal species').

reluctant or even actively opposed to the formation of special purpose restoration districts that may encourage the creation of new and potentially restrictive local land-use regulatory regimes.

### *10.2.5 Group four: non-governmental organizations*

Non-governmental organizations, particularly land trusts, make large contributions to restoration on private lands in some countries. These organizations are motivated to engage in restoration to meet stewardship obligations expected by their members and donors. Like the big personas, many land trusts have flexibility and autonomy in their restoration planning because they own the land. These trusts lack boards of directors or large donors who may influence priorities for restoration. Environmental NGOs, such as the U.S. Nature Conservancy and Bush Heritage Australia, devote increasing efforts to purchasing land for restoration and conservation as private sanctuaries, and to assist individual landowners to take similar steps, often in perpetuity through placement of conservation covenants on their properties.

In many cases, these groups will specifically acquire lands that will further landscape-level restoration goals. For example, after an initial acquisition of 525 acres from The Nature Conservancy, the Mountain Restoration Trust in the Santa Monica mountains of California acquired an additional 975 acres with the commitment of restoring and maintaining the lands as a reserve.[35] In other cases, NGOs will reach out to undertake restoration on private land that they do not own. For example, because three-quarters of the remaining wetlands in the United States are on private land, Ducks Unlimited has participated in private land restoration projects through the provision of expert advice on wetland restoration.[36] A third role that NGOs play in restoration on private lands involves coordination and facilitation of multi-stakeholder restoration networks. For purposes of typology, this group of 'private restoration actors' is described below as a separate group of 'blended' actors. NGOs who are focused as part of their organizational purpose on restoration will be key institutional players in facilitating possible restoration districts. In many of the blended restoration efforts described below, NGOs have been key in initiating the ongoing projects.

### *10.2.6 Group five: blended restoration efforts*

Because of the different skills, resources, property tenures, and constituencies involved, landscape-level restoration efforts are sometimes pursued by a confederation of actors under the auspices of a lead NGO or government agency.

35 Land Acquisition, 'Mountain Restoration Trust', www.mountainstrust.org/restoration/restoration.html.

36 Ducks Unlimited, 'Developing a Wetland Mitigation Project on your Land', www.ducks.org/conservation/du-wetland-mitigation/developing-a-wetland-mitigation-project-on-your-land.

These projects vary in their genesis and execution but tend to rely on some form of collaborative governance. The motivation for such projects is a combination of intrinsic and extrinsic drivers. The actors who initiate the network are often motivated to take action based on concerns for environmental integrity. For later actors who join the collaborative effort, there may also be intrinsic motivation based on increased knowledge about the nature of the restoration effort or an extrinsic motivation based on external pressure, real or perceived, to participate in the network.

While there has not been empirical work done to determine whether these collaborative efforts are more effective in delivering restoration outcomes than efforts by other groups in the typology of private restoration actors, these multi-stakeholder efforts should, in theory, have long-term impact because they change the relationships of multiple groups across a specific place. Out of a fragmented landscape, these types of projects have the potential to create a unitary vision. The Cape to City project on the North Island of New Zealand is illustrative of a 'blended' project, with stakeholders including the Hawke's Bay Regional Council, civil society groups, Maōri iwi, and individual landowners. The project is a response to ecologists' findings that 75 percent of indigenous vegetation has been cleared, 70 percent of forest remnants that are not presently in conservation status are on private lands, and only 2 percent of wetlands remain.[37] Focused on promoting native planting, habitat corridors, and pest eradication, the stakeholders have agreed to a long-term voluntary implementation of a biodiversity strategy that builds on existing synergies.[38] The parties intend to 'unite the past with the present to restore biodiversity in Hawke's Bay and set a precedent for the future in New Zealand'.[39]

These collaborations can take a variety of approaches to how stakeholders leading the initiatives leverage restoration efforts from other stakeholders. In most cases, leaders can be identified who drive both the facilitation process and the mechanisms to incentivize restoration. In the United States, one of the mechanisms for furthering restoration has been government-designed restoration partnerships, where the Fish and Wildlife Service provides both financial and technical assistance to private landowners undertaking certain types of habitat restoration projects on private lands. In return, landowners agree to enter contracts to maintain the project for at least ten years and contribute personal funding to the project.[40]

37 Hawke's Bay Biodiversity Strategy, November 2015, at 5, www.hbrc.govt.nz/assets/Document-Library/Strategies/biodivstratNovember2015v3.pdf

38 Ibid, 7, observing that while a lot of great work is currently under way to help restore Hawke's Bay's biodiversity, even more could be achieved if our efforts were better coordinated. We need to align and harmonise our biodiversity programs so these can be more efficient and effective.

39 Cape to City, 'Species Reintroduction', http://capetocity.co.nz/our-work/species-reintroduction.

40 United States Fish and Wildlife Service (2003).

In contrast to the Fish and Wildlife Service partnerships that are a government-led initiative, the Yellowstone to Yukon (Y2Y) Conservation Initiative is led by civil society groups. While the large landscape conservation and restoration initiative that spans 3,200 kilometres has partner commitments from conservation groups, local landowners, businesses, government agencies, Native Americans, First Nations, and scientists, the initiating civil society groups play a core role in asserting pressure on government to take specific actions supportive of restoration. In September 2016, the Y2Y project leaders and members of the Canadian Parks and Wilderness Society urged the Alberta government to implement a wildlands forest restoration project to restore degraded landscape and boost a 'restoration economy'.[41] A similar model to the Y2Y initiative is the Scotland's Trees for Life, which involves a single non-profit charity relying upon volunteers to work with public and private landowners to replant 1.5 million trees over 230,000 hectares.[42] These two models, the Partners for Fish and Wildlife and the Y2Y initiative/Scotland's Trees for Life, illustrate that blended governance is a malleable concept that takes on certain characteristics based on who is in leadership.

While there is great promise for blended governance based on collaboration among a variety of actors that inhabit and regulate the landscape, there are also many challenges in building the needed relationships between stakeholders to support restoration work. Some of these challenges involve under-representation by some stakeholders, competition among different collaborations, and risk aversion to making difficult decisions in light of a desire to maintain consensus among the community.[43] A critical challenge for increasing restoration of private lands through blended governance that needs to be addressed is the reality that many collaborations will fall apart over time, particularly where there is a lack of intermediate results.[44] How to maintain the momentum in restoration collaboration over the time that might be necessary to measure success is a key challenge.[45] The concept of a special purpose restoration district may provide one institutional solution to coordinating a diverse set of actors.

41 Yellowstone to Yukon Conservation Initiative, 'New Report Highlights Opportunity for Restoration in the Castle' (7 September 2016), https://y2y.net/news/media-releases/new-report-highlights-opportunity-for-restoration-in-the-castle.

42 Trees for Life, http://treesforlife.org.uk.

43 Margerum (2016), 27–53.

44 Ibid, 44.

45 Recognizing the need to maintain momentum for the long run is an important part of the early organizational stages of a restoration alliance. In the case of the Hawke's Bay Biodiversity Strategy, the partners agreed to a long-term commitment to invest in restoration. As the strategy document provides:

> The unifying goal flowing from the overarching vision is to ensure that important biodiversity habitats and populations of native species are enhanced, healthy and functioning by 2050 [...] 2050 might seem a long way off but we are realistic about the challenge. It has taken more than 200 years to create the biodiversity problems we have today, so it's going to take a while to make progress towards fixing them.
>
> (Hawke's Bay Regional Council 2015, 7)

### *10.2.7 Group six: residual private landowners*

The previous five groups in the typology have demonstrated some commitment to restoring lands under their control through their investments in a variety of eco-restoration efforts. Yet the efforts of these groups are atypical of private land-owners. The final category of this typology is the 'catch-all' group of 'ordinary' private landowners subject to government land-use planning and environmental regulation, who do not engage in restoration activities on their own lands.

There are three primary and often related reasons that this group of landowners may not actively participate in any restoration efforts.[46] First, landowners may lack knowledge about the need for eco-restoration on private lands, or lack understanding of how their land might function within a landscape restoration project. For some landowners, as long as their property is green, there is a presumption that the land must be healthy. Or other landowners, such as house owners, may assume that their landholding is too small to make any difference to a landscape-level restoration project, or that their individual property decisions will not have an impact beyond their own property boundaries.

Second, this residual group of landowners may have knowledge of the need for landscape restoration but not have adequate means to implement restoration. Some restoration projects, such as reconfiguring a stream bank or daylighting a stream, are technically complex and potentially expensive to execute. For individual landowners who lack confidence in their ability to sustain a restoration effort, there is unlikely to be any strong motivation for reaching out to neighboring landowners to coordinate landscape-level efforts for a greater community good.

With uncertain outcomes being inherent in many restoration projects due to the complexity of planning and implementing landscape-level restoration, the status quo bias explains why many members of the residual landowners are unlikely to be actively engaged in restoration.[47] Even when a landowner recognizes that there is a problem, there is a psychological tendency not to act if the problem does not appear to be getting worse. As long as the status quo is manageable, then many landowners are content with not making changes. Importantly, in light of the proposal discussed below, a status quo situation is not simply a matter of a particular landscape. Status quo can also refer to social relationships. If social relationships within a community were to begin to change, then individuals who have knowledge of problems, and ideas on how to solve those problems, may be catalysed into some level of action.

### *10.2.8 Summary*

Table 10.1 summarizes the various private actors with a prediction about their general motivation to restore landscapes and participate in institutional frameworks such as special purpose restoration districts.

46 Lindenmayer et al (2016); Rosenberg and Margerum (2008).
47 Telesetsky (2017)

*Table 10.1* Typology of private restoration actors

| | *Intrinsic or extrinsic motivation to restore* | *Level of control over restoration activities* | *Likely to lead efforts to create restoration special purpose districts?* |
|---|---|---|---|
| Group one: Eco-barons – philanthropists and highly committed individuals | Intrinsic | High level of control and autonomy | Yes |
| Group two: Indigenous peoples | Intrinsic | High level of control on lands owned by Indigenous group; lower levels of control on lands with culturally significant resources unless prior agreement | Yes |
| Group three: Corporate actors | Intrinsic (corporate integrity) and extrinsic (corporate social responsibility) | High level of control over corporate-owned lands | Maybe |
| Group four: Non-governmental organisations | Intrinsic (NGO mission) and extrinsic (donors) | High level of control over lands owned by NGO; low level of control on lands not owned by NGO unless private landowner is a party to a conservation contract limiting activities | Yes |
| Group five: Blended, multi-stakeholder groups | Intrinsic and extrinsic | High autonomy for private land owners unless private landowner is a party to a conservation contract | Maybe |
| Group six: Residual landowners | Unknown | High autonomy | Unknown |

The foregoing taxonomy and discussion suggests that there is a diversity of private actors who are likely to engage in eco-restoration on lands that they own or over which they have some contractually recognized interest. Overall, several of these groups play a significant role in recovering characteristics and aspects of landscapes that might otherwise continue to decline without active interventions. Yet, even with all of these commendable efforts by groups and individuals, most private landowners are part of the group of residual landowners who are unlikely to be engaged in any intentional efforts to further the project of landscape-level eco-restoration for a variety of reasons, including a lack of knowledge of the nature of the problem, lack of resources, or no compelling reason to disturb the status quo.

Reflecting on the need to engage more private landowners in landscape-level restoration activities, it seems we need more education, more financial resources, and more reasons to care about restoration activities. Section 10.3 proposes a potential governance intervention in the form of special purpose districts that may motivate more participation from residual landowners as a grassroots, landowner-led model.

## 10.3 Restoration special purpose districts

This second half of the chapter explores how special districts as a place-based governance intervention might increase involvement in landscape restoration by all private landowners connected to a specific area. In particular, this section explores whether it is possible to extend the mandate of existing conservation districts or pilot new regional institutions in the form of restoration special purpose districts in order to create new bodies of collective property norms across private lands. The following sub-sections will introduce the concept of a special district as a governing institution, explain why this type of governance might be appropriate to achieve restoration objectives, and conclude with some of the challenges in implementing this idea. The following discussion recognizes that the proposals for special districts that allow for private ordering may or may not be capable of being created by legislation, depending on the political structures of a given state.

### *10.3.1 Concept of special districts in governance*

In some jurisdictions, including many of the countries discussed so far in this chapter, special districts are created through legislation in order to perform a discrete set of functions in a particular geographical area. Special purpose districts are usually formed by residents within a district to provide services that are not currently being supplied by the government to those residents, or are being supplied in a manner that does not meet the needs of the residents. These districts cover all sorts of purposes, including irrigation and reclamation, pest and weed control, flood control, economic development, fire protection, park management, and utility delivery. The motivation behind forming these districts is usually to manage a public good or coordinate a response to some community threat for the good of the public located within the district. In most cases, these districts exist to bring attention to a specific set of shared problems, and are sized to fit the contours of the problem as agreed upon by a majority of landowners within a district.

While some of the environmental management activities associated with special districts can be articulated through conventional local government land-use planning schemes and development control powers, special districts can articulate a more focused approach suitable for active restoration work. Legislation defines the particular roles and functions of a district. For example, 'land conservation districts' in Western Australia undertake projects that remedy or mitigate land degradation within various districts.[48] What constitutes an appropriate project for a district is interpreted broadly. This gives districts flexibility in how to achieve their purpose. The Commissioner in Western Australia has concluded that a district as a governance body is authorized to engage in activities ranging from land conservation advice to land conservation work with the approval of a landowner.[49] Depending on the purpose of a given special district, landowner approval may be required by the district, enabling legislation for the district to formally exist.

Special districts are usually organized around some form of governance board, appointed by public or private parties. Where a district requires landowner approval, the board is more likely to include a number of landowners. An enabling act may authorize a district to levy a tax on its members. It is not uncommon for a special district, such as a fire protection district, to levy a tax on a portion of an existing property tax to be used by the district to achieve its legislated purposes. Generally, landowner approval is required to permit this type of levy.

One common type of special district is the 'conservation district'. The role of these districts can vary greatly depending on what the statutes permit. Conservation districts may focus on protecting shared aquifers, maintaining beaches, conserving soil, or eliminating water contamination.[50] A conservation district commonly disseminates information among its members on best restoration practices for critical habitat areas. Some conservation districts will also facilitate eco-restoration efforts across a variety of partners.[51]

A 'restoration district' could build on the success of existing conservation districts by educating community members about eco-restoration and facilitating community-wide discussions of long-term plans to restore ecological integrity to damaged or degraded privately held lands. Where a conservation district does not exist, a 'restoration special purpose' or 'landscape restoration' district might be created. In either case, a district designed to promote restoration may need either some enabling legislation to come into existence, or an amendment to existing legislation to provide necessary governance powers over restoration projects. As suggested above, certain groups within a

48 Government of Western Australia, Soil and Land Conservation Act 1945, s. 24(1).

49 Land Conservation District Committees, Role and Functions, www.agric.wa.gov.au/print/node/1620.

50 Revised Code of Washington Ch. 90.72. As an example of a unique conservation district, the State of Washington created in 1985 a 'Shellfish Protection District' or a 'Clean Water District' in order to protect the state-wide shellfish industry from onshore pollution. Within 180 days after the State Department of Health closes or downgrades a shellfish growing area due to a degradation of water quality, coastal landowners are required to create a shellfish district. The function of such districts is to identify contamination sources that are endangering water quality and seek remedies.

51 The Lake DeSmet Conservation District in the State of Wyoming in the United States provides support for federal government, state government, municipal government, and private landowner efforts to restore streams through in part by removing fish barriers.

landscape who are already engaged in eco-restoration, such as NGOs, highly committed individuals, or Indigenous groups, may assume leadership roles in promoting and lobbying for the creation of districts or the extension of the purpose of existing districts.

At the outset, a district designed to coordinate restoration across a large area might employ an eco-restoration consultant to provide a landscape-based report on the baseline for the area. Such a report might provide recommendations for district-wide projects that might be undertaken by either the landowners themselves on their landholdings, or by a third-party contractor. Funding generated within the district could either pay for external services, or fund materials to be used by landowners within the district to accomplish restoration objectives.

### *10.3.2 Justifications for the formation of a restoration special purpose district or extension of the mandate of existing conservation districts*

Where special districts can be formed under existing laws, they offer certain innovations to existing restoration efforts. Most importantly, they provide a governance mechanism that can implement ecological efforts across existing political jurisdictional boundaries. Because these districts usually rely on engaging landowner involvement, once these districts are formed, they can provide a level of landscape connectivity that would be politically difficult to achieve through government regulation. If momentum can be maintained such that the district becomes part of the political landscape of a given place, these districts have the potential to redefine the role of collective stewardship in achieving eco-restoration.

The introduction of special purpose districts to govern restoration efforts, or the extension of the mandate of existing conservation districts, offer two distinct advantages for private landowners seeking coordination of restoration activities across a region. First, these districts create a formal structure for landowners to interact, build relations, foster trust, create connectivity, and network. Second, by creating a peer group focused on achieving a mandate of place-based restoration, the districts are capable of locking in for a certain number of years the financial resources to enable some ongoing investments into restoration efforts. Through the district model, individual landowners can share costs across a wider pool of stakeholders. As Benjamin Richardson explains (Chapter 3 in this book), the spatial context of eco-restoration at the landscape scale requires governance arrangements that transcend the artificial boundaries of property tenure or traditional government agency jurisdiction, an agenda to which special districts can contribute if designed appropriately.

After these two justifications for restoration districts are discussed, this section will conclude with a couple critiques of the district model as a governance approach designed for residual landowners. At the end of the chapter, some model statutory language is proposed for the adoption of a restoration special purpose district.

*Districts as structures for relationship building and private ordering*

In terms of building people-to-people relations, districts offer a governance space within which to build a private order to govern private resources that meet not just private interests but also general public needs for restoration. While governments can regulate activities on private lands, these types of interventions may be resisted in countries with a strong individualistic private property tradition. Landowners are much more likely to respond positively to work programs that align with their beliefs and values and over which they can assert some control. Special purpose districts allow landowners to maintain control over their own lands while coordinating efforts with other landowners. The districts establish the relationships between landowners through a type of private ordering where private parties regulate themselves.

Here, private ordering ensures that stakeholders, in this case landowners, are able to give input on what rules will govern their relationship with their own land as well as their relationship with other landowners. The application of private ordering for landscape-level environmental outcomes has already proven somewhat successful with the introduction of coordinated conservation easement programs in some areas of the United States.[52]

Because restoration districts rely on private ordering to create the rules that will bind the landholders together, such districts may appeal to a broader set of private landowners who understand that they have a choice in whether to opt-in to a district. When they opt-in to a district, landowners will be able to continue to assert control over the types of projects that the district will pursue. Empirical evidence suggests that landowners are more positively engaged in specific projects when undertaking voluntary projects than when they are being externally regulated.[53]

The restoration district as a governance mechanism is far more than just a legal framework for private ordering. It also provides a community structure for connecting groups who otherwise may not yet share a common cause. Here, groups and individuals that are already involved in eco-restoration, such as the NGOs and big personas who promote the formation of a restoration district, may be able to use the concept of a district to demonstrate to other landowners both the need for restoration and the urgency for coordination. Residual landowners who lack knowledge of the need for eco-restoration or who may have been content with the status quo may agree to participate in a district in order to conform to the actions of nearby landowners. Inviting neighboring landowners to see conservation work in action and to be part of a community-wide conservation project, community leaders may be able to change landowners' perspectives about both the value of conservation and the people involved in the conservation.[54] Restoration districts

52 See Land Trust Alliance, www.findalandtrust.org (Identifying acreages of protected land by hundreds of U.S. based land trusts composed of place-based actors).

53 Carr et al (2014); Cooke et al (2012).

54 Fischer and Bliss (2008), 280. Undertaking a study of 36 family forest owners in Oregon and noting that 'most of the landowners said they trust and learn from other landowners' and believed that grassroots landowner-led efforts were 'the ideal mode for private-lands conservation'.

provide a formal vehicle for individual landowners to build long-term relationships based on trust and shared concerns.

### *Districts as long-term investors*

In addition to providing a platform for private ordering, restoration special purpose districts could serve as hubs for long-term investment in landscape restoration efforts. Depending on the enabling legislation for a particular district, the district board may have the authority to levy taxes from members for use in achieving district functions.

Long-term eco-restoration projects involve costs not just in the initial restoration efforts of removing invasive species and reintroducing native biodiversity, but also in maintenance efforts. The ability to pool financial resources across members has the potential to increase strategic investments in which no single landowner would be willing to invest on his or her own. It may be possible to use funds raised through a district levy to compensate landowners who bear disproportionate burdens from the implementation of a restoration strategy. For example, an individual who owns lands being restored to recover a breeding or feeding area for a particular species may have to remove previously productive land from use, and may need financial compensation before they will agree to participate in the project. A neighboring landowner may also be unintentionally damaged by a landowner undertaking restoration, as has been the case in coastal wetland restoration projects inadvertently flooding adjacent farmlands, and so also will need compensation to offset the damages.[55]

Existing conservation districts have successfully provided restoration funding for various projects, such as stream restorations that cut across multiple landowners' property. A landscape-based restoration special purpose district could build on the successes of these conservation districts by seeking to coordinate more connectivity between projects rather than simply funding individual efforts. In Dakota County in Minnesota, individual landowners can apply for incentive payments to offset wetland restoration investments, including funds for establishing restoration and for completing and certifying the restoration.[56] The success of the existing program depends on individual members of the conservation district deciding whether or not to participate in projects requiring 10 to 15 years of commitment. As currently structured, the long-term achievements of the program as a landscape-level restoration program depend on what appears to be uncoordinated participation among individual landowners. A restoration special purpose district or a conservation district with a new mandate would, in theory, put greater emphasis on coordinating individual efforts across individual property lines.

One of the greatest advantages to the district model is the possibility of building into the district governance structure an ongoing financial mechanism for the districts to undertake monitoring of projects. One of the recurring critiques of

55 Sayles (2018), 67.
56 Dakota County (2010).

restoration ecology projects has been a lack of ongoing monitoring, which leads to projects that are ultimately considered to be failures.[57] Depending on how the district manages its funds, this issue could be addressed by having a line item in the district's budget dedicated to ongoing monitoring of restoration projects.

### *10.3.3 Challenges of instituting special purpose districts for ecological restoration*

Assuming that the formation of a restoration special purpose district, or the enlarged mandate of a conservation district, results in a functional and well-funded network of landowners that also includes residual landowners, conflicts might still arise between eco-restoration 'experts' and landowners about the long-term vision. The former may propose restoration targets on the basis of certain pre-European, historical landscape features or to recover certain rare or endangered landscapes, while the landowners may prioritize functional outcomes (e.g. aesthetic or economic benefits) or stress cultural considerations, as occurs particularly with Indigenous landowners. Given that the appeal of political control over decision-making is likely to be one of the major factors in the formation of a restoration special purpose district, legitimate concerns may arise over whether a district will be engaged in 'restoration' as understood by restoration ecologists or engaged in some other enterprise. Just because a landowner participates in a community project in order to be a member in good standing within a community of landowners does not mean that the project will further long-term objectives of eco-restoration.

For example, in the Gratiot Conservation District in Michigan, landowners work in cooperatives to support the Michigan Pheasant Restoration Initiative. Individually, landowners are replanting habitat in grasslands and wetlands to help increase pheasant numbers.[58] Pheasants are, however, not native to the United States but were introduced by settlers for recreational hunting. The 'common pheasant', which is the focus of the habitat restoration project in Michigan, is a Chinese Ringnecked Pheasant introduced to the United States in 1881.[59] While eco-restoration practitioners might be more committed to recovering species such as the 'greater prairie chicken' that once existed in Michigan before European settlement, members of a restoration district may not support this project and might even actively oppose it. While the greater prairie chicken is not currently listed under the U.S. Endangered Species Act, it could become listed in the future and the presence of such a bird on private lands would restrict landowners' activities.[60]

Even when landowners willingly participate in restoration activities, they may be reluctant to invest in the precautionary standards set by eco-restoration experts.

57 Woodworth (2017), 271.
58 See www.gratiotconservationdistrict.org/pheasant-restoration.html.
59 History of Pheasants in the United States, www.pheasant.com/facts.
60 Fischer and Bliss (2008), 278 (interviewing landowners who expressed concern that if they created oak habitat that it might attract endangered species leading to new 'opportunity costs on land uses').

For example, in a survey sponsored in part by the Snohomish Conservation District in the U.S. State of Washington, researchers learned that among a group of landowners who have planted a riparian buffer, or are willing to plant such a buffer, almost half of the landowners were unwilling to plant a 100-foot riparian buffer based on the recommended ecological standard.[61] While the 100-foot width for riparian buffers had been recommended for stream restoration to meet natural resource goals for salmon recovery, landowners were unwilling to allow this change in land use. Offers of financial compensation for planting the buffer did not change the desire of survey respondents to plant the recommended buffer.[62]

These examples from existing conservation districts suggest that there will be ongoing negotiations among members and advisers within any proposed restoration special district regarding its objectives and work plan. A district's capacity to achieve restoration outcomes as understood by restoration ecologists will depend on the ability of the district members to create a shared vision and to sustain political will over many years to make measurable progress towards restoration targets. Sustaining political will in turn depends on the resolve of the group to persist in the face of project implementation difficulties. Because restoration ecology is not an exact science, and often requires multiple iterations before there is an ecologically desirable outcome, district leadership that is aligned with eco-restoration outcomes will be particularly critical at the outset of any special district formation designed to further restoration efforts. The ability of groups such as big personas and NGOs to maintain momentum over the course of decades will be essential to the success of restoration districts in creating a private order that parties are willing to maintain.

In some cases a less formal regime for promoting cooperation among landholders towards shared environmental stewardship and restoration objectives may be more politically feasible. In Australia, 'Landcare' and 'Land for Wildlife' schemes have flourished in many regions as voluntary frameworks for landholders to collaborate or at least work towards shared environmental goals without prescriptive regulation.[63] Their governance flows from peer pressure, public seed funding, and community support, and they help build social legitimacy for eco-restoration as a bottom-up initiative driven by community needs.

## 10.4 Conclusion: shifting from private to collective rules

Certain private actors are strong proponents of eco-restoration. Yet the majority of private actors, referred to in this chapter as residual landowners, are not engaged in restoration activities for a variety of reasons, ranging from a lack of capacity to a lack of interest. These residual landowners represent a large amount of acreage

61 Carr et al (2014), 10.
62 Ibid.
63 See e.g. https://landcareaustralia.org.au; www.wildlife.vic.gov.au/protecting-wildlife/land-for-wildlife.

that could be managed at a landscape level if there was some coordination among the landowners. This chapter proposes statutorily created restoration special purpose districts, or the enlargement of the purpose of existing conservation districts to include restoration, in order to coordinate restoration efforts across a given landscape. The introduction of restoration-oriented special purpose districts to coordinate restoration would have the dual advantages of creating a shared governance network for restoration, and supplying a reliable source of private financing to restore a landscape. The challenge for any district would be creating a shared vision that has the ongoing support of the landowners while still delivering meaningful ecological outcomes.

By localizing governance at the level of the private landowner, the proposal to introduce restoration special purpose districts also has the potential to shift how private property ownership is understood socially, an issue also canvassed by Robyn Bartel and Nicole Graham (Chapter 5 in this book). In some regions, such as North America, private property is considered an inviolable right. In political conversations in places where private property is fiercely protected by individual landowners, there is little rhetoric of the responsibilities that accompany property ownership. Some property law scholars argue for a wider recognition that private property ownership includes social obligations not just to other owners but also to non-owners.[64] On the basis of recognizing such innate social obligations as an incident of land title, collective community interests associated with a physical property might take priority over individual interests.[65] A special district facilitates what John Rawls refers to as the 'social division of responsibility' between society and an individual where an individual should be prepared as a matter of justice to invest in fair institutions that benefit society.[66] Special districts developing collective rules to be implemented by private landowners offer a potentially just approach to landscape management.

The proposal in this chapter for introducing restoration districts offers an institutional vehicle for fulfilling the social objectives of private property ownership. In practice, these districts acknowledge individual control by owners over a piece of property, but more importantly exist to recognize the eco-relationships of property across a landscape and the relational duties of landowners. A well-functioning restoration special purpose district would reflect a shift from a physical landscape governed and fragmented by private property tenure to a landscape governed instead by collective property norms. The latter allows for coordination as property owners act through a shared vision that is less likely to emerge in a system dominated by private property rules.

A restoration special purpose district, or a conservation district with an enlarged mandate, offer practical approaches for achieving socio-ecological stewardship by landowners who are not engaged in active restoration on lands that would benefit from such restoration. The formality of a district offers one approach for joint

64 Alexander (2013), 453
65 France-Hudson (2017), 108–109.
66 Rawls (1999), 371.

planning, and then ideally joint implementation of the planning. With private lands posing a major challenge to landscape-level eco-restoration, new governance ideas that reimagine how landowners interact at the landscape level should be given some consideration. This chapter concludes with model text for creation of a restoration district that reflects the foregoing ideas.

## Annex 1: Model language for a creation of a restoration district

**Restoration Commission**: The commission is established as an agency of the state to perform all functions conferred upon it by law. The original restoration commission shall consist of ____ appointed members to serve terms of ____ years. The majority of the appointed members should be landowners within the region where a restoration district is proposed. The members of the commission will receive ____ compensation. After a restoration district has been formed, the commission shall consist of ____ appointed members and ____ elected members to serve terms of ____ years.

**General Duties of Commission**: The commission has the following duties and powers:

1 To propose the creation of a restoration district to be adopted by the district voters.
2 To develop, with the involvement of a majority of the district's landowners, a set of district-wide restoration objectives with a timetable for achieving the objectives.
3 To facilitate coordination between district work projects and programs and activities of other jurisdictions as related to the restoration objectives.
4 To coordinate a special financial assessment across district landowners in a voter-approved district to fund district projects.
5 To seek and accept grants from external public or private sources to fulfill the purposes of the commission.

**Formation of a Restoration District**: The commission may propose to the voters the organization of a restoration district. The proposal should describe the general objectives of forming the district and define the geographical boundaries of the proposed district. The commission must deliver copies of the proposal to all registered landowners within the proposed district. The proposal should include an estimate of an annual levy on landowners to achieve certain restoration goals. ____ percent of eligible voters within the area to be affected must approve the formation of the district.

**Supervisors of Restoration District**: If voters approve a restoration district, the commission shall appoint ____ supervisors and the voters shall elect ____ supervisors. The majority of the supervisors shall be landowners within the district. The supervisors will constitute the governing board of the district with oversight over the commission and will serve without compensation other than reimbursement of their expenses.

**Enlargement of District**: A restoration district can be enlarged through a petition process approved by a vote of ____ percent of eligible voters in the affected area.

**Request to Withdraw from Restoration District**: After the decision to approve a restoration district, its individual landowners may submit a request to withdraw from the district. The request must be submitted to the commission who will organize an independent hearing to determine whether to approve or reject the withdrawal depending on a number of factors including, but not limited to, the quantity of land owned by the landowner and the significance of the land to the eco-restoration effort. Judicial review is available for any rejected request.

**Powers of Restoration District**: A restoration district shall constitute a governmental subdivision and shall exercise the following powers:

1 To make an inventory of all natural resources within the district that could be used to achieve restoration objectives.
2 To conduct projects on any lands within the district with the consent of the affected landowners.
3 To enter into agreements with landowners to provide financial or other aid to assist in achieving the objectives of the district.
4 To obtain by purchase, exchange, lease, gift, or other legal means any real or personal property that contributes to achieving the district's restoration objectives.
5 To make available to landowners equipment and material necessary for achieving restoration objectives.
6 To coordinate across properties projects necessary to achieve the district-wide restoration objectives.
7 To streamline, in consultation with other agencies, permits and other regulatory requirements for restoration projects within the district.
8 To regularly monitor and report the outcome of work projects.
9 To sue or be sued in the name of the district.

### *Special assessment and exemptions*

A system of assessments shall classify lands in the restoration district into suitable classifications according to ecological benefits conferred or to be conferred by the activities of the restoration district. Lands deemed not to receive any benefits from the activities of the restoration district shall be placed into a separate classification and shall not be subject to the special assessments. An annual assessment rate shall be stated as either uniform annual per acre amount or an annual flat rate per parcel. The maximum annual per acre special assessment rate shall not exceed ____ cents per acre. The maximum annual per parcel rate shall not exceed ____ dollars. Public lands within the district shall be subject to special assessments on the same basis as privately owned lands. Special assessments will be collected for ____ years with the possibility of renewal.

A landowner within a restoration district may apply for an exemption from the special assessment if they can demonstrate to the commission that they have already implemented best eco-restoration practices on their lands.

### *Rates*

Government authorities may approve revenues to a restoration district by fixing additional property tax rates beyond the special assessment.

### *Dissolution of Restoration District*

At any time after ____ years from the organization of a district, ____ percent of the voters in the district may file with the commission a petition to request that district be dissolved. A public hearing will be held within ____ days of the submission of the petition. The commission will organize for a vote within ____ days of the public hearing. If a majority of the landowners within the district vote for dissolution, the district will be dissolved.

## References

### *Primary sources*

Endangered Species Act, 16 U.S.C.§§ 1541–1544.

Surface Mining Control and Reclamation Act, 30 U.S.C. 1201–1328.

United Nations Declaration on the Rights of Indigenous People, UNGA 61/295 (2008).

### *Secondary sources*

Gregory Alexander (2013) 'Ownership and Obligations: The Human Flourishing Theory of Property' 43 *Hong Kong Law Journal* 451.

Matthew Brown (2014) 'Bison Treaty Signed by Alberta, Montana Tribes', *CBC News*, 23 September, www.cbc.ca/news/canada/calgary/bison-treaty-signed-by-alberta-montana-tribes-1.2775965

Christy Carr et al (2014) *Landowner Perspectives on Voluntary Conservation Incentive Programs*, American Farmland Trust.

Scott Cendrowski (2015) 'How a Less-Spotted Billionaire is Saving Alaska's Wildlife', *Fortune*, 18 May, http://fortune.com/2015/05/18/how-a-lesser-spotted-chinese-billionaire-is-saving-alaskas-wildlife.

Benjamin Cooke et al (2012) 'Social Context and the Role of Collaborative Policy Making for Private Land Conservation' 55(4) *Journal of Environmental Planning and Management* 469.

Nancy Daiyi and Linda Ford (2002) 'Life in Country: Ecological Restoration on Aboriginal Homelands', June, *Cultural Survival Quarterly*.

Dakota County (2010) *Dakota County Wetland Restoration Program*, Soil and Water Conservation District Fact Sheet, www.dakotaswcd.org/pdfs/ag_fs_wetrest.pdf.

Paige Fischer and John Bliss (2008) 'Behavioral Assumptions of Conservation Policy: Conserving Oak Habitat on Family-Forest Land in the Willamette Valley, Oregon' 22 *Conservation Biology* 2.

Ben France-Hudson (2017) 'Surprisingly Social: Private Property and Environmental Management' 29(1) *Journal of Environmental Law* 101.

Hawke's Bay Regional Council (2015) *Hawke's Bay Biodiversity Strategy 2015–2050*, Hawke's Bay Regional Council.

Tony Hiss (2014) 'Can the World Really Set Aside Half of the Planet for Wildlife?', *Smithsonian Magazine*, September, www.smithsonianmag.com/science-nature/can-world-really-set-aside-half-planet-wildlife-180952379.

Edward Hume (2009) *Eco Barons: The Dreamers, Schemers, and Millionaires Who Are Saving Our Planet*, Harper Collins.

David Lindenmayer et al (eds) (2016) *Wildlife Conservation in Farm Landscapes*, CSIRO Publishing.

Bruce Lindsay (2016) 'Legal Instruments in Private Land Conservation: The Nature and Role of Conservation Contracts and Conservation Covenants' 24 *Restoration Ecology* 5.

Richard Margerum (2016) 'Theoretical Perspectives on the Challenges of Collaboration' in Richard Margerum and Cathy Robins (eds) *The Challenges of Collaboration in Environmental Governance*, Edward Elgar.

Vanessa Masterson et al (2017) 'The Contribution of Sense of Place to Social-ecological Systems Research: A Review and Research Agenda' 22(1) *Ecology and Society* 48.

John Rawls (1999) 'Social Unity and Primary Goods' in S. Freeman (ed.) *J. Rawls, Collected Papers*, Harvard University Press, 359.

Leighton Reid (2016) 'South Africa 3: Town and Country: Aiming for Ecological Restoration at the Landscape', https://mbgecologicalrestoration.wordpress.com/tag/working-for-water/.

Benjamin J. Richardson (2017) 'Restoring Layered Geographies: Ecology, Society and Time' 26 *Griffith Law Review* 2.

Benjamin J. Richardson and Ted LeFroy (2016) 'Restoration Dialogues: Improving the Governance of Ecological Restoration' 24 *Restoration Ecology* 5.

Stacy Rosenbergand Richard D. Margerum (2008) 'Landowner Motivations for Watershed Restoration: Lessons from Five Watersheds' 51(4) *Journal of Environmental Planning and Management* 477.

Jesse Sayles (2018) 'Effects of Social-Ecological Scale Mismatches on Estuary Restoration at the Project and Landscape Level in Puget Sound, USA' 36 *Ecological Restoration* 1.

Eric Staats (2016) 'Chinese Billionaire Wenliang Wang Has Agreed to Donate Money for Long-Delayed Restoration of Dying Mangroves', *Naples Daily News*, 11 January.

Anastasia Telesetsky (2017) 'Eco-restoration, Private Landowners and Overcoming Status Quo Bias, 26 *Griffith Law Review* 2.

United States Fish and Wildlife Service (2003) *Policy 64 FW1*, Partners for Fish and Wildlife Program, www.fws.gov/policy/640fw1.pdf.

Paddy Woodworth (2017) 'Can Ecological Restoration Meet the Twin Challenges of Global Change and Scaling Up Without Losing its Unique Promise and Core Values?' 102(2) *Annals of the Missouri Botanical Gard*en 102.

# 11 Reconstructing the environment

## Perception and change in Australia's Murray-Darling Basin

*Francine Rochford*

### 11.1 Introduction

Ecological restoration is not simply an action or series of actions on a passive system. The system is both active and reactive, and a participant in its own processes. The insight of eco-semiotics is the conceptualisation of 'ecosystems as communicative systems'[1] and, moreover, recursive communicative systems.[2] Through this insight we can critique the processes by which restoration of ecosystems is attempted, and perhaps identify some of the barriers to success and how they might be overcome. This chapter will apply these insights to the processes of eco-restoration represented by the implementation of the Murray-Darling Basin Plan ('the Plan').

Semiotic scholarship[3] as a formal cataloguing of objects against the signifier of the object can be seen in the mapping of meaning in geography, geology or landscape studies. Some scholarship maps 'individual units of meaning',[4] but in relation to living systems, Maran and Kull note that eco-semiotics can be the study of 'the role of environmental perception and conceptual categorisation in the design, construction, and transformation of environmental structures'.[5] Cobley,[6] following Kull,[7] notes that the objects in the domain of semiotics include recognition, memory, categorisation, mimicry, learning and communication. In the restoration of the Murray-Darling Basin ('the Basin') through the implementation of the Plan, the objects in the domain of semiotics extend to recognition and memory of lived experiences and construction of meaning according to those experiences.

Eco-restoration in Australia and elsewhere has tended to focus on projects with discrete physical and temporal boundaries.[8] The restoration of a mine site, the

1 Maran and Kull (2014), 41.
2 Bateson (2000).
3 For an introduction to the eco-semiotic approach see Nöth (1998), Kull (1998), Kull (2007), Nöth and Kull (2001).
4 Lindström, Palang and Kull (2013), 101.
5 Maran and Kull (2014), 41.
6 Cobley (2010), 227.
7 Kull (2007), 2.
8 Richardson (2016).

revegetation of a coastal dune, or the re-snagging of a river tributary are projects with specific end (and start) points, clear parameters and budgets, and accordingly a clear narrative of value, temporal and geographical scope. However, eco-semiotics can be deployed more usefully in the case of landscape-level eco-restoration projects, such as the reallocation of water to the environment in the Basin in Australia, prohibitions on land clearing, 'rewilding' or re-establishing free flooding zones. These are large-scale, long-term projects affecting multiple communities and a range of ecological processes, and effecting significant change to landscapes. It is to be expected that processes of communication, differences in degree of learning, categorisation of what is harmful or beneficial, and attributions of value,[9] all insights from eco-semiotics and biosemiotics, can be usefully considered in analysis of these projects. Transcending mapping of individual units of meaning, eco-semiotics can provide a critique of processes by which the social and community impacts of eco-restoration projects are assessed. This is particularly important because of the politicisation of large-scale restoration processes. The extent of the proposed restoration and the significance of its effects on communities require a careful conversation on equal terms between those affected and those proposing the change.

This should also occur in recognition of the place of human communities as part of the ecosystem. Cobley, utilising the concept of biosemiotics, addresses the artificiality of the divide between the semiosis of human animals and other organisms and 'to erode, or at the very least, to make more porous, the boundary between living nature and culture, the sciences and the humanities'.[10] In the case of landscape-level ecological restoration, such as that occurring in the Basin, the interface between the human aspect of the reform and the remainder of the ecology is critical to the success of the project.

The first part of this chapter will 'map' the meaning of water in political and economic terms against the geography of the Basin. It will consider the large-scale attempt at eco-restoration – the Water Act 2007 (Cth) and the Plan and its associated legislative instruments – which is currently redeploying water from human economic and social uses to environmental uses. The purpose of this program is set out in the objects of the Act – 'to protect, restore and provide for the ecological values and ecosystem services of the Murray-Darling Basin'. This has resulted in a reduction in water availability for human consumptive use and caused planned inundation of floodplains. Using the analytical insights of eco-semiotics, the chapter will demonstrate the re-assignment of values through this process, leading to human-influenced environments, and an oppositional symbolic semiosis creating and imposing a meta-level revaluation of the environment. It will consider the techniques of remapping meaning using legal and policy – systemic – processes.

This analysis raises questions about the modern centrality of law – or more precisely regulation – as a mechanism for mapping meaning. Legislative instruments, which are the primary force behind large-scale ecological interventions,

9 Cobley (2010), 228.
10 Ibid, 225.

prescribe definitions, thus fixing meaning to specified words. This can suppress or override previous meanings and thus can be experienced as dissonance in conversation or 'communication' with affected communities. Legislative selection of certain ecological characteristics over others results in elevation of those characteristics, creating the impression of imbalance. Processes of legislatively prescribed 'consultation' in these circumstances are experienced as remapping of meaning – consultation on fundamental aspects of the reform is an assertion of new meaning. Moreover, the package of legislation mandating the Plan, and the Plan itself, manifest a particular type of value attribution – the audit, measurement and trade of certain characteristics of the ecosystem.

## 11.2 The Murray-Darling Basin

The Basin, in southeast Australia, is geographically extensive and politically diverse. It covers around 1 million square kilometres, but the mouth of the Murray in South Australia is the only outlet to the sea. The Basin itself receives comparatively little rainfall – in some parts of the Basin the annual evaporation rate exceeds the rainfall – and the rivers themselves carry little water by international standards.[11] Most of the rivers in the Basin arise in the Great Dividing Range – the Murray rising in the Snowy Mountains and the Darling in southeastern Queensland. The confluence of these two rivers is in Wentworth, New South Wales.[12] As a result of low inflows, high evaporation rates and seepage exacerbated by slow flows – due to its flat gradient – the system has massive transmission losses. Under natural conditions the tendency is for a decrease in flows downstream.[13] However, in wet periods the Basin is subject to overland flooding covering thousands of square kilometres. With climate change, scientists forecast that the Basin will become even drier.[14]

The political diversity of the Basin – traversing four states and one territory in Australia's federal system of government – also signals differences in dates of settlement and development. Traditional owners had their own mappings of meaning across the landscape, incorporating locations of ceremonial and sacred significance, hunting, fishing and harvesting[15] and cultural flows.[16] These were mapped over (although the existing mapping continued to exist) after European settlement, but not consistently or in a single process. Since it was first settled by Europeans from the early nineteenth century, the Basin has experienced diverse

11 Adamson, Mallawaarachchi and Quiggan (2006), 4.
12 Walker (1985), 11.
13 Walker (1985), 117.
14 MDBA, 'Climate', www.mdba.gov.au/discover-basin/landscape/climate.
15 MDBA, 'Traditional owners map, land and water use in northern Murray-Darling' www.mdba.gov.au/media/mr/traditional-owners-map-land-water-use-northern-murray-darling.
16 MDBA, 'MDBA welcomes significant step towards establishing cultural flows' www.mdba.gov.au/media/mr/mdba-welcomes-significant-step-towards-establishing-cultural-flows.

patterns of infrastructure development and settlement (or depopulation later) in each state. Soil type and natural water availability informed the 'closer settlement' of regions, and the requirements of settlers as mandated by legislation varied according to the perceived needs of the new colony or nation at the time of settlement, or as later priorities prevailed.

Narratives accompanying colonial settlement were explicit and sometimes elevated to legislative status. The 'nation-building'[17] motif was the basis of a series of Acts setting out the terms of settlement on terms considered best for the emerging nation. Other principles also influenced the developing landscape. For instance, the Amending Land Act 1865 (Vic) (the Grant Act) was grounded in social justice, intending to make settlement possible for less affluent farmers. The Land Act 1898 (Vic) sought to create the conditions for self-reliant homesteading, and in the case of water infrastructure development the Water and Conservation District Act 1880 (Vic) sought to 'preserve life', 'increase agricultural yield' and provide 'water security'.[18] Legislation governing settlement and infrastructure in each state echoed the social and economic needs of each colony or state at the time, as influenced by the prevailing agricultural requirements – particularly water availability. Massive infrastructure works delivered water to arid regions, enabling irrigated agriculture that increased the productive capacity of the land. Some of these infrastructure works themselves – such as the Snowy River hydro-power project of the 1950s – have iconic status and represent nation-building and immigration narratives (or 'mythmaking').[19]

Just as Traditional owners had their own meanings mapped across the landscape, settlers and their descendants in each part of the Basin have their own settlement narrative. Many groups of settlers share a chronology, having been part of a 'closer settlement' scheme.[20] The normative vision is not only shared by the community, but formed the (colonial) community. The motifs of settlement and consequent utilisation of the water resource for productive purposes have been supported by the contribution of irrigated agriculture to the national economy – which is high:

> relative to other sectors of the Australian economy [...] and high relative to the average of the agricultural sectors in other Organisation for Economic Cooperation and Development (OECD) countries. [...] This is achieved despite farming, largely unsubsidised [...] on fragile soils and in one of the most variable climates in the world.[21]

However, with the declining political importance of agriculture in Australia's economy, the narrative of the Federal government – and, to an extent, state governments – has shifted. The increasing importance of the sustainability of

17 Butcher (2008), 9.
18 Ingle Smith (1998), 151.
19 Griffin (2003).
20 Connors (1970).
21 Carberry et al (2010).

extraction of water from the Basin, the developing understanding of the inter-activity of sub- and surface-water extractions, the increased pressure on water resources arising from urban development and, not least, the political impact of episodic droughts required a Grand Plan – a rewriting of the narrative of water extraction in the Basin.

The federal Water Act 2007 (Cth) and the subsequent Plan were the instruments designed to achieve a semiotic remapping of the ecosystem. Of growing political and legal salience in this transformation has been the resurgence of Aboriginal Australian interests; formerly dispossessed of their lands in the Basin during the nineteenth century, in recent decades they have had some success with land claims pursuant to state legislation and the federal Native Title Act 1994 (Cth). Aboriginal people are also asserting demands for a greater share of water resources in the Basin to be managed by their communities for their cultural and economic needs, and they are collaborating through the Murray Lower Darling Rivers Indigenous Nations alliance.[22]

## 11.3 Transformation of the legal policy framework

The return of water to the environment through the Plan has occurred through a transformation of the legal and policy landscape. Two significant features of the policy framework are first, the marketisation of water; and second, the reallocation of water shares between consumptive and non-consumptive (environmental) purposes. The Federal government sought to achieve this reversion through the Water Act 2007 (Cth), a political compromise beaten out in the face of the direct impediment of s.100 of the Constitution, which placed water extraction and management in the hands of the states. The political and legislative strategies required to enable passage of the Act relied both on Commonwealth–state agreement and on a range of Commonwealth constitutional powers. The most significant of these is the external affairs power in section 51(xxix) of the Constitution, coupled with Australia's accession to the Convention on Wetlands of International Importance (Ramsar Convention)[23] and the Convention on Biological Diversity.[24] Sections 3, 4 and 21 of the Water Act 2007 (Cth) ensure the centrality of the Conventions to the validity of the legislation.

The Act requires the Murray-Darling Basin Authority to prepare a Basin Plan[25] to provide for integrated management of the Basin's water resources. This includes both surface and ground water, water in overland flows, in watercourses, wetlands or aquifers.[26] The Plan must provide 'for limits on the quantity of water that may be taken from the Basin water resources as a whole, and from the water resources of each water resource plan area'.[27] As a consequence of the

22 Winter (2015).
23 Convention on Wetlands of International Importance Especially as Waterfowl Habitat (1971).
24 Convention on Biological Diversity (1992).
25 Water Act 2007 (Cth) s 43A.
26 Ibid, s 4.
27 Ibid, s 19.

implementation of the Plan, irrigators in the Basin will have access to water diminished. The limits are expressed as 'long-term average sustainable diversion limits'.[28] These limits 'must reflect an "environmentally sustainable level of take"', defined in the Act to mean:

> The level at which water can be taken from that water resource which, if exceeded, would compromise:
> a) key environmental assets of the water resource; or
> b) key ecosystem functions of the water resource; or
> c) the productive base of the water resource; or
> d) key environmental outcomes for the water resource.[29]

The limits on water extraction had to take account of principles of ecologically sustainable development, on the basis of the best available scientific knowledge and socio-economic analysis, having regard to the consumptive and other economic uses of water resources, and to social, cultural, indigenous and other public benefit issues.[30] As originally developed, the long-term, average sustainable diversion limits required recovery of 2,750 gigalitres for the environment. Taking into account water recovery that had already occurred, this target left 1,203 gigalitres to be recovered, primarily from farm irrigators. The 'return' of this water to the environment is intended to result in eco-restoration not of a single site, but of the system as a connected network, thus qualifying it as a process of eco-restoration. The Commonwealth Environmental Water Holder now controls 6 percent of average system inflows.[31] State programs also hold significant water: the Victorian Environmental Water Holder now controls 19 percent of surface water entitlements in Victoria.

The implementation of the Plan and the limits on water extraction have, however, contributed to 'unavoidable socio-economic impacts'.[32] The highly politicised debate has most recently resulted in another political compromise to 'save' the Plan by allowing water recovered from *infrastructure works* to be counted to the 2,750 gigalitre environmental water target in the southern Basin and a further 70 gigalitres to be redirected to the northern Basin using the same method.[33] The prolonged and fraught process of negotiating and then implementing the Plan has had significant effects on the livelihoods of irrigators and rural communities. Indeed, the process of 'consultation' and the legislative and bureaucratic processes resulting in, and implementing, the Plan were intended to have the effect of remapping the established semiotic relations between irrigators, water and the environment. It could be argued that what has resulted is not a new or deeper understanding of the environment or environmental water, but a diminished

28 Ibid, s 22(1).
29 Ibid, s 4.
30 Ibid, s 21(4).
31 Webb et al (2018).
32 MDBA (2018).
33 Baxendale (2018).

conceptualisation of the environment. Water is now understood to be a fungible market commodity, even when it is associated with environmental outcomes, and the environment itself is now considered to have equivalent interests to other stakeholders, rather than having prior claims.[34]

## 11.4 The biography of water

Kaplan[35] uses the insights of Kopytoff[36] to describe the multiple 'natures or careers'[37] of water over time. Transcribed into the Australian perspective, the long traversal of water through the Basin takes it through myriad (and multiple) consumptive and non-consumptive uses. It begins – or begins again – as precipitation. We see its pooling or puddling, draining or flooding, seepage or storage, utilisation (and re-utilisation) for domestic consumption, industry or agriculture, or/and eventual evaporation or drainage to the mouth of the Murray. Sequentially and simultaneously the same physical water can have several ecological, legal and cultural manifestations. In ecological terms, water multiplies the potential for interaction, since 'availability of water is perhaps the most important single element in the complex of factors controlling variation in ecological systems'.[38] Water has additional significance in Australia as the planet's second-driest continent (after Antarctica).

Water symbolises life, and the water in a landscape is identified with the liveability of the land. Pre-Christian water rituals were geographically situated – life-giving springs were associated with sacred beings. In Roman religion, water was powerfully associated with the divine. Edlund-Berry notes that

> wherever these natural water sources appear, we are likely to find remains of sacred areas, ranging from isolated votive deposits to monumental sanctuaries. Along roads, at river crossings, and along mountain passes, the presence of water reinforces the link between nature and the divine.[39]

Association of water with life appears throughout the Christian Bible, from Genesis 1:21, water 'brought forth abundantly […] every living creature that moveth', to water as a representation of the sacred in Matthew 3:11, where water is the medium of baptism into repentance. But the more prosaic representations of water are also evident in sacred texts. With the eye of a farmer or gardener, Lot 'beheld all the plain of Jordan, that it was well watered every where'. In the Quran the relationship between water and food is also emphasised: '[i]t is He Who sends down water from the sky, and with it we bring forth vegetation of all kinds' (Quran 6:99). The antithetical association – lack of water with death – is also a

34 O'Donnell (2018).
35 Kaplan (2007).
36 Kopytoff (1986).
37 Kaplan (2007), 685.
38 Hughes and Lampert (1980), 52.
39 Edlund-Berry (2006), 162.

deeply wrought image in sacred literature – the 'dry ground, the "thirsty land"' (Isaiah 44:3), the 'dry and weary land where there is no water' (Psalm 63:1). Water is deeply embedded other lifeworld contexts, such as rites associated with birth, cleansing and hospitality. The use of water to grow food in a garden setting, or to water animals or maintain hygiene, would naturally fall within the lifeworld context. The sacredness of water arises from its essentiality, not vice versa.

Water has fundamental value to indigenous Australians, with 'ongoing, complex cultural and economic relationships with both fresh and salt water'.[40] For instance, for the Arabana people of the Kati Thanda–Lake Eyre region, water is considered to be 'the locus for cultural identity, as a source of survival in a very dry place, as markers of travel routes and as places where historical, recreational and cultural activities take place'.[41] The relationship between individuals, communities and water was not made up of discrete, quantifiable transactions; analysis should stress the 'interpretive significance of human/habitat interactions by emphasising their connectivity rather than their independence'.[42] Indeed, the relationship perfectly describes the interrelatedness of the semiotic process: 'Aboriginal people, lands and waters are in a continual and interconnected process of mutual transformation'.[43] Aboriginal art, songs and stories link the present to the past and continue the process of creation and re-creation of meaning.[44]

In the Australian environment, the periodic scarcity of water is evident both in settler narratives and in early Australian film, with *The Breaking of the Drought* (1920) contrasting catastrophic bushfire with the 'cleansing' power of rain. Thus,

> Individuals, communities, and societies attach culturally influenced, symbolic, economic, and ecological meaning to natural resources like irrigation water. According to phenomenological and constructivist approaches to natural resources, this perception process determines (potential) human action with regard to using and living with the respective natural resource.[45]

The European settlement of Australia vested cultural meaning in water and water resources. This meaning varied depending on the interaction between the community and the resource; water was a pre-requisite for mining, for settlement, for agriculture and for conveyance. It also evoked emotional reactions. Early colonial art and literature provide many examples of real and symbolic interaction with water.[46] 'Meaning, although interpreted individually, is [...] a shared cultural product.'[47] Art has a particularly evocative role in attempts to find and re-create meaning. Cultural forms such as literature help to create and re-create meaning, so

40 Butterly and Richardson (2016), 3.
41 Nursey-Bray and Arabana Aboriginal Corporation (2016), 13.
42 Toussaint, Sullivan and Yu (2005), 61.
43 Ibid.
44 Somerville (2013).
45 Oberkircher and Hornidge (2011), 396–7.
46 Bonyhady (2000).
47 Drenthen (2016).

there is a degree of alignment in the perception and categorisation of the environment, at least amongst contemporaries. For instance, poet A.B. 'Banjo' Patterson's childhood in the Southern Tablelands of New South Wales informs his writing, but his self-consciously 'wild erratic fancy' visualises the experience of a drover following water in heroic, romantic terms: Clancy has

> [g]one a-droving 'down the Cooper' where the Western drovers go;
>
> as the stock are slowly stringing,
>
> Clancy rides behind them singing,
>
> For the drover's life has pleasures that the townsfolk never know.[48]

Cooper Creek in the Lake Eyre Basin is a symbol of a largely unknown inland for predominantly urban Australia, traversing Queensland and South Australia, and is the site of the deaths of explorers Burke and Wills. Indigenous groups lived in the Channel country, and for these groups the Cooper represented both trade and relative abundance.[49] For a drover, the ephemeral channels of the Cooper provided water, and the creek's intermittent flooding, once subsided, provided feed. The 'roaming' occupation of the drover matches the ephemeral river system, and for the city-dweller in Patterson's imagination it symbolises adventure. Water in the inland is part of the grand, nation-building narrative; the 'conservation' and distribution of water is a bold experiment, necessary to ensure the viability of inland farming.

Conversely, the colonial, settled farmer's interaction with water is small-scale, cyclical and fatalistic. Burdened with stock and crops and buildings, the farmer cannot follow the water, as a drover will, but must rely on the rain. In a parody of farmers' ongoing preoccupation with water, poet Patrick Hartigan's 'Said Hanrahan' counters the stanza reacting to drought with a stanza reacting to flood:

> And so around the chorus ran
>
> 'It's keepin' dry, no doubt.'
>
> 'We'll all be rooned,' said Hanrahan,
>
> 'Before the year is out.'And:
>
> It pelted, pelted all day long,
>
> A-singing at its work,
>
> Till every heart took up the song
>
> Way out to Back-o'Bourke.
>
> And every creek a banker ran,

48 Paterson (1889).
49 Hughes and Lampert (1980).

And dams filled overtop;
'We'll all be rooned,' said Hanrahan,
'If this rain doesn't stop.'[50]

Access to water in Australian settlement is also charged with political meaning and is part of the 'class' narrative of early colonial culture. Clancy, for instance, is a drover, an itinerant, alongside the 'Jolly Swagman'[51] as the anti-establishment 'other'. There is a continual struggle between the tendency for land and water 'grabs' symbolised by the squatter and the need to facilitate population growth. This is also a representation of the tension between sovereignty and nativism – the squatter with political power is often literally English nobility, and in consequence has the financial resources to set up the infrastructure required to clear and profit from the land. Fitzpatrick records the manipulations of Victoria's Land Act by the 'bank-backed squatter' who, 'knowing the location of the water holes on his run, put up his dummies [straw creatures such as infants and friends in whose names applications for property were made] to select the watered areas. The eyes having been picked out of the run by this peacocking, the rest of it was rendered useless to the farmer or small grazier.'[52] The squattocracy – the class and money distinctions bound up in water availability – is explicit in Rudd's portrayal of settlement:

> when Dave wanted to know why Dad didn't take up a place on the plain, where there were no trees to grub and plenty of water, Dad would cough as if something was sticking in his throat, and then curse terribly about the squatters and political jobbery.[53]

In this representation, water defined every day's labour:

> Whenever there came a short drought the waterhole was sure to run dry; then it was take turns to carry water from the springs—about two miles. We had no draught horse, and if we had there was neither water-cask, trolly, nor dray; so we humped it—and talk about a drag! By the time you returned, if you hadn't drained the bucket, in spite of the big drink you'd take before leaving the springs, more than half would certainly be spilt through the vessel bumping against your leg every time you stumbled in the long grass. Somehow, none of us liked carrying water. We would sooner keep the fires going all day without dinner than do a trip to the springs.[54]

50 Hartigan (1921).
51 Paterson (1885).
52 Fitzpatrick (1947), 10.
53 Rudd (1899), 5.
54 Ibid.

The politicisation of water and water conservation was reflected in art. Henry Lawson, who considered himself the 'government irrigation poet',[55] allied himself with the cause of water certainty: 'I wrote of the unlocked rivers, in the days when my heart was full, / And I pleaded for irrigation where they sacrifice all for wool'.[56] The artist Piguenit used the visual artistic form: his representations of the Darling, Nepean and Hawkesbury Rivers are significant in Australian landscape painting. Bonyhady argues that Piguenit, with his brother-in-law Gerald Halligan, was an advocate for irrigation.[57]

The multiple meanings of water over time and space mean that the Plan, a landscape-level eco-restoration project, requires a process to transform a heteroglot dispute across multiple geologies into a single narrative. What this means is that established mappings between multiple meanings and values are disrupted and replaced with a single articulation of asserted correspondence. In the case of water, the established meanings include those deeply embedded in the lifeworld. As Oberkircher and Hornidge note, '[w]ith regard to water use [...] the individual's lifeworld forms the fundamental basis for understanding, attachment of meaning, and finally communication, behavior, and action in the sense of a personal encyclopedia of water use'.[58] The farmer attaches meaning to water through this catalogue of experiences. Water to a farmer is a signifier of political power, of productivity, and of success. More than that, however; the 'life world' of the farmer is entirely engaged with water. Water is a signifier of other aspects of life, of personal history and memory and of future viability and family legacy. As Keane notes, 'semiotic ideology directs attention to the full range of possible sign vehicles and the sensory modalities they might engage, including sound, smell, touch, muscular movement, pain, affect, and other somatic phenomena'.[59] The smell of pasture during irrigation is a powerful sensory memory. The abrupt tearing of the silent night by the motorbike on an irrigator's trip around the water, the churning sound of the Dethridge wheel, the splash of water against gumboots of an irrigator – these make up a summer season. The irrigator going around the water knows each bay as intimately as a gardener knows a garden – its vegetation and soil type, its fall, its history of watering and likely saturation point – how quick it will be to reach the end of a bay and therefore when the water should be shut off. The line of ibis grazing the irrigation bay measures the run of the water. The erratic line of irrigation channels, hugging the high ground, is a geographical constant and a demarcation of possibilities. These sounds and smells and other sensory modalities are the lifeworld experience of water extraction. For irrigation communities, the green pasture dependent on irrigation is an amenity and a critical aspect of the visual environment. The replacement of green pastures with dryland agriculture evokes a different affective response: it has been shown that

55 Bonyhady (2000), 300.
56 Lawson (1906).
57 Bonyhady (2000), 301–305.
58 Oberkircher and Hornidge (2011), 397.
59 Keane (2018), 65.

'colour is a key variable'[60] in perception of landscape, and that the colour green in particular evokes positive feelings. These responses have been explained as 'being due to an evolutionary predisposition to prefer landscape aesthetics that are associated with productive seasons'.[61]

## 11.5 Framing devices

### *11.5.1 Semiotics of water*

Remapping the multiple meanings of water to a new set of semiotic correspondences has necessitated a lengthy process utilising semiotic framing devices. A degree of 'semiotic regimentation' was required to strategically shift the culturally embedded meanings of water. Taddei notes that 'historical narratives on climate, coupled with political and economic genres, structures and processes that developed through time, are strategically used as framing devices in discussions and decision making processes related to the environment'.[62] He notes that:

> the political game consists of the use of strategy in order to make a preferred take on reality more widely accepted or authorized than it really is, and to make use of strategies in order to sustain such a state of affairs for as long as possible or profitable. [...] one important strategy in such a game is to try and manipulate the contextual factors that define the way in which messages are interpreted, rather than simply controlling the content of a message: this is what Silverstein calls the *metapragmatic* dimension of language use [...] it is also part of what Goffman refers to as *framing*. Institutions and institutionalized rituals play a major role in such processes.[63]

Various framing concepts were deployed to 'regiment' the semiotics of water; the first was the harnessing of 'efficiency' terminology to accepted environmental outcomes. This required the creation of methods of measurement – an initial audit, quantitative measurement and a corresponding measurement of environmental value – the articulation of 'icon' sites. The second was to associate the Plan and its processes with distributional justice mechanisms – in this case, communication practices and transparency; and the third was to create a reallocation ritual.

One technique for escalating pressure to accede to this new conceptual framing was to elevate the sense of crisis by assuring the existence of a 'new normal' in relation to water scarcity. In the Basin we see these framing devices deployed to characterise drought as 'the new normal' – for instance in explicit manipulation of terminology. 'There needs to be a new national approach to *living with dryness*, as

60 Villagra Islas (2010).
61 Ibid.
62 Taddei (2005), iv.
63 Ibid, 17, 19.

we prefer to call it, rather than *dealing with drought*', according to the Federal government.[64] The use of this framing device was particularly useful during the so-called 'millennium drought' in the early 2000s, which was the context and major contributing factor for political agreement to enter into the Murray-Darling Basin Plan Agreement. The continued deployment of framing devices meta-pragmatically risks realignment of meaning over time, so the 'health' of concepts has to be nurtured to ensure that they retain their new meaning. Institutional rituals used to ensure that the concepts retain the reframed meaning include ongoing consultation processes, consistent word choices, and newsletters, emails, webpages and advertisements that associate the process of reform with positive environmental outcomes (such as river flows) and images of participating communities.

Conversely, a series of negative framing techniques have been deployed to diminish 'off-message' concepts. This process has characterised the lengthy history of the Plan and every renegotiation. Semiotic ideology is evident in some literature and discussions relating to the Plan. This is particularly potent as a semiotic counter-strategy when it overlooks or diminishes the importance of cultural non-instrumental values placed on water. Semiotic counter-strategies characteristic of the remapping of values to the Basin is most evident in the narrative of the 'corporate farmer' other, the profit motivation of which sets it against the traditional 'small' irrigator and water user.[65] Just as early colonial narratives created a political narrative around the 'squattocracy', there is a clear semiotic counter-strategy developing around the upstream 'large corporate farmers'. These are frequently large stations capturing overland flows in on-farm dams to irrigate cotton crops, but they may also be river diverters. There has been a clear delineation of smaller irrigators as supportive of the Plan and as 'winners' in the process. For instance, during drought the small irrigators survived not through production (because many did not have sufficient water to produce), but through the sale of 'temporary water'. However, in this narrative large corporate irrigators have been identified as 'profiteering', sometimes by engaging in exactly the same market process – selling water rather than using it.[66]

The semiotic importance of the 'compliance' narrative has also been utilised to undermine the Plan,[67] necessitating scaffolding action to reinforce the integrity of the verification techniques.[68] One of the conditions of the recent reintroduction of a Bill to amend the provisions relating to the return of water to the environment was a commitment to 'greater compliance and integrity in the basin to prevent water theft'.[69] The depiction of contestation between small and large irrigators also revisits the state-versus-state theme framed in particular by South

64 Commonwealth of Australia (2008).
65 See e.g. Neales (2017); Brewster (2018).
66 ABC (2017). See also ANAO (2017); Sutton (2017); McNally (2018).
67 ABC (2017).
68 ANAO (2017).
69 Baxendale (2018).

Australia as an 'institutionalized ritual'.[70] A proposed amendment to the Sustainable Diversion Limit under the Plan was used by the South Australian government to justify the creation, in 2018, of a Royal Commission 'to investigate the operations and effectiveness of the Murray-Darling Basin system'.[71] Another counter-strategy can be seen in the creation of the national register of foreign ownership of water entitlements established by the Foreign Acquisitions and Takeovers Act 1975 (Cth) and the Foreign Acquisitions and Takeovers Regulations 2015 (Cth).

### *11.5.2 Efficiency and the market mechanism*

The nomination of the market mechanism as the preferred option to manage water resources had been articulated by the federal Productivity Commission many years prior to the millennium drought – the National Competition Policy was endorsed by the Council of Australian Governments (CoAG) in April 1995, and made provision for payments to States and Territories which implement competition reforms. The CoAG Water Reform Framework – now the National Water Initiative – was one of those reforms. At that point, however, it was framed as a preferred allocation mechanism on an economic basis and was premised on the creation of a 'grid' for water to enable it to shift from lower-value uses to higher-value uses. Thus, the use which could yield a higher money value per megalitre of water could enable the user to afford to purchase water from a lower-value use. In irrigator terms, this would generally mean the movement of water from, for instance, irrigated pasture to horticulture. At that time there was little political appetite for extending the 'grid' to the purchaser able to demand the highest-value use – urban domestic use.

The later framing of the market as the preferred mechanism to manage environmental outcomes was carefully constructed. This framing device aligned efficiency with environmental benefit and neutral or positive socio-economic impacts.[72] The creation of environmental water managers with the capacity to trade in water was a necessary concomitant to extracting environmental value from efficiency mechanisms, which may otherwise be deployed for other consumptive uses. Partly by participation in the water market, however, state and federal environmental water holders have extensive water rights. Water policy is now entirely systematised. Expert analysis now audits, traces and measures water availability and compliance with highly stylised economic principles and rules. Water is geographically detached from its source, and no longer subject to natural laws – modernisation of water delivery diminishes reliance on gravity and instead utilises pumps so that a 'grid' for water can be created. Trade in water is facilitated by a set of bureaucracies, corporatised entities benchmarked against 'sustainable revenue streams', recovery of costs, expenditure, and return on assets and investments. Water providers are required to provide incentives for sustainable water use

70 Taddei (2005), 9.
71 MDBRC SA (2018a).
72 Ernst and Young (2018).

*by providing appropriate signals* to consumers, providing incentives for efficiency improvements, and providing information to ensure rational economic decisions.

The social change engineered by the Plan was characterised by the reorganisation of the symbolic nature of water. This reorganisation mandated a transformational process - the creation of new institutions and rituals which transformed meanings.[73] Previous correspondences between water and its meanings have been severed and replaced with a constructed set of valuations derived from instrumental verificational techniques. Numbers are used iconically, indexically or symbolically to represent water in its various manifestations. Verran argues that 'in shape-shifting between the form one/many and whole/parts, and in moving between symbolic/indexical and indexical/iconic modes of semiosis, number works in inventive ways in mediating a frontier embedded in knowing and working Australia's water resources as enumerated'.[74] Verran notes the use of 'numbers as working indexically or symbolically. They are being used to represent the ecological health of Australia's creeks and rivers, lakes and billabongs. With an imperceptible shift such numbers can be used iconically to constitute a water market.'[75]

This process is facilitated by a suite of techniques, including the elevation of certain ecological elements over others, narrative and strategic consultation processes. The outcome is not a single articulation of meaning, but a semiotic re-regulation of the environment and a disruption of the relationship between inhabitants and the environment of the Basin. In the case of water entitlement holders, a series of existing correspondences between meaning and value existed. These were both engendered and translated by legal understandings of the relationship between water and land, and cultural understandings of the value of food and fibre production. These correspondences have been severed and replaced with a constructed set of valuations derived from instrumental verification techniques.

The construction and utilisation of the numerical form to mediate the social change required by the Plan was inevitable. The complex and multiple semiotic relationships between individuals, states and groups, the land and the water, resisted a single narrative structure. Multiple understandings of value forms and relativities meant that there was no universal imperative for change. At the outset of the change process there was no agreed position even on the amount of water in the Basin, let alone the value of that water. States, and regions in states, deployed localised regulatory forms. Thus, in Victoria for instance, different regulatory techniques applied to surface water, groundwater, drainage water, flood water, stock and domestic water, rural and urban water, and water in regulated and unregulated streams. Access to water in each of these categories attracted different regulatory regimes with significantly different ongoing rights. There was no comparison, for instance, between a secure 'right' to surface water attached to land in an irrigation region, a seasonal 'sales' entitlement, and an ephemeral

73 Taddei (2005).
74 Verran (2010), 173.
75 Verran (2010).

groundwater or drainage pumping right. Conversely, 'gaps' in regulation existed where water was intercepted by farm dams in the upper catchment before it reached a waterway, or where commercial tree plantations resulted in water extraction. There were also temporal differences in regulation – the regulation of water differed across irrigation seasons so that, for instance, water could not be 'carried over' into a different regulatory period in Victoria – in contrast to New South Wales. The value of water on-farm was entirely dependent on timing as well – the growth stage of crops could mean that water could be very valuable one week and worth little or nothing the next.[76]

These distinctions were amplified across state government boundaries, so that there was no shared common meaning to the language of water 'rights' or the articulation of their volume or value. In New South Wales and Queensland, management of water was less conservative than in Victoria, so rights of access to water were less valuable because in some years there would be reduced or no water attached to those licences. Other significant differences attached to the geography of the regions. Cyclical 'overland flows' are a feature of the Basin; during these flooding events water will spread hundreds of kilometres across a floodplain. The 'interception' of these flows in massive on-farm storages was permitted on quite different legal bases than the 'abstraction' of water from a river with defined boundaries. The abstraction of this water was opportunistic and not reliable, so again the volume and value of the resource was highly variable and suitable only for enterprises that could mitigate the damage from lack of water in the event of drought – so, for instance, it could not support horticulture or dairying that had ongoing water needs to maintain stock.

From the perspective of environmental values, there was little or no correspondence between these values and the environmental values of the river system. The two sets of value propositions (agricultural productive use and environmental value) did overlap to a degree – for instance in drainage regulation and indicators of quality such as salinity or nutrient load – but productive and environmental discourses in relation to water extraction and use were high-level principles. An exhaustive analysis of the lengthy process of semiotic re-alignment is beyond the scope of this chapter, but there are several stages in this process that are characteristic instrumental techniques.

First, there was an audit process, then the creation of a single unit of value, then a 'participatory' mediated reassignment of meaning. The project to create a single value proposition for these multiple physically and temporally variable values commenced with the technical processes of index and audit. *An Audit of Water Use in the Murray-Darling Basin*[77] was conducted to provide a set of baseline parameters for decision-making by assessing diversions in volumetric terms. Verran notes that:

76 Claughton (2014).
77 MDBMC (1995).

> We see that while appearing merely as a means of assembling information, numbering Australia's water resources serves two separate and distinct purposes: first, environmental monitoring which we can understand as a form of cadastral accounting characterized by a moment of audit and achieved in arithmetic calculation; second, constitution of an archive to constitute an emergent Australian water market and the possibility of the complex calculation that far exceeds the simple arithmetic involved in cadastral numbering.[78]

The 'cadastral' process appears to align agreed units of measurement with their value. In origin, a cadastral process deployed legal units of measurement with legally articulated valuations. Extension of this process to the natural water resources of an entire system required high levels of stated equivalency – in other words, estimates of value. In the development of the Plan, the evolution of a single 'language' descriptor for different water extraction types was a precursor to the process of creating a single index of value.

Whereas water released from reservoirs could be estimated with a reasonable degree of certainty, and water extracted through metered outlets could be measured with a higher degree of certainty, measurement of overland flows prior to water reaching a stream, plantation interception, evaporation, seepage, and other natural processes, both in and outside a watercourse, involved high levels of complexity and inaccuracy. At the outset, the lack of baseline information required caveats on the measured resource. Terminology such as 'consumptive', 'usable' and 'system' signalled the distinction between actual water and measurable water. 'Conveyance water', 'system losses' and 'return flows' were audited based on expert estimates. Conversely, system water measurement relied on streamflow.

More significantly, narratives constructed around 'return' of water focused on system losses within irrigation infrastructure, rather than natural losses of potentially exponentially higher degrees of magnitude. The primary mechanism by which a balance between environmental, social and economic outcomes was to be achieved was through the use of 'infrastructure funding'. This massive project involved a suite of technological 'fixes' to known water losses, such as diminishing seepage and evaporation from irrigation channels by lining channels or replacing them with pipes. Estimated 'measurement error' of Dethridge wheels justified their replacement by modernised outlets. 'Deemed' water losses were 'recovered' by 'investing' in modernised infrastructure, both off- and on-farm.[79] Programs providing funding for new or 'modified' infrastructure are therefore aligned with water recovery and used as a mechanism to 'mediate' change:

> It was always known that change of the magnitude required to save the Basin system would have some unavoidable socio-economic impacts. That is why unprecedented funding of $13 billion was allocated to its implementation—

78 Verran (2010), 175.
79 Department of Sustainability and Environment (2012).

with $8 billion earmarked specifically for infrastructure programs to recover water with minimal socio-economic harm to communities.[80]

Alternatively, water recovery could occur through purchase of water entitlements in the market for return to environmental flows. These techniques were of greatest benefit in irrigation systems with developed water products with a high level of certainty in measured extraction. Thus, in the Victorian irrigation system, with historically conservative water management policies, purchase of a megalitre of permanent water resulted, in most years, in a megalitre of water being returned to the environment. Conversely, in New South Wales, where water management had been less conservative (so that irrigators could in some years obtain well over a normal entitlement and in other years receive nothing at all), the purchase of a megalitre of permanent water would, in many years, return less than a megalitre to the environment.

The lengthy process of remapping meaning across the Basin required a variety of semiotic techniques. Along with the measurement or 'audit' of water in the Basin, the process required a mechanism for measuring success. This had to be aligned with accepted environmental values. However, ecosystems are reflexive communication systems, constantly creating and re-creating themselves in response to the actions and interactions of component parts – including human animals. Environmental values not only diverge across different states and parts of states, but also over time. One irrigator's understanding of a healthy landscape might include a warm pasture with shelter belts with high livestock-carrying capacity, free-draining loam and low water table. Land of such quality could be supplemented with nutrients or 'improved' pasture which would, over time, increase carrying capacity. A vegetable grower or orchardist might look for different characteristics in land, but for each the understanding of value is allied with productive capacity. Conversely, some Australian land in its natural state might be considered 'hungry', having been leached of nutrients by flooding. It might lack tree cover, or be 'heavy' and poor in drainage. Other land might be heavily treed but therefore thirsty, unable to grow pasture. Poorly drained soil, prone to hold water, is a loss of productive capacity.

Conversely, environmental values can map entirely different virtues. For instance, swamps, or 'wetlands' are valued ecosystem features. Optimal salt loads are necessary to achieve certain environmental outcomes in the lower lakes. Soils with high nutrient content may deliver negative outcomes to native plants accustomed to poor soils. Bushfires, which can have catastrophic effects on water quality, are necessary for some plant regeneration. In other words, an ecosystem highly adapted to variable climate and poor soils does not necessarily thrive in artificial conditions of plenty.

80 MDBA, 'MDBA Analysis Shows Effect of Basin Plan on Communities', 3 May 2018 www.mdba.gov.au/media/mr/mdba-analysis-shows-effect-basin-plan-communities.

### *11.5.3 Icon sites*

Associating the 'audit' culture with the environmental narrative was a difficult process of semiotic remapping. Entirely different valuations of ecosystem features require that commonality be established by a series of manufactured processes. The articulation of 'icon sites' had the effect of constraining the Basin-wide negotiation of environmental value to measurable outcomes. Visual symbols characterised the re-alignment of value by the creation of discrete 'signs' corresponding with the value of the Basin. These served to shift value designations from the productive uses of water to environmental uses of water, whilst at the same time aligning ongoing environmental health with continued productive uses.

The articulation of 'icon' sites, the mapping of value to particular environmental assets, is a semiotic technique. The Plan utilises the creation of ecological icons through a 'forced narrative', a process by which the manipulation of symbols strategically realigns the relationship between humans and the environment. The River Murray channel itself has been conferred with the status of an 'icon site' under The Living Murray program.[81] Other 'icons' include the Barmah-Millewa Forest, the Gunbower-Koondrook-Perricoota Forest, the Hattah Lakes, the Chowilla Floodplain and the Lindsay-Wallpolla Islands, and the Lower Lakes, the Coorong and Murray Mouth. These sites were selected for the 'high ecological and economic value and their cultural and heritage significance to Aboriginal people and the whole community'[82] but also because there is already significant environmental information about them, so improvement (or lack of improvement) could be measured effectively. This audit function is buried in the 'icon' narrative, which borrows religious and cultural force. By this technique, '[t]exts mediate human activity at a distance and help enlist and align people to larger social institutions and practices, and text genres provide means of recognizing social relations, obligations, and interactions embodied within communications'.[83]

The 'mouth of the Murray' itself is a powerful visual tool used to signify the health (or lack of health) of the Basin.[84] A key visual comparison of the 'closed' mouth and the 'open' mouth of the Murray was extensively deployed during the millennium drought, often without the historical context (that the mouth of the Murray would naturally close during drought). The icon site designation also acts as a reference point for reiterations of value in the Plan itself. It reduces assertions of value to particular acts, and this becomes a measure of success.[85] The compilation of information is then strategically utilised to support additional monitoring, evaluation and review activities.[86]

81 MDBMC (2002); MDBA (2018).
82 MDBA (2011), 14.
83 Bazerman, Little and Chavkin (2003), 456.
84 Fyfe (2003); MDBA (undated) 'A Dwindling River', http://riverstories.mdba.gov.au/heartbreak/article/a-dwindling-river .
85 MDBA (2018).
86 Ibid, 13.

### *11.5.4 Distributional justice and consultation*

A meta-discourse of participatory democracy was deployed to strengthen the re-assignment of values according to these processes. The consultation device is both a framing device emphasising communal benefit ('water sharing', 'shared targets', 'irrigation communities', 'regional communities', 'stakeholder consultation') and a strategy for reinforcing that framing device over time. Highly organised communication strategies accompanied the Australian National Water Initiative[87] and the *Sharing the Murray*[88] and *Living Murray*[89] processes. Legislative authority is provided for consultative processes; for instance, section 202 of the Water Act 2007 (Cth) establishes the Basin Community Committee with functions that include advising the Murray-Darling Basin Authority about engaging the community in the preparation of each draft Basin Plan; and community matters relating to the Basin water resources. Indeed, consultation is so frequent that consultants themselves have identified 'participation fatigue'.[90] Despite the dominance of techniques of communication, it is arguable that alternative mechanisms would have resulted in a greater (and more rapid) return of water to the environment.[91]

Taddei notes that

> [t]he political convenience of a participatory approach, in such a conflictive environment, besides meeting a requirement imposed by the World Bank, is that it allows for the possibility to exercise a great deal of semiotic manipulation of how things are framed.[92]

Taddei describes the parallel process of aligning values in the Brazilian context, involving 'participatory water allocation meetings'. At these meetings, knowledge is delivered to users in using technocratic language, and approaches to allocation are based on 'rationalization and efficiency'. The overarching conceptual framing is of water as a scarce economic resource which must be utilised to achieve the highest economic value. Their audience, however, has

> radically different ways of thinking about, and relating to water; differences which have developed throughout the region's local history, and which naturally transform the water allocation meetings into battlefields focused on the representation of natural resources, the role of science and technicality in society, economic development and politics.[93]

87 Daniell et al (2010).
88 MWEC (1997).
89 MDBMC (2002), 29.
90 Ernst and Young (2018), 221.
91 Crase, Dollery and Wallis (2005).
92 Taddei (2005), 326.
93 Ibid, 7.

These observations are highly relevant to the Basin context, even taking into account the differences in culture and financial and legal literacy. The legislative and scientific instruments accompanying the Plan and its implementation are highly technical, both in language and substantive context. Although the implementation of the Plan is dependent on the development of localised 'Water Resource Plans' (WRPs), the development and accreditation of WRPs is highly regulated and managed by a bureaucratic process.[94] Development of a WRP requires understanding of technical concepts such as 'risk identification' and 'risk assessment', 'mitigation' and 'management' of risk. The process also requires ongoing monitoring of implementation and effectiveness.

### *11.5.5 The [re]allocation ritual*

Taddei describes the water allocation 'ritual' in Brazil.[95] It involves, he notes, the 'technical-scientific terminology of the language game in which water management is decided upon', a terminology which 'alienates' some users from 'those who can play the language game well – primarily technicians from state and municipal governments, and from the agribusiness sector – [who] maintain control over water negotiations'.[96] The process also involves the attachment of 'economic rationality and capitalistic values to the technicalities of water management. In this environment, the committee meetings are ritual performances in which these discourses are legitimised and become a structuring element that shape the practice of water allocation.'[97]

Whilst the participatory allocation process in that context does not describe the process in Victoria (or Australia generally) for allocation of water, a description of similar 'ritual' could be applied to the process of determining the 'limits on the quantity of water that may be taken from the Basin water resources as a whole, and from the water resources of each water resource plan area'[98] – the 'long-term average sustainable diversion limits'[99] and the subsequent creation of WRPs, some of which will also include a water-sharing plan. The development, assessment and accreditation of these plans is a highly bureaucratised process involving state and Federal government agencies[100] and designated stakeholders as well as community consultation sessions.

## 11.6 Conclusions

Restoration of the eco-system of the Basin is premised on the return of water to the 'system'. This enormous project relies to a large extent on the disruption of

94 MDBA, *Water Resource Plans,* www.mdba.gov.au/basin-plan-roll-out/water-resource-plans.
95 Taddei (2005).
96 Ibid, 327.
97 Ibid.
98 Water Act 2007 (Cth) s 19.
99 Ibid, s 22(1).
100 MDBA (2017) 'Water Resource Plans: What They Are and How They are Developed', www.mdba.gov.au/publications/policies-guidelines/water-resource-plans-what-they-are-how-they-are-developed.

established understandings and relationships between human inhabitants of the Basin and the use of water in the Basin, including the economic, amenity, cultural and subsistence uses. The process of semiotic remapping of the relationship between water and water users has been consciously directed towards the overwriting of multiple associations between water and human users with a single understanding of water as an economic good. Simultaneously, the multiple sequential or simultaneous natures or careers[101] of water over time have been replaced by bureaucratic processes with alienating 'rituals' of accreditation. Human participants in the previous 'ecosystem' of water consumption have been disembedded from the environment, experiencing 'participatory' processes as symbolically compliant with democratic processes but distanced from lived communication. The semiotic framing of the process of reform involved the utilisation of 'efficiency' terminology in association with positive environmental outcomes, the incorporation of language and techniques of audit and verification, and the harnessing of traditional distributional justice techniques. The superstructure of semiotic framing, alongside the infrastructure of detailed regulatory systems, creates an irresistible momentum. All elements of the ecosystem now sit in a different relation with other elements as a result of this enterprise. The disruption in semiotic understanding will settle back into familiar associations along waterways and lakes – water in recreational and environmental contexts will find new or associated attributions to a degree, but the dissociation rendered by the intervening market and regulatory signifiers will change the relationship between landowners and water, hastening a purely instrumental attribution of value.

## References

### *Primary sources*

Convention on Biological Diversity (1992, 1760) UNTS 79; 31 ILM 818
Convention on *Wetlands* of International Importance Especially as Waterfowl Habitat (1971) TIAS No. 11084, 996 UNTS 245
Foreign Acquisitions and Takeovers Act 1975 (Cth) (Australia)
Foreign Acquisitions and Takeovers Regulations 2015 (Cth) (Australia)
Water Act 2007 (Cth) (Australia)

### *Secondary sources*

ABC (2017) 'Pumped: Who's Benefitting from the Billions Spent on the Murray-Darling?', *ABC*, 24 July, www.abc.net.au/4corners/pumped/8727826.
David Adamson, Thilak Mallawaarachchi and John Quiggin (2006) 'State-Contingent Modelling of the Murray Darling Basin: Implications for the Design of Property Rights', presented at the 50th Annual Conference of the Australian Agricultural and Resource Economics Society, Sydney, 8–10 February.

101 Kaplan (2007), 685.

ANAO (2017) *New South Wales' Protection and Use of Environmental Water in the Murray-Darling Basin*, Australian National Audit Office.

Gregory Bateson (2000) *Steps to an Ecology of Mind*, The University of Chicago Press.

Rachel Baxendale (2018) 'David Littleproud Strikes Deal with Labor to Save Murray-Darling Basin Plan', *The Australian*, 7 May, www.theaustralian.com.au/national-affairs/david-littleproud-strikes-deal-with-labor-to-save-murraydarling-basin-plan/news-story.

Charles Bazerman, Joseph Little and Terri Chavkin (2003) 'The Production of Information for Genred Activity Spaces' 20 *Written Communication* 455.

Tim Bonyhady (2000) *The Colonial Earth*, Melbourne University Press.

Kerry Brewster (2018) 'Murray Darling Basin Authority powerless to act against farmers harvesting floodwaters', *ABC News*, 14 February, www.abc.net.au/news/2018-02-13/murray-darling-authority-powerless-against-floodwater-harvesting/9426138.

John Butcher (2008) 'The Unfinished Project of Nation-building' in John Butcher (ed.) *Australia Under Construction*, ANU E-Press.

Lauren Butterly and Benjamin J. Richardson (2016) 'Indigenous Peoples and Saltwater/Freshwater Governance' 8(26) *Indigenous Law Bulletin* 3.

Peter S. Carberry et al (2010) *Technological Innovation and Productivity in Dryland Agriculture in Australia*, ABARE-BRS and CSIRO.

David Claughton (2014) 'Minister Rejects Namoi Water Call for Relaxation of Water Rules', *ABC News*, 21 January, www.abc.net.au/news/rural/programs/nsw-country-hour/2014-01-22.

Paul Cobley (2010) 'The Cultural Implications of Biosemiotics' 3(2) *Biosemiotics* 225.

Commonwealth of Australia (2008), *It's About People: Changing Perspective. A Report to Government by an Expert Social Panel on Dryness*, Report to the Minister for Agriculture, Fisheries and Forestry, Commonwealth of Australia, Drought Policy Review Expert Social Panel.

Tom Connors (1970) 'Closer Settlement Schemes' 42(1) *The Australian Quarterly* 72.

Lin Crase, Brian Dollery and Joe Wallis (2005) 'Community Consultation in Public Policy: The Case of the Murray-Darling Basin of Australia' 40(2) *Australian Journal of Political Science* 221.

Katherin Daniell et al (2010) 'Co-engineering Participatory Water Management Processes: Theory and Insights from Australian and Bulgarian Interventions' 15(4) *Ecology and Society* 11.

Peter Davies and Susan Lawrence (2014) 'A "Mere Thread of Land": Water Races, Gold Mining and Water Law in Colonial Victoria' 16 *Journal of Australian Colonial History* 165.

Department of Sustainability and Environment (2012) *Water Savings Protocol – Technical Manual for the Quantification of Water Savings in Irrigation Water Distribution Systems*, Government of Victoria.

Martin Drenthen (2016) 'Understanding the Meaning of Wolf Resurgence. Ecosemiotics and Landscape Hermeneutics' in M. Tønnessen, K. Oma and S. Rattasepp (eds) *Thinking about Animals in the Age of the Anthropocene*, Lexington Books, 109.

Ingrid Edlund-Berry (2006) 'Hot, Cold, or Smelly: The Power of Sacred Water in Roman Religion, 400–100 BCE' in C.E. Schultz and P.B. Harvey (eds) *Religion in Republican Italy*, Cambridge University Press, 162.

Ernst and Young (2018) *Analysis of Efficiency Measures in the Murray-Darling Basin*, Independent Report to the Murray-Darling Basin Ministerial Council.

Brian Fitzpatrick (1947) 'The Big Man's Frontier and Australian Farming' 21(1) *Agricultural History* 8.

Melissa Fyfe (2003) 'Nearly Closing Time Again for Mighty Murray' *The Age*, 23 April.
Grahame Griffin (2003) 'Selling the Snowy: The Snowy Mountains Scheme and National Mythmaking' 27(79) *Journal of Australian Studies* 39.
Patrick Hartigan (1921) 'Said Hanrahan' in *Around the Boree Log and Other Verses*, Angus and Robertson, 80.
P.J. Hughes and R.J. Lampert (1980) 'Pleistocene Occupation of the Arid Zone in Southeast Australia: Research Prospects for the Cooper Creek-Strzelecki Desert Region' 10 *Australian Archaeology* 52.
D. Ingle Smith (1998) *Water in Australia*, Oxford University Press.
Martha Kaplan (2007) 'Fijian Water in Fiji and New York: Local Politics and a Global Commodity' 22(4) *Cultural Anthropology* 685.
Webb Keane (2018) 'On Semiotic Ideology' 6(1) *Signs and Society* 64.
Igor Kopytoff (1986) 'The Cultural Biography of Things: Commoditization as Process' in A. Appadurai (ed.) *The Social Life of Things: Commodities in Cultural Perspective*, Cambridge University Press, 64.
Kalevi Kull (1998) 'Semiotic Ecology: Different Natures in the Semiosphere' 26 *Sign Systems Studies* 344.
Kalevi Kull (2007) 'A Brief History of Biosemiotics' in M. Barbieri (ed.) *Biosemiotics: Information, Codes and Signs in Living Systems*, Nova Science, 1.
Henry Lawson (1906) 'The Heart of Australia' in *When I was King and Other Verses*, Angus Robertson.
Kati Lindström, Hannes Palang and Kalevi Kull (2013) 'Semiotics of Landscape' in P. Howard, I.H. Thompson and E. Waterton (eds) *The Routledge Companion to Landscape Studies*, Routledge, 97.
Timo Maran and Kalevi Kull (2014) 'Cosemiotics: Main Principles and Current Developments' 96(1) *Geografiska Annaler: Series B, Human Geography* 41.
Lucy McNally (2018) 'Alleged Barwon-Darling Water Thieves to be Prosecuted after ABC Investigation', *ABC News*, 8 March, www.abc.net.au/news/2018-03-08/nsw-water-theft-barwon-darling-government-prosecuting/9527364.
MDBA (2011) *The Living Murray Story – One of Australia's Largest River Restoration Projects*, Report 157/11, Murray-Darling Basin Authority.
MDBA (2018) *Icon Site Condition*, Report 06/18, Murray-Darling Basin Authority.
MDBMC (1995) *An Audit of Water Use in the Murray-Darling Basin*, Murray-Darling Basin Ministerial Council.
MDBMC (2002) *The Living Murray: A Discussion Paper on Restoring the Health of the River Murray*, Murray-Darling Basin Ministerial Council.
MDBRC SA (2018a) *Issues Paper 1*, Murray-Darling Basin Royal Commission South Australia.
MDBRC SA (2018b) *Issues Paper 2*, Murray-Darling Basin Royal Commission South Australia.
MDBRC SA (2018c) *Terms of Reference – Explanatory Memorandum*. Murray-Darling Basin Royal Commission South Australia
MWEC (1997) *Sharing the Murray: Proposal for Defining People's Entitlements to Victoria's Water from the Murray*, Murray Water Entitlement Committee.
Sue Neales (2017) 'Farmers Fear the Darling's Down and Out' *The Australian*, 26 June.
Melissa Nursey-Bray and Arabana Aboriginal Corporation (2016) 'Lore, law and later governance: Insights into managing water for country, Australia' 8(27) *Indigenous Law Bulletin* 12.
Winfried Nöth (1998) 'Ecosemiotics' 26 *Sign Systems Studies* 332.

Winfried Nöth and Kalevi Kull (2001) 'Introduction: Special Issue on Semiotics of Nature' 29(1) *Sign Systems Studies* 9.

Erin O'Donnell (2018) 'Legal Rights for Rivers: More Power, Less Protection?', 23 April, www.internationalwaterlaw.org/blog.

Lisa Oberkircher and Anna-Katharina Hornidge (2011) '"Water is Life"—Farmer Rationales and Water Saving in Khorezm, Uzbekistan: A Lifeworld Analysis' 76(3) *Rural Sociology* 394.

A.B. 'Banjo' Paterson (1885) 'Waltzing Matilda' published 1903 to music composed by Christina Macpherson in 1995.

A.B. 'Banjo' Paterson (1889) 'Clancy of the Overflow', *The Bulletin*, 21 December.

Benjamin J. Richardson (2016) 'The Emerging Age of Ecological Restoration Law' 25(3) *Review of European, Comparative and International Environmental Law* 277.

Steele Rudd (Arthur Hoey Davis) (1899) 'On Our Selection', *Bulletin Newspaper*.

Margaret Somerville (2013) *Water in a Dry Land: Place-Learning Through Art and Story*, Routledge.

Malcolm Sutton (2017) 'Murray-Darling Basin: "Striking Variations" in Compliance as Review Criticises NSW and Qld', *ABC News*, 25 November, www.abc.net.au/news/2017-11-25/mdb-review-scathing-assessment-of-nsw-and-qld/9193362.

Renzo Romano Taddei (2005) '"Of Clouds and Streams, Prophets and Profits": The Political Semiotics of Climate and Water in the Brazilian Northeast', PhD thesis, Department of Anthropology, Columbia University.

Sandy Toussaint, Patrick Sullivan and Sarah Yu (2005) 'Water Ways in Aboriginal Australia: An Interconnected Analysis' 15(11) *Anthropological Forum* 61.

Helen Verran (2010) 'Number as an Inventive Frontier in Knowing and Working Australia's Water Resources' 10(1/2) *Anthropological Theory* 171.

Paula Villagra Islas (2010) 'Re-growth of Vegetation, Colour and Landscape Complexity: The Role of Design in the Interpretation of Natural Systems in Botanical Gardens' *Proceedings of the 4th Global Botanic Gardens Congress*, June.

K.F. Walker (1985) 'A Review of the Ecological Effects of River Regulation in Australia' 125 *Hydrobiologia* 111.

Angus Webb et al (2018) 'It Will Take Decades, but the Murray Darling Basin Plan is Delivering Environmental Improvements', *CSIROscope*, 2 May, https://blog.csiro.au/it-will-take-decades-but-the-murray-darling-basin-plan-is-delivering-environmental-improvements.

Caroline Winter (2015) 'Indigenous Groups Call for a New Water Partnership with Government on the Murray-Darling Basin System', *ABC News (Rural)*, 20 February, www.abc.net.au/news/rural/2015-02-20.

# 12 Reforming restoration law to support climate change adaptation

*Phillipa C. McCormack*

## 12.1 Introduction[1]

The editors of this book, and other contributions to legal academic literature on ecological restoration, have argued that environmental laws focus too heavily on the future.[2] They suggest that the primary focus of environmental law on preventing future loss and remediating immediate past harm is an impediment to ambitious restoration of the vast biological wealth that has been lost due to human activities. While supporting the goal of increasing ambition in ecological restoration laws, this chapter takes a different perspective. It argues that environmental laws are, in one very important respect, insufficiently focused on future threats. Despite increasingly detailed information about the challenge that rapid climate change poses to biodiversity in every biome on Earth, environmental laws, including for restoration, are currently ill-equipped to anticipate and respond to this challenge.[3]

This chapter takes the emerging concept of 'renewal ecology'[4] as a lens through which to analyse whether restoration laws and policies can enhance conservation in a period of rapid, anthropogenic environmental change. Renewal ecology emphasises the need to take adaptation-oriented approaches to restoring ecological health and function. While continuing to emphasise the importance of conserving the natural world, renewal ecology accommodates concepts of ecological novelty, and accepting a potential role for humans as well as non-human 'novel' species and interactions, in the task of renewing landscape-scale ecological functions.[5] This chapter demonstrates that Australia's legal frameworks for restoration, by contrast, are typically reactive, focused on a stationary and simplistic view of nature that assumes that harm can be 'undone' over relatively short timeframes.[6] The

1 This research was generously supported by a grant from the Institute for the Study of Social Change, for which I am very grateful.
2 For example, Akhtar-Khavari and Richardson (2017).
3 See generally, Ruhl (2010).
4 Bowman et al (2017).
5 Ibid; see discussion of a spectrum, especially in areas that cannot be restored to historical states, in Zedler, Doherty and Miller (2012).
6 For instance, rather than acknowledging long histories of dynamic environments and the certainty of increasingly rapid and complex environmental and climatic change.

chapter argues that the concept of 'renewal' provides a useful way to reconceive of the task of restoration. In particular, the concept of renewal has the potential to support new legal mechanisms for helping biodiversity to thrive, despite the dramatic challenge that climate change represents to life on Earth.

This chapter takes Australia as a case study, and proposes legal reform to support climate adaptation and ecological 'renewal' in terrestrial environments. Australia provides an interesting case study for analysing the challenge of climate change for ecological restoration. The Australian continent is 'megadiverse' but also has one of the highest rates of biodiversity decline in the world.[7] More than two centuries of landscape-scale fragmentation and habitat loss have had both immediate effects on species' survival and ecosystem functioning, and long-lasting legacy effects on environmental health and the adaptive capacity of remnant biodiversity.[8] Australian species and ecosystems are also particularly susceptible to climate change. A large proportion of Australia's biota cannot be found anywhere else on Earth and has evolved to survive in narrow, specialised environmental conditions. As a result, many Australian species have limited capacity to adapt independently to rapid climate change.[9] As such, Australian lawmakers and conservation managers will face the challenge of climate-driven species losses and redistributions sooner, and to a larger degree, than in many other jurisdictions.[10]

Ongoing biodiversity decline and the emerging threat of climate change render the task of improving Australia's laws for restoring ecological health, function and climate-adapted habitat increasingly urgent.[11] The case study approach taken here allows a detailed analysis of the relevant legal frameworks in Australia, and the practical challenges that climate change poses for traditional ecological restoration goals. However, recommendations made in this chapter for reforming Australia's legal frameworks for restoration are likely to resonate with, and may inform similar considerations in, other jurisdictions.

This chapter proceeds in four sections. Section 12.2 briefly describes the imperative to account for climate change in ecological restoration law and practice. Section 12.3 provides an overview of important trends in restoration scholarship. It highlights growing recognition of the importance of anthropogenic landscapes, human values and human engagement in restoration success. This section also introduces renewal ecology, to inform legal reform that facilitates biodiversity adaptation as the climate changes. Section 12.4 summarises the existing legal

7 Woinarski, Burbidge and Harrison (2015).

8 Cresswell and Murphy (2017).

9 Australian Department of the Environment and Energy (2004); Steffen et al (2009), 8–9, 93; on the implications of climatic sensitivity and barriers to adaptation for Australian conservation laws more generally, see McCormack (2018a).

10 A comparatively large proportion of the Australian biota will reach the limits of their environmental tolerance and adaptive capacity over coming decades, with potentially rapid increases in species extinctions and ecosystem collapse, Steffen et al (2009), 94–98.

11 For instance, more than 1,700 species of animals and plants are listed by the Australian Government as being at risk of extinction, Australian Department of the Environment and Energy (n.d.); Geyle (2018).

framework for facilitating restoration in Australia, highlighting particular challenges for climate adaptation-oriented restoration. Section 12.5 then argues in favour of supplementing restoration laws with new, renewal-oriented legal mechanisms to support conserving, and in some cases creating, novel habitats to shelter species and ecological communities from the worst effects of climate change.

## 12.2 Australian biodiversity and climate change: the imperative to restore and create climate refugia

Land clearing is currently the most significant driver of ecological decline and species extinction in Australia.[12] Large, intact and ecologically complex areas of vegetation provide habitat for species, sustain ecological communities and ecosystem processes, and facilitate climate change adaptation.[13] However, even regrowth, mixed native and non-native vegetation communities, and small patches of vegetation can provide important habitat, landscape connectivity and climate 'buffering' for biodiversity, especially in otherwise heavily cleared landscapes.[14]

Historical clearing practices have left just 25 percent of the original extent of Australia's native vegetation intact.[15] Those intact areas are distributed unevenly across vegetation and community types, with less than 1 percent of the original extent remaining for some ecological communities, and many more heavily fragmented and close to collapse.[16] In some cases, lost vegetation types will never be able to be restored within their original distribution because that land has been converted to other uses, such as cities or agriculture. These legacy effects of historical habitat clearing underpin an urgent need to restore lost vegetation and ecosystems. However, renewing ecological functions, such as habitat provision, will also be crucial for mitigating climate-driven habitat loss (such as from more severe and frequent bushfires) and climate changes that render existing habitat unsuitable, including drying and warming trends.

The most significant changes to Australia's climate for the purposes of this chapter include rising temperatures, shifting rainfall patterns, rising sea levels, and more frequent and severe weather events such as bushfires, droughts and floods.[17] The implications are more complex as they reverberate within and between ecological and social systems. For example, rising temperatures and shifting rainfall patterns will break down the 'niche' habitat conditions in which species have evolved.[18] As a result, preserving species in their historical location and conditions will often be impossible.

12 Evans et al (2011); although other threats are substantial, especially invasive predators such as feral cats, Woinarski, Burbidge and Harrison (2015).
13 Watson et al (2016).
14 Bowen et al (2007).
15 Cresswell and Murphy (2017), 3.
16 Ibid.
17 See generally, Steffen et al (2009); Reisinger et al (2014).
18 Steffen et al (2009), 138.

Instead, climate change is already triggering species redistributions to new areas.[19] In some cases, these redistributions will cause species to migrate well outside of the historical ranges in which they are considered native, with implications for the application of conservation laws that are typically founded on concepts of 'nativeness'.[20] Other species and ecological communities will retreat to small pockets of remnant habitat. Under such conditions, many species will become extinct.[21] Furthermore, those species with the capacity to move independently to new areas will do so at different rates, triggering ecological 'shuffling' and ecosystem collapse. These movements will trigger demand for new forms of habitat – including in locations where the necessary climatic conditions to support those species may not yet exist. It will also create new challenges for conserving ecosystems into which new species are migrating, and for conserving their component 'native' species.

A recent report by Australia's Commonwealth Scientific and Industrial Research Organisation found that by 2070, most of Australia's bioregions could be entirely dissimilar to those that exist today.[22] That report used medium- and high-impact emissions scenarios from the Intergovernmental Panel on Climate Change to generate modelling on the significance of future environmental change in Australia.[23] It found that – for a large proportion of Australia's biodiversity – the combinations of temperature, rainfall and vegetation that species and ecological communities currently rely on to survive, will no longer exist anywhere on the continent.[24] Given these findings, the goal of restoring any environment to an ecological condition with a historical reference point – whether from the recent or more distant past – may be futile.[25] Importantly, even focusing on current environmental conditions may be unhelpful for informing the restoration or creation of habitat to sustain biodiversity over the medium to long term.

This chapter investigates the potential value of emphasising ecological renewal, rather than restoration, to support conservation in the context of such significant environmental change. For example, focusing on ecological renewal could support restoration activities to *create* 'climate refugia' – habitat that can shelter species from the most severe impacts of climate change.[26] Climate refugia could be created for climate adaptation-oriented conservation *in situ* (to benefit species and ecological communities in the areas in which they are considered native) and/or

19 Pecl et al (2017).
20 McCormack and McDonald (2014); McCormack (2018b).
21 IPCC (2014); Moritz and Agudo (2013).
22 Dunlop et al (2012), 22.
23 Ibid, 14–17; the 'medium' emissions scenario (A1B) anticipates the introduction of renewable energy sources to meet growing energy demands, while the 'high' emissions scenario (A1F1) anticipates high dependence on fossil fuels. Dunlop et al note that the high scenario 'is actually closer to "business as usual" for the near future', as greenhouse gas emissions had already reached or exceeded the trajectory of that scenario in the decade to 2012, 14.
24 Ibid.
25 See Hagerman and Satterfield (2014).
26 Reside et al (2013).

outside species' native distributions, in areas that will be both accessible and climatically suitable as the climate changes.

Early research on the location of existing climate refugia in Australia indicates that refugia are not distributed evenly across the continent. One particularly important landscape feature for climate adaptation is rapid elevation change, that is, mountains.[27] This is because as temperatures increase, species can access cooler climates by shifting a short distance upslope, rather than needing to travel potentially vast distances across land.[28] Australia is a relatively flat continent but greater elevation changes can be found in the south and especially the southeast. Figure 12.1 illustrates the topography of the Australian continent, with those areas of highest elevation in black. This figure demonstrates the uneven distribution of areas of high altitude across the continent, and the significance of south and southeast Australia for potential climate refugia.[29] This figure also reveals substantial potential adaptation challenges for biodiversity currently located in the north and northwest of the continent.

Data produced and analysed by Dr April Reside and colleagues recently highlighted the areas of the continent that will be subject to the lowest rates of temperature change and levels of species emigration (that is, species 'moving out' to find more suitable habitat conditions). They found that, again, these areas are heavily biased to the south and southeast of Australia. Reside and colleagues also demonstrated that many Australian species will be unable to access existing climate refugia.[30] This is because the rate and scale of climate change will be too large, and the distance from existing habitat to climate refugia will be too great, for most Australian species to migrate in time.[31] 'Hard barriers' such as cities, deserts and water bodies will also limit the accessibility of climate refugia.[32] Alarmingly, this research demonstrates that for some Australian biodiversity, climate refugia do not exist.[33] Those species and ecological communities that cannot access climate refugia – or for which climate refugia do not exist – face a far greater risk of extinction.

This increased risk of extinction may be able to be reduced through human intervention.[34] One such intervention that has received little attention in the climate adaptation literature to date is to create climate-adapted ecological characteristics in degraded areas, supporting *new* climate refugia where existing environments cannot provide climate-ready habitat. Given the scale of the

27 Reside et al (2014).
28 While specific features of climate refugia will differ, depending on specific habitat and life-cycle requirements, elevation is a consistent indicator, with other features of climate refugia include high levels of vegetation cover and non-seasonal or ephemeral water sources such as rivers and soaks, Morelli et al (2016), 2–4.
29 Reside et al (2013).
30 Ibid; Reside et al (2014).
31 Reside et al (2013).
32 For instance, the southern-most state of Tasmania will likely be the location of many important climate refugia but, for most species, the state will be out of reach without assistance to cross the Strait that separates the state from mainland Australia.
33 Reside et al (2013), 1, 2.
34 For example using conservation introductions, McCormack (2018b).

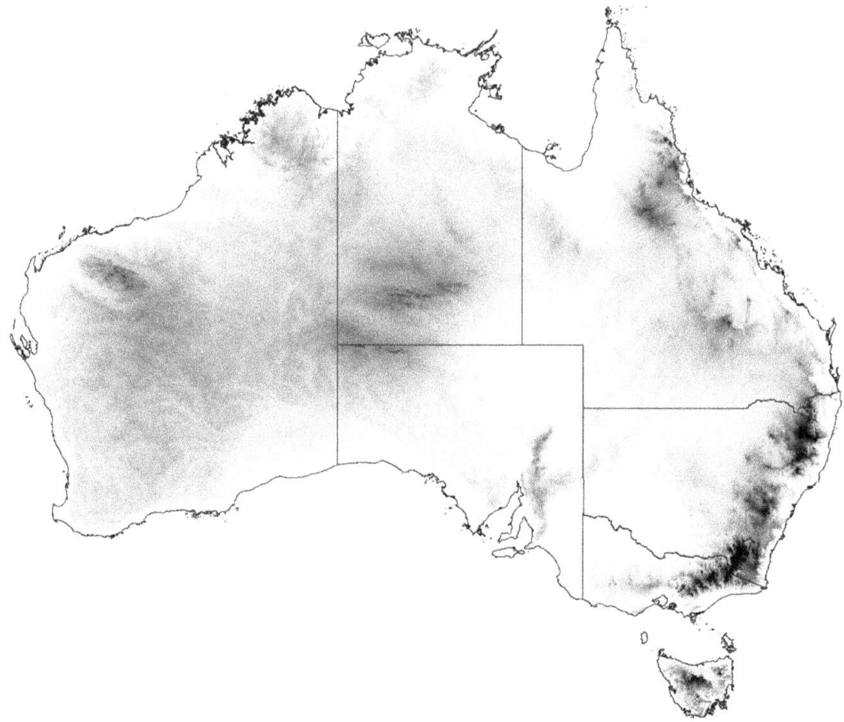

*Figure 12.1.* Topographical diversity of the Australian continent, using data from Geosciences Australia 2018

ecological threat posed by climate change, new approaches to restoration that optimise climate-adapted outcomes are essential.[35] Given the extent of vegetation and habitat loss across the Australian continent, actively creating climate refugia may become crucial for avoiding catastrophic, climate-driven biodiversity loss in future.

It is noteworthy that it is precisely the anthropocentric 'domination' of Australian landscapes, especially through broadscale clearing for agriculture and urban development,[36] that has created the need for more assertive human intervention for conservation. It is also perhaps paradoxical that an important response to the harm caused by human manipulation of the environment might be greater and more purposeful human intervention. Human intervention has been a feature of Australian landscapes for some 50,000 years under Aboriginal stewardship, mostly far more ecologically benign than that of the past two centuries of European settlement, so the issue to reckon with is not the presence of human intervention as such, but rather its purpose and quality. Importantly, engineering new habitat will

35 Breed et al (2013), Williams et al (2014) and Havens et al (2015) in Prober et al (2015), 1.

36 Australian Department of the Environment and Water (n.d.).

certainly not diminish the urgent need to reduce ongoing, native habitat loss; to invest appropriately in recovering populations of threatened species and ecological communities; and to fully implement and enforce existing conservation laws.[37] Nevertheless, the remainder of this chapter argues that, among other responses, lawmakers can and should expand legal frameworks for ecological restoration to facilitate the creation and conservation of new forms of habitat, providing climate refugia for as much Australian biodiversity as possible.

## 12.3 Restoration ecology: evolving concepts and practices

Scientific scholarship on restoration theory and practice has evolved a great deal over recent decades. While continuing to recognise the role of historical ecological baselines, including for informing restoration goals,[38] restoration ecology has expanded in focus to accommodate new concepts, theories and paradigms.[39] This section describes a selection of examples of this broader focus, providing context for the legal analysis that follows.

'Classical' approaches to restoration emphasised 'recovering' environments to historical reference points or trajectories, although restoring an area to a precise historical baseline has rarely been the sole objective in practice.[40] Historical information about the characteristics and components of ecosystems can be valuable, including for adaptation, in demonstrating the resilience of a local ecosystem in periods of past, rapid environmental change.[41] Indeed, Principle 1 of the *National Standards for the Practice of Ecological Restoration in Australia* (the 'Australian National Standards') adopts the idea of 'history as information and reference',[42] suggesting that ecological restoration practice should be 'based on an appropriate local indigenous reference ecosystem'.[43] Classical approaches have been criticised for failing to account for long histories of Indigenous use and connection with 'natural' landscapes, and the implications of anthropogenic climate change, among other things.[44] Restoration sciences now typically acknowledge that reinstating historical reference ecosystems is often undesirable, including because historical conditions will be difficult or impossible to recreate as temperatures, rainfall and species compositions shift and change.[45]

New paradigms have emerged in restoration ecology scholarship and practice to accommodate rapid environmental change and acknowledge the inseparability of

37 For a discussion of legal reforms more broadly, see APEEL (2017a).
38 Higgs et al (2014); McDonald et al (2016), 12; SERA (2017), 19, 27.
39 Perring et al (2015), 131.
40 For example, McDonald et al (2016); Rohwer and Marris (2016); Higgs et al (2014), 499.
41 Higgs et al (2014), 501–503.
42 Ibid, 504.
43 SERA (2017), 4–6.
44 For example, Rohwer and Marris (2016); Hobbs et al (2014).
45 For example, SERA (2017), 5; but see Higgs et al (2014), 499 arguing that 'discounting the importance of history in [restoration ecology] is both premature and unwarranted'.

human and non-human nature. For example, ecologists are increasingly targeting restoration projects towards improving ecological resilience to climate change and other environmental disruptions,[46] while also maximising ecological diversity, functions and health rather than pursuing fidelity to historical baselines.[47]

Second, there is growing recognition of the integral role of humans in ecological restoration. Examples of such scholarship include agroecology,[48] eco-stewardship,[49] reconciliation ecology,[50] and social-ecological systems thinking.[51] Research in these areas contributes new perspectives that challenge the idea of ecological restoration being divorced from human influence; and that seek to engage human cultures and values in setting restoration goals and priorities.[52] For example, the most recent version of the Australian National Standards acknowledges that 'social aspects are critical to successful ecological restoration'.[53]

One area of slow progress has been integrating ecological knowledge, experience and land-management expertise held by Indigenous and First Nations peoples into restoration planning and practice.[54] Some examples of successful policy responses can be seen in Australia's Caring for Country[55] and Indigenous Rangers programmes,[56] where traditional land management techniques such as periodic burning are being implemented to restore to health Indigenous-owned and -managed lands, known as 'Country'.[57] Researchers are also sharing new ways of bridging traditional knowledge systems and Western science to inform policy-making as well as practical, on-ground management.[58] Such considerations already form part, for example, of the work of the Intergovernmental Science-Policy Platform on Biodiversity and Ecosystem Services (IPBES) on land degradation and restoration.[59] However, strong critiques remain, including of the strict nature–culture dichotomies in some restoration literature; privileged 'colonial visions' of

46 Prober et al (2015), 2.
47 For example, Wardell-Johnson et al (2015).
48 For example, Perring et al (2006); and see Bartel and Graham, Chapter 5 in this volume.
49 For example, Chapin et al (2010); there are some limited examples of the concept of stewardship in environmental law, such as the Environmental Management and Pollution Control Act (1994) (Tas), s 23A, and Land Act 1994 (Qld), s 199 and Environment Protection Amendment Bill 2018 (Vic), which each impose a form of 'general environmental duty'.
50 Rosenzweig (2003).
51 Eisenberg (2017).
52 Martin et al (2014), 746ff; Wiens and Hobbs (2015); Rohwer and Marris (2016).
53 SERA (2017), principle 6.
54 Godden and Cowell (2016); Evans and Davis (2018).
55 Weir, Stacey and Youngetob (2011).
56 Australian Department of Prime Minister and Cabinet (2017).
57 Traditional owners have a fundamental role to play in Australia's conservation land management and ecological restoration planning, as more than 40 percent of the Australian National Reserve System is on Indigenous owned land, Australian Department of the Environment and Energy (2016).
58 Tengö et al (2017).
59 Ibid; IPBES (2018).

ecological restoration; and marginalising cultural goals that stem from some definitions of ecological restoration.[60]

As described in section 12.2, environmental degradation by humans is a key driver of the need for ecological restoration. Human adaptations to climate change, such as retreat from exposed coastal inundation, and 'tree changes' to escape urban heat islands, have the potential to both undermine existing restoration efforts and dramatically affect remnant and intact habitats.[61] As different organisms respond to climate change in different ways and at different rates, restoration planning will need to anticipate and manage adaptive and, potentially, ecologically maladaptive human behaviours and values.

A third and more controversial development in restoration ecology recognises potential value in novel combinations of organisms for providing ecological functions and processes in 'anthropogenic landscapes'.[62] Relevant scholarship includes intervention ecology and research on ecosystem realignment, managed relocation and novel ecosystems.[63] This scholarship takes a particular interest in environments that are subject to pervasive human influence, heavily degraded ecosystems and the implications of rapid environmental change.[64] For example, proponents of these approaches assert 'that future conservation actions must be extended to ecosystems directly shaped by humans and must embrace and leverage varying intensities of human activity'.[65]

Intervention-oriented scholarship is not applied to highly functional, intact and remnant ecosystems. Rather, proponents argue that '[c]onservation goals in areas directly shaped by humans need not be less ambitious than those in "natural areas"'.[66] For the purposes of this chapter, intervention-oriented paradigms are not highlighted to advocate the 'benefits' of anthropogenic environmental degradation, nor are they adopted as an excuse for giving up on ambitious ecological restoration goals.[67] These approaches are highlighted here for their contribution to ecological restoration theory and practice in landscapes that could never be restored to historical standards of health, 'nativeness' or 'pristine' ecological interactions.

In fact, intervention-oriented paradigms may have particular value in the context of climate-driven change. They explicitly adopt purposeful, future-oriented planning and experimental methods, and they are guided by adaptive forms of management designed to respond in conditions of uncertainty. Accepting that climate change will trigger cascading ecological and social

60 Evans and Davis (2018); Godden and Cowell (2016).

61 Ruhl (2010), 388–389.

62 Zedler, Doherty and Miller (2012).

63 Perring, Audet and Lamb (2014); Hobbs et al (2011) argue that the capacity to reinstate ecological *functions* and *processes* should be the goal of restoration; while Seastedt et al (2008) focus on 'desired outcomes and trajectories' for new species combinations; and see Martin et al (2014), 746.

64 For example, reconciliation ecology, urban ecology, social-ecological systems thinking (or 'human ecology'), intervention ecology and novel ecosystems.

65 Martin et al (2014), 746.

66 Young, Petersen and Clary (2005).

67 Murcia et al (2014); Miller and Bestelmeyer (2016).

transformations, these paradigms seek to integrate emerging science, including for climate adaptation and ecological restoration, with accountability for informed and pragmatic decision-making.

Finally, a recent paper describes a new sub-discipline called 'renewal ecology'.[68] The authors propose to reconceive ecological restoration theory and practice in the context of anthropogenic climate change. They argue for a 'solutions-focused discipline' that advocates innovation in restoring, creating and managing habitat and ecosystems for the future, 'to maximise both biodiversity and human well-being in the face of rapid environmental change'.[69] While renewal ecology continues to support avoiding species extinctions and promoting healthy environments, its emphasis is not on maintaining ecological functions that rely on species assemblages in 'native' environments. Rather, it emphasises the need to sustain ecosystem functions and ecological health under rapidly changing climatic and environmental conditions – without requiring fidelity to native species combinations or assemblages.

As the latest development in a discipline that is continually evolving, renewal ecology represents a 'growing acceptance that the future of biodiversity and the provision of ecosystem services will depend on more radical interventions than have been previously countenanced, *including the intentional creation of novel ecosystems*'.[70] The authors propose renewal ecology as a new scholarly focus, acknowledging resistance to the concept of novel ecosystems and objections to perceived managerialism or environmental dominion inherent in the idea of engineering climate-adapted habitat. However, they argue, given the enormity of the changes we face, 'a new approach is required' that can help to 'minimize the collateral damage of human adaptation and development in an anticipatory and collaborative way'.[71]

To further define the concept, the authors provide a series of 'concept examples' that demonstrate co-benefits for human and non-human adaptation and well-being. These concept examples include constructing coastal reef, mangrove and saltmarsh ecosystems to reduce erosion and mitigate storm damage instead of hard-engineering solutions such as concrete seawalls; focusing urban design on city 'greening' to cool cities and connect people to nature; and capitalising on abandoned mining and marginal agricultural land by using those areas to enhance connectivity and construct critical habitat.[72]

Renewal ecology is a very new concept and needs further work to define its full scope and operation. For example, the concept's authors highlight the importance of 'minimising losses and maximising opportunities' for both humans and non-human biodiversity,[73] but they do not provide any guidance about balancing competing interests outside of the concepts listed above. For example, abandoned

68 Bowman et al (2017).
69 Ibid, 675.
70 Ibid, 678, emphasis added.
71 Bowman et al (2017), 676; Heller and Hobbs (2014).
72 Bowman et al (2017), 676–678.
73 Ibid, 678.

mine sites have long been the site of proposed restoration projects, but there is no guidance for 'renewal' in the context of new threats such as biodiversity that is lost as humans relocate to escape sea-level rise. Similarly, there is no guidance for balancing trade-offs in the allocation of emergency services resources or recovery grants where bushfires concurrently affect both human communities and World Heritage natural values.[74]

An appropriate focus for 'purposeful human intervention', as described in this section, may be easy to identify in some cases.[75] However, decision-making will be more complex (and controversial) for ecosystem-based interventions such as creating climate refugia, where trade-offs between human well-being and other biodiversity are unavoidable. This is not unique to 'high-intervention' conservation strategies. Debates about triage and the allocation of limited funding in conservation decision-making are ubiquitous.[76] Policy-makers, researchers and concerned communities will need to continue to grapple with these complex questions, regardless of whether renewal ecology is used to inform adaptation-oriented restoration laws.

In a time that has been described as both the 'sixth mass extinction' and a period of 'biodiversity crisis',[77] these developments in restoration ecology, highlighting opportunities to repair some of the damage caused by human activities, may offer hope for the future. Ecological restoration is proposed, in these emerging paradigms, as a tool for active intervention, and a decisive response to some of the most intractable environmental challenges that we face.[78]

## 12.4 Australia's legal frameworks for restoration: increasingly at odds with restoration science

Despite the evolution in scientific scholarship and practice, there has not been an equivalent evolution in restoration law. Australian law imposes few legal obligations to halt existing harmful activities, and even fewer obligations or incentives to actively recover species, habitats or ecosystems. The limited legislative and policy provisions for restoration in Australia focus almost exclusively on reinstating current (or recent historical) environmental conditions.[79] These provisions typically seek to return environments to 'normal' after anthropogenic harm, framing restoration as a process of returning to pre-disturbance environmental conditions. The absence of more holistic and comprehensive approaches to ecological restoration (or even to more simplistic

74 For example, Hobday et al (2016).
75 Current examples of purposeful interventions already exist in Australia, including engineering habitat for endangered handfish (Australian Department of the Environment and Energy 2015) and for endangered seahorses (Millington and Notzon 2018).
76 For example, Bottrill et al (2009).
77 Hoag (2010); Ceballos et al (2015), e1400256.
78 Perring et al (2015); Suding et al (2015).
79 But for a shift in international legal frameworks, supporting 'ecological' restoration and acknowledging climate change, see Conference of the Parties to the Convention on Biological Diversity (2014).

ideals of rehabilitation) are at odds with the dramatic scale of biodiversity loss experienced across the Australian continent. Nevertheless, limited provisions for restoration in Australia's domestic environmental laws do arise, in the following contexts.[80]

First, statutory planning processes may require habitat restoration, including in threatened species-recovery planning or protected area-management plans. Restoration obligations in threatened species-recovery plans have traditionally taken a classical approach to historical baselines for restoration.[81] However, since the introduction of the latest Australian National Standards for ecological restoration, recovery plans no longer require restoration activities to adhere to the concept of an 'original range' or to reinstate historical habitat conditions.[82] Despite this (informal) shift, fewer than 40 percent of federally listed threatened species and ecological communities have recovery plans, with state and territory governments also falling short of statutory requirements to make and implement plans. Where recovery plans exist, very few place any prescriptive limits on future habitat loss, let alone impose specific obligations to restore or create new habitat to support these species.[83] Recovery plans also have limited capacity to influence land management for biodiversity wholly or partially distributed on private land – especially in the absence of a formal conservation agreement such as a covenant. As a result, and without dramatic increases in implementation and funding, threatened species-recovery planning does not represent a promising legal mechanism for facilitating adaptation-oriented ecological restoration in practice.

Rehabilitation provisions in protected area-management plans can also require, in some cases, the installation of new habitat features such as large woody debris or fabricated tree hollows for habitat.[84] These developments indicate that existing legal and policy frameworks have some potential to facilitate adaptation for biodiversity, including through the creation of new, climate-adapted habitat. However, restoration provisions in protected area-management plans are typically restricted to using 'local provenance'. That is, re-vegetation and restoration activities must rely on native species and even be restricted, in some cases, to species 'from within the boundaries of the park'.[85] Such exclusions may have a protective effect, for example by prohibiting the introduction of invasive alien species to protected areas. However, they also limit the capacity for land managers to engage in adaptation-oriented restoration activities, especially as the distribution of species that

80 See also Richardson (2016) for a similar typology.

81 Although, exceptions include the Australian Department of the Environment and Energy (2015), *Recovery plan for three handfish species*, which details habitat-related threats and promotes introducing engineered or artificial spawning habitat to improve recruitment and population survival, 4, 19.

82 Applies to conservation advice and threatened species-recovery plans drafted since 2016 under the federal Environment Protection and Biodiversity Conservation Act 1999 (Cth), personal communication, December 2017, member of the Commonwealth Threatened Species Scientific Committee.

83 ACF, Birdlife Australia and Environmental Justice Australia (2015).

84 For example, *Kara Kara National Park Management Plan 2013* (Parks Victoria 2013), 13, 18.

85 For example, Tasmanian Parks and Wildlife (2002), 25; Parks Victoria (2003) 10; McCormack (2018b).

are native 'within the boundaries of the park' shift or contract as a result of new climate conditions.

Second, an obligation to restore environmental processes or functioning may be imposed as a penalty, either for harm to the environment or for a breach of environmental laws. For example, a regulatory authority may order a person who has polluted the environment to take action to clean up the pollution and remediate the area of the spill.[86] These provisions are very common but do not go so far as to require the remediation of ecological functions. On the contrary, a respondent who intentionally cleared part of a Ramsar wetland on his property to plant wheat was ordered to restore the wetland to its former condition, but the fact that the wetland was not in pristine condition prior to its destruction was held to be a *mitigating* factor in sentencing.[87] Climate change will undermine the 'pristine' nature of some, and perhaps most, protected areas, and harm that increases the climate vulnerability of an area or reduces its adaptive capacity may need to be an aggravating, not a mitigating, sentencing factor.

Third, 'make good' obligations, especially in forestry and mining legislation, may require some form of rehabilitation.[88] That is, upon completion of certain permitted activities, legislation typically imposes an obligation to rehabilitate or revegetate the activity site.[89] In addition to the limited success of many statutorily mandated rehabilitation projects, these provisions also suffer from limited implementation and poor regulatory enforcement.[90]

Fourth, biodiversity offsetting schemes may support habitat restoration to create climate refugia, to offset a development removing or damaging habitat elsewhere. Offsets are gaining popularity as a market mechanism that promises to ensure biodiversity losses in one location are compensated by comparable biodiversity gains in another.[91] However, offsetting policies have been heavily criticised, including for allowing habitat to be destroyed without the necessary safeguards to ensure that the loss is, in fact, effectively compensated by equivalent, or indeed any, environmental gains.[92] Biodiversity offsets are not a preferred method for creating and conserving climate refugia. This is primarily because any activity that permits destruction or damage to existing habitat, especially habitat of a high ecological value, may in fact be the cause of current habitat and future climate refugia being lost, rather than gained.

86 Protection of the Environment Operations Act 1997 (NSW) s 91; Environmental Management and Pollution Control Act 1994 (Tas) s 44(3)(c).
87 *Minister for the Environment and Heritage v Greentree* (No. 3) (2004) LGERA 136.
88 Restoration, with its focus on the recovery of ecological health and function, is often distinguished from 'remediation' or 'rehabilitation', eg, SERA (2017), 1–2; Richardson (2015).
89 For example, Forest Practices Act 1985 (Tas); Mining Act 1992 (NSW) Part 11 Division 3 'Environmental, rehabilitation and other directions'.
90 ABC (2017); Parliament of Australia (2017).
91 Bull et al (2013).
92 Maron et al (2012); Gibbons and Lindenmayer (2007).

Fifth, restoration may be cited as a goal or object in overarching statutory purpose provisions. For example, the main object of the Nature Conservation Act 2014 [Australian Capital Territory (ACT)] is to 'protect, conserve and enhance the biodiversity of the ACT'.[93] Similarly, the federal Water Act 2007 aims to 'protect, restore and provide for the ecological values and ecosystem services of the Murray-Darling Basin'.[94] The Water Act, in particular, emphasises conserving and improving ecological health rather than reinstating specific historical assemblages in particular locations. However, while these provisions are often ambitious and ecologically holistic, they are so rarely attached to substantive, enforceable legal obligations that their practical utility is limited.[95]

Finally, some state jurisdictions have enacted general enabling provisions for restoration or rehabilitation in natural resource management legislation, other than as a penalty for environmental harm. For example, the Natural Resources Management Act 2004 (SA) provides that the Minister may enter into management agreements that relate to, among other things, 'the conservation [...] enhancement, restoration or rehabilitation of any natural resources'.[96] Along similar lines, section 6 of the Environmental Restoration and Rehabilitation Trust Act 1990 (NSW) creates a financial trust 'to encourage and support restoration and rehabilitation projects'. However, statutory provisions that seek to generate environmental 'gains' in this way are extremely rare in Australian law.[97]

Shifts in restoration practice in Australia hold promise for ecological restoration as a tool for facilitating climate adaptation for biodiversity, despite the limitations of environmental laws. For example, the recently released Australian National Standards recommend prioritising local, 'native' reference systems as the starting point for restoration activities, but they also explicitly acknowledge the implications of climate change for restoration activities, and encourage revisions to the restoration goals and outcomes of any project in light of 'observable or likely changes'.[98] Restoration practitioners are also increasingly acknowledging the need to do more than re-create past ecological conditions, for example by incorporating climate projections into restoration decision-making,[99] and combining native and warm-adapted seed stock to improve the adaptive capacity of newly planted areas.[100]

93 Nature Conservation Act 2014 (ACT) s 6(2); see also Catchment and Land Protection Act 1994 (Vic) s 4(ii); protected area legislation may also specify management objectives for restoring the environment, eg National Parks and Reserves Management Act 2002 (Tas) Sch 1, cl 1-10.

94 Water Act 2007 (Commonwealth) s 3(d)(ii), emphasis added.

95 McCormack (2018a).

96 Natural Resources Management Act 2004 (SA) s 205; see also Conservation and Land Management Act 1984 (WA) s 33(1)(cc).

97 Cf, Canada's National Parks Act (SC 2000, c32), s 8(2) provides that 'the maintenance or restoration of *ecological integrity* [...] shall be the first priority of the Minister administering the Act', referenced in APEEL (2017b), 40.

98 SERA (2017), 6.

99 Wardell-Johnson et al (2015).

100 Prober et al (2015).

No restoration law in Australia supports the creation and conservation of engineered habitat as climate refugia, or anticipates the need to focus on future threats or challenges to long-term restoration goals. Although there have been instances dating from the early twentieth century where threatened species have been relocated huge distances to safer environments, such as placement of koalas on Kangaroo Island and lyrebirds in Tasmania, these interventions did not involve the creation of new habitat. Formal legal frameworks, including legislation, regulations and government policies, have failed to evolve alongside developments in restoration scholarship. A response to this legal failure is urgently required, despite some promising non-government developments such as the 'Alps to Atherton' restoration and connectivity initiative.[101] Without explicit legal and policy guidance for ecological restoration, there remains little incentive to prioritise government and private funding for restoration, and to take necessary, coordinated actions across state and territory borders to achieve landscape or continental-scale restoration goals.

## 12.5 Designing restoration laws for climate adaptation and ecological 'renewal'

Recognising climate change as an important driver of habitat loss and decline, and as an exacerbating factor for historical fragmentation and habitat loss, is an important prerequisite for adaptation-oriented restoration laws. However, recognising the threat of climate change to Australian biodiversity will not be sufficient. Conservation decision-makers must also be equipped to 'confront and respond, quickly and effectively, to new challenges and issues as they arise'.[102] Improving legal responsiveness in this way will require a shift from reactive and *ad hoc* restoration laws, to purposeful and holistic frameworks including restoration-specific provisions in legislation, supporting policy and appropriate funding.[103] This section draws on recent developments in the restoration ecology literature, described in section 12.3, to propose an adaptation-oriented approach to ecological restoration.

First, clear legislative goals will be required, to support substantive obligations or incentives to identify, restore and conserve climate refugia. Given that 'native' habitat conditions will no longer exist for some Australian species and ecological communities, restoration legislation (or restoration provisions in general conservation laws) will need to redefine goals for protecting 'native' habitat and species in their 'natural' ranges, to accommodate climate-driven species shifts. The form of such provisions may simply require the introduction of a qualification, that 'native habitat' is habitat in which a species 'occurs *or is likely to occur*'.[104]

101 Worboys et al (2008).
102 Hobbs et al (2011); Trouwborst (2009), 424.
103 Trouwborst (2009), 424.
104 Drawing on the Nature Conservation Act 2014 (ACT) s 64(7), which provides that a 'native species' includes species that occur or are likely to occur in the Territory; see discussion of 'climate envelopes' in Thomas et al (2011).

Restoration legislation should also support habitat restoration to *create* future climate refugia, even if the required habitat type does not yet exist in the area. For experimental or habitat 'engineering' projects such as these, a precautionary approach should be taken to selecting the most appropriate site. New risk-assessment procedures could balance the risks to any biodiversity resident in an area proposed to be restored with the risk of future extinctions if the proposed climate refugia was not created. Areas that may be appropriate for anticipatory and experimental restoration activities may include former mine sites, abandoned agricultural areas and brownfield sites.[105]

The state of Western Australia has a policy framework for proactively identifying potential habitat in the analogous context of conservation translocations. Western Australia's translocation policy[106] creates a category of land defined as 'Fauna Reconstruction Sites': '"Fauna Reconstruction Site" means an area where the Department of Conservation and Land Management (CALM) is proposing to reconstruct, or is reconstructing, the vertebrate fauna as far as is possible through predator control, habitat management and translocations.'[107]

Clause 4.1 of the policy provides that CALM will 'Designate areas as Fauna Reconstruction Sites and Species Recovery Sites, and publish a list of such sites annually.'[108] Research to identify and designate areas as climate refugia over coming decades, in anticipation of species requirements, could employ a similar policy approach. This approach would support strategic planning opportunities, across private, public and Indigenous owned-land, at a lower cost than identifying climate refugia on a case-by-case basis. Establishing a list of 'climate refugia sites' in anticipation of future habitat requirements may also mitigate the risk of particularly valuable future climate refugia being allocated to urban or resource developments.

As a necessary corollary to the reforms proposed in this chapter, once created, climate refugia must also be able to be *protected* under conservation laws. Protection mechanisms may include designating new climate refugia as 'critical habitat' for species that may persist there as the climate changes. At the very least, these areas should be able to be recognised as areas of high ecological value, worthy of conservation effort, funding and protection from resource extraction and other development interests.[109]

Laws and policies for conserving protected areas could be useful in this regard, but in 2013, 'approximately half of the intact habitat currently occupied by [...] threatened species [was] not protected [...] and this [was] much the same for

105 For example, employing the ecosystem design framework described by Drake et al (2010); Bowman et al (2017), 676–678.
106 Policy Statement 29: Translocation of Threatened Flora and Fauna (Western Australia, July 1995).
107 Ibid, 1.
108 Ibid, 5.
109 For example, as 'areas of outstanding biodiversity value', under the Biodiversity Conservation Act 2016 (New South Wales) s 3.2, including areas that make a significant contribution to the persistence of 'irreplaceable biological distinctiveness' or 'ecological processes or ecological integrity' – concepts that are sufficiently broad to apply to climate refugia.

projected future habitats'.[110] Some of these areas may represent valuable potential climate refugia. Government policy and conservation strategies should 'complete' the National Reserve System (NRS) as an urgent priority, by incorporating a comprehensive, representative and adequate selection of sites from each of Australia's unique bioregions into the NRS.[111] This task will require a significant increase in extent of privately owned land conserved in the NRS, including under conservation covenants or other conservation agreements.[112] It will also require improved financial incentives and supportive market mechanisms, potentially coupled with new forms of stewardship and land-management obligations on private landholders.[113]

As the climate changes, it will not be enough for climate refugia to exist or even to be protected under law. Species must also have the capacity, opportunity and, in some cases, human-assisted translocation to ensure that they can *access* climate refugia. Section 12.2 outlined some of the barriers to species accessing climate refugia, particularly where the 'niche' environmental conditions for a species' survival will no longer exist anywhere near their current distribution and habitat. Legal frameworks for adaptation-oriented planning and ecological 'renewal' must include a focus on accessibility, including where necessary through direct human intervention. To this end, strategies such as enhancing landscape-scale connectivity and engaging in conservation introductions will become increasingly important.[114]

Complicating the question of accessibility, highly accessible refugia may face greater risk of invasion by destructive invasive species such as feral cats, and pathogens such as the deadly amphibian chytrid fungus. Invasive species and pathogens may undermine the capacity of such areas to offer refuge from climate change. As a result, the capacity to *reduce* landscape-scale connectivity to some important climate refugia may improve the viability of such areas, and the populations of native species that they can support.

Finally, the question of funding for new conservation initiatives is a persistent one, given Australia's high rates of biodiversity loss and its status as one of the 40 most underfunded nations in the world for biodiversity conservation.[115] Funding sources and opportunities for ecological restoration are expertly addressed by Froukje Platjouw in Chapter 7 of this book, so this chapter will not investigate the matter in detail. Suffice to say, while conservation funding has historically been the primary domain of governments in Australia, future conservation activities are increasingly likely to be funded by the private sector, through NGO land managers and with philanthropic funding. For example, NGOs such as the Australian Wildlife Conservancy and Bush Heritage have achieved some success in restoring

110 Maggini et al (2013), 39.
111 Taylor (2017).
112 For example, ibid; Fitzsimons (2015).
113 See further Bartel and Graham (Chapter 5 in this volume) on eco-restoration governance on private land in Australia; and stewardship obligations and duties on private landholders.
114 McCormack and McDonald (2014).
115 Adjusted for gross domestic product, among other things, Waldron et al (2013).

species assemblages and ecological functions to large, private protected areas, especially in desert and outback regions.[116] That success has seen governments such as the New South Wales government partnering with NGOs to implement government policies, including species reintroductions and ecological restoration policies, in publicly owned protected areas.[117]

## 12.6 Conclusion

Climate change represents a new and compounding threat to biodiversity all over the world. There is an urgent imperative to protect healthy and diverse ecosystems while also reversing ecological damage to improve the health and function of remnant ecosystems. This chapter demonstrates that, in the Australian context, existing environmental laws provide piecemeal opportunities for some restoration tasks. However, the rate and scale of climate change will undermine most efforts to return environmental conditions to a historical baseline, or even to protect and sustain existing environmental conditions, as is the focus of most Australian environmental law. Thus, effective restoration laws to conserve biodiversity in the future must be closely integrated with climate change adaptation goals. The climate is changing now, and conservation and restoration efforts must include a focus on conserving climate refugia to support species and communities as their existing habitats collapse or disappear. As Emily Barritt argues in Chapter 4 of this book on ecological stewardship, this chapter proposes purposeful human intervention, not as a form of mastery, but as a response to potentially catastrophic losses of the rich diversity of life on Earth. Adaptation-oriented restoration will be crucial for sustaining species and ecological communities, and we must do what we can to facilitate their survival.

## References

### *Primary sources (law materials)*

Biodiversity Conservation Act 2016 (New South Wales).
Catchment and Land Protection Act 1994 (Victoria).
Conservation and Land Management Act 1984 (Western Australia).
Environment Protection Amendment Bill 2018 (Victoria).
Environment Protection and Biodiversity Conservation Act 1999 (Commonwealth of Australia).
Environmental Management and Pollution Control Act 1994 (Tasmania).
Forest Practices Act 1985 (Tasmania).
Land Act 1994 (Queensland).

116 For example, Australian Wildlife Conservancy, 'Sanctuaries: Newhaven Restoration Project', www.australianwildlife.org/sanctuaries/newhaven-sanctuary/newhaven-restoration-project.aspx,Bush Heritage, 'Restoration', www.bushheritage.org.au/what-we-do/science/themes/restoration.
117 *Saving our Species* programme (New South Wales Office of Environment and Heritage n.d.).

Mining Act 1992 (New South Wales).
*Minister for the Environment and Heritage v Greentree (No. 3)* (2004) LGERA 136.
National Parks Act, SC 2000, c 32 (Canada).
National Parks and Reserves Management Act 2002 (Tasmania).
Natural Resources Management Act 2004 (South Australia).
Nature Conservation 2014 (Australian Capital Territory).
Protection of the Environment Operations Act 1997 (New South Wales).
Water Act 2007 (Commonwealth of Australia).

## *Secondary sources*

ABC (2017) 'Mining Report Finds 60,000 Abandoned Sites, Lack of Rehabilitation and Unreliable Data', *ABC Online*, 16 February, www.abc.net.au/news/2017-02-15.

ACF, Birdlife Australia and Environmental Justice Australia (2015) *Recovery Planning: Restoring Life to Our Threatened Species*, Australian Conservation Foundation.

Afshin Akhtar-Khavari and Benjamin J. Richardson (2017) 'Ecological Restoration and the Law: Recovering Nature's Past for the Future' 26(2) *Griffith Law Review*, 147.

APEEL (2017a) *Blueprint for a New Generation of Environmental Laws*, Australian Panel of Experts on Environmental Law.

APEEL (2017b) *The Foundations of Environmental Law: Goals, Objects, Principles and Norms*, Technical Paper 1, Australian Panel of Experts on Environmental Law.

Australian Department of the Environment and Energy (2004) *Threatened Species and Ecological Communities in Australia*, Australian Government.

Australian Department of the Environment and Energy (2015) *Recovery Plan for Three Handfish Species*, Australian Government.

Australian Department of the Environment and Energy (2016) 'Collaborative Australian Protected Area Database (CAPAD)', Australian Government, www.environment.gov.au/land/nrs/science/capad/2016.

Australian Department of the Environment and Energy (n.d.) 'Species Profile and Threats Database', Australian Government, www.environment.gov.au/cgi-bin/sprat/public/publicthreatenedlist.pl.

Australian Department of the Environment and Water (n.d.) 'Native Vegetation in Australia', Australian Government, www.environment.gov.au/land/vegetation/index.html.

Australian Department of Prime Minister and Cabinet (2017) *Indigenous Rangers – Working on Country*, Australian Government.

Australian Wildlife Conservancy (n.d.) 'Sanctuaries: Newhaven restoration project', www.australianwildlife.org/sanctuaries/newhaven-sanctuary/newhaven-restoration-project.aspx.

Madeline C. Bottrill et al (2009) 'Finite Conservation Funds Mean Triage is Unavoidable' 24(4) *Trends in Ecology and Evolution* 183.

Michiala E. Bowen et al (2007) 'Regrowth Forests on Abandoned Agricultural Land: A Review of their Habitat Values for Recovering Forest Fauna' 140(3) *Biological Conservation* 273.

David Bowman et al (2017) 'Renewal Ecology: Conservation for the Anthropocene' 25(5) *Restoration Ecology* 674.

M.F. Breed et al (2013) 'Which Provenance and Where? Seed Sourcing Strategies for Revegetation in a Changing Environment' 14 *Conservation Genetics* 1.

Joseph W. Bull et al (2013) 'Review: Biodiversity Offsets in Theory and Practice' 47(3) *Oryx* 369.
Bush Heritage (n.d.) 'Restoration', www.bushheritage.org.au/what-we-do/science/themes/restoration.
Gerardo Ceballos et al (2015) 'Accelerated Modern Human-induced Species Losses: Entering the Sixth Mass Extinction' 1(5) *Science Advances* e1400253, 1.
F. Stuart Chapin et al (2010) 'Ecosystem Stewardship: Sustainability Strategies for a Rapidly Changing Planet' 25 *Trends in Ecology and Evolution* 241.
Conference of the Parties to the Convention on Biological Diversity (2014) *Decision of the COP in its Twelfth Meeting, Held in Pyeongchang from 6–17 October 2014 – Ecosystem Conservation and Restoration*, UN Doc UNEP/CBD/COP/DEC/XII/19, 17 October.
Ian Cresswell and Helen Murphy (2017) *Australia: State of the Environment 2016*, independent report to the Minister for the Environment and Energy, Australian Government.
J.A. Drake et al (2010) 'A Review of Landscape Rehabilitation Frameworks in Ecosystem Engineering for Mine Closure', presented at the International Conference on Mine Closure, Perth, Western Australia.
Michael Dunlop et al (2012) *The Implications of Climate Change for Biodiversity Conservation and the National Reserve System: Final Synthesis*, CSIRO Climate Adaptation Flagship, Australian Government.
Ann M. Eisenberg (2017) 'Alienation and Reconciliation in Social-ecological Systems' 47 (1) *Environmental Law* 127.
Megan Evans et al (2011) 'The Spatial Distribution of Threats to Species in Australia' 61 (4) *BioScience* 281.
Nicole M. Evans and Mark A. Davis (2018) 'What About Cultural Ecosystems? Opportunities for Cultural Considerations in the "International Standards for the Practice of Ecological Restoration"' 26(4) *Restoration Ecology* 612.
James A. Fitzsimons (2015) 'Private Protected Areas in Australia: Current Status and Future Directions' 10 *Nature Conservation* 1.
Hayley Geyle et al (2018) 'Quantifying Extinction Risk and Forecasting the Number of Impending Australian Bird and Mammal Extinctions' 24 *Pacific Conservation Biology* 157.
Philip Gibbons and David B. Lindenmayer (2007) 'Offsets for Land Clearing: No Net Loss or the Tail Wagging the Dog?' 8(1) *Ecological Management and Restoration* 26.
Lee Godden and Stuart Cowell (2016) 'Conservation Planning and Indigenous Governance in Australia's Indigenous Protected Areas' 24(5) *Restoration Ecology* 692.
Shannon Hagerman and Terre Satterfield (2014) 'Agreed But Not Preferred: Expert Views on Taboo Options for Biodiversity Conservation, Given Climate Change' 24(3) *Ecological Applications* 548.
K. Havens et al (2015) 'Seed Sourcing for Restoration in an Era of Climate Change' 35 *Natural Areas Journal* 22.
Nicole Heller and Richard J. Hobbs (2014) 'Development of a Natural Practice to Adapt Conservation Goals to Global Change' 28(3) *Conservation Biology* 696.
Eric Higgs et al (2014) 'The Changing Role of History in Restoration Ecology' 12(9) *Frontiers in Ecology and the Environment* 499.
Hannah Hoag (2010) 'Confronting the Biodiversity Crisis' 38(4) *Nature Reports Climate Change* 51.

Richard J. Hobbs et al (2011) 'Intervention Ecology: Applying Ecological Science in the Twenty-first Century' 61(6) *BioScience* 442.

Richard J. Hobbs et al (2014) 'Managing the Whole Landscape: Historical, Hybrid, and Novel Ecosystems' 12(10) *Frontiers in Ecology and the Environment* 557.

Alistair Hobday et al (2016) 'Was Tasmania's Summer of Fires and Floods a Glimpse of its Climate Future?', *The Conversation*, 19 April, https://theconversation.com/was-tasmanias-summer-of-fires-and-floods-a-glimpse-of-its-climate-future-58055.

IPBES (2018) *Sixth session, Held in Medellin, Colombia from 18–24 March 2018 – Summary for Policymakers of the Thematic Assessment of Land Degradation and Restoration*, Plenary of the Intergovernmental Science-Policy Platform on Biodiversity and Ecosystem Services, www.nespnorthern.edu.au/wp-content/uploads/2018/04/K1800679-IPBES-6-L-9.pdf.

IPCC (2014) 'Summary for Policymakers' in C.B. Field et al (eds) *Climate Change 2014: Impacts, Adaptation and Vulnerability*, Cambridge University Press for Intergovernmental Panel on Climate Change.

Ramona Maggini et al (2013) *Optimal Habitat Protection and Restoration for Climate Adaptation, Final Report*, National Climate Change Adaptation Research Facility.

Martine Maron et al (2012) 'Faustian Bargains? Restoration Realities in the Context of Biodiversity Offset Policies' 155 *Biological Conservation* 141.

Laura J. Martin et al (2014) 'Biodiversity Conservation Opportunities Across the World's Anthromes' 20 *Diversity and Distributions* 745.

Phillipa C. McCormack (2018a) 'The Legislative Challenge of Facilitating Climate Change Adaptation for Biodiversity' 92(7) *Australian Law Journal* 546.

Phillipa C. McCormack (2018b) 'Conservation Introductions for Biodiversity Adaptation under Climate Change' 7(2) *Transnational Environmental Law* 323.

Phillipa C. McCormack and Jan McDonald (2014) 'Adaptation Strategies for Biodiversity Conservation: Has Australian Law Got What it Takes?' 31 *Environmental and Planning Law Journal* 114.

Tein McDonald et al (2016) *International Standards for the Practice of Ecological Restoration – Including Principles and Key Concepts*, Society for Ecological Restoration.

James R. Miller and Brandon T. Bestelmeyer (2016) 'What's Wrong with Novel Ecosystems, Really?' 24(5) *Restoration Ecology* 577.

Ben Millington and Nancy Notzon (2018) '"Seahorse Hotels" Bring an Endangered Species Back from the Edge of Extinction', *ABC News*, 8 July, www.abc.net.au/news/2018-07-08/seahorse-hotels-bringing-endangered-species-back/9924454.

Toni L. Morelli et al (2016) 'Managing Climate Change Refugia for Climate Adaptation' 11(8) *PLoS ONE* e0159909 1.

Craig Moritz and Rosa Agudo (2013) 'The Future of Species under Climate Change: Resilience or Decline?' 341(6145) *Science* 504.

Carolina Murcia et al (2014) 'A Critique of the "Novel Ecosystem" Concept' 29 *Trends in Ecology and Evolution* 48.

New South Wales Office of Environment and Heritage (n.d.) *Saving our Species*, New South Wales Government, www.environment.nsw.gov.au/topics/animals-and-plants/threatened-species/saving-our-species-program.

Parks Victoria (2003) *Cape Liptrap Coastal Park Management Plan 2003.*

Parks Victoria (2013) *Kara Kara National Park Management Plan 2013.*

Parliament of Australia (2017) *Inquiry into the Rehabilitation of Mining and Resources Projects as it Relates to Commonwealth Responsibilities*, Australian Government.

Gretta T. Pecl et al (2017) 'Biodiversity Redistribution Under Climate Change: Impacts on Ecosystems and Human Well-being' 355(6332) *Science*, doi:10.1126/science.aai9214.

Michael P. Perring et al (2006) 'Biodiversity in Agricultural Landscapes: Saving Natural Capital Without Losing Interest' 20 *Conservation Biology* 263.

Michael P. Perring, Patrick Audet and David Lamb (2014) 'Novel Ecosystems in Ecological Restoration and Rehabilitation: Innovative Planning or Lowering the Bar?' 3(1) *Ecological Processes* 8.

Michael P. Perring et al (2015) 'Advances in Restoration Ecology: Rising to the Challenges of the Coming Decades' 6(8) *Ecosphere* 13.

Suzanne M. Prober et al (2015) 'Climate-adjusted Provenancing: A Strategy for Climate-resilient Ecological Restoration' 1(65) *Frontiers in Ecology and Evolution* 1.

A. Reisinger et al (2014) 'Australasia' in V.R. Barros et al (eds) *Climate Change 2014: Impacts, Adaptation, and Vulnerability, Part B: Regional Aspects*, Contribution of Working Group II to the Fifth Assessment Report of the IPCC, Cambridge University Press, 1371.

April Reside et al (2013) *Climate Change Refugia for Terrestrial Biodiversity: Defining Areas that Promote Species Persistence and Ecosystem Resilience in the Face of Global Climate Change*, National Climate Change Adaptation Research Facility.

April Reside et al (2014) 'Characteristics of Climate Change Refugia for Australian Biodiversity' 39(8) *Austral Ecology* 887.

Benjamin J. Richardson (2015) 'Reclaiming Nature: Eco-restoration of Liminal Spaces' 2 (1) *Australian Journal of Environmental Law* 1.

Benjamin J. Richardson (2016) 'The Emerging Age of Ecological Restoration Law' 25(3) *Review of European, Comparative and International Environmental Law* 277.

Yasha Rohwer and Emma Marris (2016) 'Renaming Restoration: Conceptualizing and Justifying the Activity as a Restoration of Lost Moral Value Rather than a Return to a Previous State' 24(5) *Restoration Ecology*, 674.

Michael L. Rosenzweig (2003) 'Reconciliation Ecology and the Future of Species Diversity' 37(2) *Oryx* 194.

J.B. Ruhl (2010) 'Climate Change Adaptation and the Structural Transformation of Environmental Law' 40 *Environmental Law* 363.

Timothy R. Seastedt, Richard J. Hobbs and Katharine N. Suding (2008) 'Management of Novel Ecosystems: Are Novel Approaches Required?' 6(10) *Frontiers in Ecology and the Environment* 547.

SERA (2017) *National Standards for the Practice of Ecological Restoration in Australia*, Standards Reference Group, Society for Ecological Restoration Australasia.

Will Steffen et al (2009) *A Strategic Assessment of the Vulnerability of Australia's Biodiversity to Climate Change*, independent report to the Australian Government.

Kathrine Suding et al (2015) 'Committing to Ecological Restoration' 348(6235) *Science* 638.

Tasmanian Parks and Wildlife (2002) *Mount Field National Park, Marriotts Falls State Reserve and Junee Cave State Reserve Management Plan 2002.*

M.F.J. Taylor (2017) *Building Nature's Safety Net 2016: The State of Australian Terrestrial Protected Areas 2010–2016*, WWF-Australia.

Maria Tengö et al (2017) 'Weaving Knowledge Systems in IPBES, CBD and Beyond—Lessons Learned for Sustainability' 26/27 *Current Opinion in Environmental Sustainability* 17.

Chris D. Thomas et al (2011) 'A Framework for Assessing Threats and Benefits to Species Responding to Climate Change' 2(2) *Methods in Ecology and Evolution* 125.

Arie Trouwborst (2009) 'International Nature Conservation Law and the Adaptation of Biodiversity to Climate Change' 21 *Journal of Environmental Law* 419.

Anthony Waldron et al (2013) 'Targeting Global Conservation Funding to Limit Immediate Biodiversity Declines' 110(29) *Proceedings for the National Academy of Sciences* 12144.

Grant Wardell-Johnson et al (2015) 'Integrating Rehabilitation, Restoration and Conservation for a Sustainable Jarrah Forest Future During Climate Disruption' 21 *Pacific Conservation Biology* 175.

James E.M. Watson et al (2016) 'Catastrophic Declines in Wilderness Areas Undermine Global Environment Targets' 26(21) *Current Biology* 2929.

Jessica Weir, Claire Stacey and Kara Youngetob (2011) *The Benefits Associated with Caring for Country*, Australian Institute of Aboriginal and Torres Strait Islander Studies (AIATSIS).

Western Australian Government (1995) *Policy Statement 29: Translocation of Threatened Flora and Fauna.*

John A. Wiens and Richard J. Hobbs (2015) 'Integrating Conservation and Restoration in a Changing World' 65(3) *BioScience* 302.

A.V. Williams et al (2014) 'Next Generation Restoration Genetics: Applications and Opportunities' 19 *Trends in Plant Science* 529.

John C.Z. Woinarski, A.A. Burbidge and P.L. Harrison (2015) 'Ongoing Unraveling of a Continental Fauna: Decline and Extinction of Australian Mammals Since European Settlement' 112(15) *Proceedings of the National Academy of Sciences* 4531.

G.L. Worboys et al (2008) 'The Australian Alps to Atherton (A2A) Connectivity Conservation Area: A National Response to Climate Change', paper for the Australian Protected Area Congress 2008, 24–28 November, Sunshine Coast.

P.T. Young, D.A. Petersen and J.J. Clary (2005) 'The Ecology of Restoration: Historical Links, Emerging Issues and Unexplored Realms' 8(6) *Ecology Letters* 662.

Joy Zedler, James M. Doherty and Nicholas Miller (2012) 'Shifting Restoration Policy to Address Landscape Change, Novel Ecosystems, and Monitoring' 17(4) *Ecology and Society* 36.

# Index

For Product Safety Concerns and Information please contact our EU
representative GPSR@taylorandfrancis.com
Taylor & Francis Verlag GmbH, Kaufingerstraße 24, 80331 München, Germany

www.ingramcontent.com/pod-product-compliance
Ingram Content Group UK Ltd.
Pitfield, Milton Keynes, MK11 3LW, UK
UKHW022219190926
475702UK00003B/437
*9781138605015*